Springer Proceedings in Complexity

Springer Proceedings in Complexity publishes proceedings from scholarly meetings on all topics relating to the interdisciplinary studies of complex systems science. Springer welcomes book ideas from authors. The series is indexed in Scopus. Proposals must include the following: - name, place and date of the scientific meeting - a link to the committees (local organization, international advisors etc.) - scientific description of the meeting - list of invited/plenary speakers - an estimate of the planned proceedings book parameters (number of pages/articles, requested number of bulk copies, submission deadline). submit your proposals to: christopher.coughlin@springer.com

More information about this series at http://www.springer.com/series/11637

Christos H. Skiadas · Ihor Lubashevsky
Editors

11th Chaotic Modeling and Simulation International Conference

 Springer

Editors
Christos H. Skiadas
ManLab, Technical University of Crete
Chania, Crete, Greece

Ihor Lubashevsky
Complex Systems Modeling Laboratory
The University of Aizu
Aizuwakamatsu, Fukushima, Japan

ISSN 2213-8684   ISSN 2213-8692   (electronic)
Springer Proceedings in Complexity
ISBN 978-3-030-15299-4   ISBN 978-3-030-15297-0   (eBook)
https://doi.org/10.1007/978-3-030-15297-0

This Springer imprint is published by the registered company Springer Nature Switzerland AG
The registered company address is: Gewerbestrasse 11, 6330 Cham, Switzerland

Committees—CHAOS 2018

Honorary Committee and Scientific Advisors

Florentino Borondo Rodríguez, Universidad Autónoma de Madrid, Instituto de Ciencias Matemáticas, ICMAT (CSIC-UAM-UCM-UCIII)

Leon O. Chua, EECS Department, University of California, Berkeley, USA, Honorary Editor of the International Journal of Bifurcation and Chaos

Giovanni Gallavotti, Universita di Roma 1, "La Sapienza", Italy and Rutgers University, USA

Gennady A. Leonov, Dean of Mathematics and Mechanics Faculty, Saint-Petersburg State University, Russia, Member (corresponding) of Russian Academy of Science

Gheorghe Mateescu, Department of Chemistry, Case Western Reserve University, Cleveland, OH, USA

Yves Pomeau, Department of Mathematics, University of Arizona, Tucson, USA

David Ruelle, Academie des Sciences de Paris, Honorary Professor at the Institut des Hautes Etudes Scientifiques of Bures-sur-Yvette, France

Ferdinand Verhulst, Institute of Mathematics, Utrecht, The Netherlands

International Scientific Committee

C. H. Skiadas, Technical University of Crete, Chania, Greece, Co-Chair

H. Adeli, The Ohio State University, USA

J.-O. Aidanpaa, Division of Solid Mechanics, Lulea University of Technology, Sweden

N. Akhmediev, Australian National University, Australia

M. Amabili, McGill University, Montreal, Canada

J. Awrejcewicz, Technical University of Lodz, Poland

E. Babatsouli, University of Crete, Rethymnon, Greece

J. M. Balthazar, UNESP-Rio Claro, State University of Sao Paulo, Brasil
S. Bishop, University College London, UK
T. Bountis, University of Patras, Greece
Y. S. Boutalis, Democritus University of Thrace, Greece
C. Chandre, Centre de Physique Theorique, Marseille, France
M. Christodoulou, Technical University of Crete, Chania, Crete, Greece
P. Commendatore, Universita di Napoli 'Federico II', Italy
D. Dhar, Tata Institute of Fundamental Research, India
J. Dimotikalis, Technological Educational Institute, Crete, Greece
B. Epureanu, University of Michigan, Ann Arbor, MI, USA
G. Fagiolo, Sant'Anna School of Advanced Studies, Pisa, Italy
M. I. Gomes, Lisbon University and CEAUL, Lisboa, Portugal
V. Grigoras, University of Iasi, Romania
A. S. Hacinliyan, Yeditepe University, Istanbul, Turkey
K. Hagan, University of Limerick, Ireland
L. Hong, Xi'an Jiaotong University, Xi'an, Shaanxi, China
G. Hunt, Centre for Nonlinear Mechanics, University of Bath, Bath, UK
T. Kapitaniak, Technical University of Lodz, Lodz, Poland
G. P. Kapoor, Indian Institute of Technology Kanpur, Kanpur, India
W. Klonowski, Nalecz Institute of Biocybernetics and Biomedical Engineering, Polish, Academy of Sciences, Warsaw, Poland
A. Kolesnikov, Southern Federal University Russia
I. Kourakis, Queen's University Belfast
J. Kretz , University of Music and Performing Arts, Vienna, Austria
V. Krysko, Department of Mathematics and Modeling, Saratov State Technical University, Russia
I. Kusbeyzi Aybar, Yeditepe University, Istanbul,Turkey
W. Li, Northwestern Polytechnical University, China
B. L. Lan, School of Engineering, Monash University, Selangor, Malaysia
V. J. Law, University College Dublin, Dublin, Ireland
I. Lubashevsky, The University of Aizu, Japan
V. Lucarini, University of Hamburg, Germany
J. A. T. Machado, ISEP-Institute of Engineering of Porto, Porto, Portugal
W. M. Macek, Cardinal Stefan Wyszynski University, Warsaw, Poland
P. Mahanti, University of New Brunswick, Saint John, Canada
G. M. Mahmoud, Assiut University, Assiut, Egypt
P. Manneville, Laboratoire d'Hydrodynamique, Ecole Polytechnique, France
A. S. Mikhailov, Fritz Haber Institute of Max Planck Society, Berlin, Germany
E. R. Miranda, University of Plymouth, UK
M. S. M. Noorani, University Kebangsaan, Malaysia
G. V. Orman, Transilvania University of Brasov, Romania
O. Ozgur Aybar, Department of Mathematics, Piri Reis University, Tuzla, Istanbul, Turkey
S. Panchev, Bulgarian Academy of Sciences, Bulgaria
G. P. Pavlos, Democritus University of Thrace, Greece

G. Pedrizzetti, University of Trieste, Trieste, Italy
F. Pellicano, Universita di Modena e Reggio Emilia, Italy
D. Pestana, Lisbon University and CEAUL, Lisboa, Portugal
S. V. Prants, Pacific Oceanological Institute of RAS, Vladivostok, Russia
A. G. Ramm, Kansas State University, Kansas, USA
G. Rega, University of Rome "La Sapienza", Italy
H. Skiadas, Hanover College, Hanover, USA
V. Snasel, VSB-Technical University of Ostrava, Czech
D. Sotiropoulos, Technical University of Crete, Chania, Crete, Greece
B. Spagnolo, University of Palermo, Italy
P. D. Spanos, Rice University, Houston, TX, USA
J. C. Sprott, University of Wisconsin, Madison, WI, USA
S. Thurner, Medical University of Vienna, Austria
D. Trigiante, Università di Firenze, Firenze, Italy
G. Unal, Yeditepe University, Istanbul, Turkey
A. Valyaev, Nuclear Safety Institute of RAS, Russia
A. Vakakis, University of Illinois at Urbana-Champaign, Illinois, USA
J. P. van der Weele, University of Patras, Greece
M. Wiercigroch, University of Aberdeen, Aberdeen, Scotland, UK
M. V. Zakrzhevsky, Institute of Mechanics, Riga Technical University, Latvia
J. Zhang, School of Energy and Power Engineering, Xi'an Jiaotong University, Xi'an, Shaanxi Province, P. R. China

Plenary–Keynote–Invited Speakers

Chris G. Antonopoulos, University of Essex, Department of Mathematical Sciences, UK
Modelling the brain: From dynamical complexity to neural synchronisation, chimera-like States and information flow capacity
Jean-Marc Ginoux, Institut Universitaire de Technologie de Toulon, La Garde, France
The paradox of Vito Volterra's predator-prey model
Nikolaos Katopodes, University of Michigan, Ann Arbor, MI, USA
Instability of flow between rotating disks
Ihor Lubashevsky, University of Aizu, Aizu-Wakamatsu, Fukushima, Japan
Do we need a new physics to describe human behaviour? Phenomenological standpoint
Wieslaw M. Macek, Faculty of Mathematics and Natural Sciences, Cardinal Stefan Wyszynski University, Warsaw, Poland and Space Research Centre, Polish Academy of Sciences, Warsaw, Poland
Complex dynamics in the generalized lorenz system
Riccardo Mannella, Dipartimento di Fisica, Universita di Pisa, Italy

Alternative approach to treatment of separatrix chaos: 10 years of development
Leszek Sirko, Institute of Physics, Polish Academy of Sciences, Warszawa, Poland
Influence of topology and absorption on properties of quantum graphs and microwave networks
Alexander V. Sosnitsky; Anatoly I. Shevchenko
Department of Computer Technologies, Berdyansk State Pedagogical University; Institute of AI Problems of the MES and NAS of Ukraine, Ukraine
Intelligence (Life) as a universal transformation mechanism of chaos into harmony
Beatrice Venturi, Department of Economics and Business, University of Cagliari, Sardinia, Italy
On the structure of the solutions of a resource optimal model
Xiaoming Wang, Shanghai Center for Mathematics Sciences, Fudan University, Shanghai, China
Numerical algorithms for approximating long-time statistical properties of turbulent systems

Preface

11th Chaotic Modeling and Simulation International Conference (5–8 June 2018, Rome, Italy)

It is our pleasure to thank the guests, participants and contributors to the 11th International Conference (CHAOS2018) on Chaotic Modeling, Simulation and Applications. We support the study of nonlinear systems and dynamics in an interdisciplinary research field and very interesting applications where presented. We provide a widely selected forum to exchange ideas, methods and techniques in the field of Nonlinear Dynamics, Chaos, Fractals and their applications in General Science and in Engineering Sciences.

The principal aim of CHAOS2018 International Conference is to expand the development of the theories of the applied nonlinear field, the methods and the empirical data and computer techniques, and the best theoretical achievements of chaotic theory as well.

Chaotic Modeling and Simulation Conferences continue to grow considerably from year to year thus making a well-established platform to present and disseminate new scientific findings and interesting applications. We thank all the contributors to the success of this conference and especially the authors of this *Proceedings Volume* of CHAOS 2018.

Special thanks to the Plenary, Keynote and Invited Presentations, the Scientific Committee, the ISAST Committee and Yiannis Dimotikalis, the Conference Secretary Mary Karadima and all the members of the Secretariat.

Chania, Crete, Greece   Christos H. Skiadas
Aizuwakamatsu, Fukushima, Japan   Ihor Lubashevsky

Contents

Contents   xiii

Gravitational Waves, Relic Photons and Higgs Boson in a Fractal Models of the Universe

Valeriy S. Abramov

Abstract Models for describing separate large-scale fractal structures of the Universe are proposed. The relationships between the parameters of gravitational waves, relic photons and the Higgs boson are established. Estimates of these parameters are given on examples of: merging two black holes, binary neutron stars; "Cold relict spot" (supervoid). The behavior of deformation fields on the fractal index for a number of quantum model systems with variable parameters is investigated. It is shown that the presence of nonlinear oscillations is characteristic for a fractal layer without a quantum dot. Stochastic behavior for the boundaries of the quantum dots cores is observed, an anisotropy effect is possible.

Keywords Fractal structures of the Universe · Higgs boson · Gravitational waves · Relic photons · Binary black holes and neutron stars

1 Introduction

The hypothesis of the presence of dark matter, dark energy in the Universe can be examined on the basis of direct gravitational effects, waves [1]. Electromagnetic radiation (photons) does not carry this direct information. For the creation of a gravitational wave detector and experimental proof of their existence, R. Weiss, K. Thorne and B. Barrish were awarded the Nobel Prize in Physics in 2017. A binary black hole merger with the energy release in the form of gravitational waves (GW) was recorded by the LIGO interferometers in Livingston and Hanford [2]. These signals represent the gravitational wave amplitude dependencies on time, and was recorded by the detectors LD (Livingston Detector) and HD (Hanford Detector). The appearance of GW from a binary neutron stars was recorded on August 17, 2017 [3]. These achievements in cosmology give impulse to the development of new theoretical models of the fractal structures of the Universe: Galaxies, superclusters of Galaxies, walls, fila-

V. S. Abramov (✉)
Donetsk Institute for Physics and Engineering named after A.A. Galkin, Donetsk, Ukraine
e-mail: vsabramov@mail.ru

ments, voids [4], supervoid or the CMB Cold Spot [5], black holes, neutron stars [6]. The hypothesis of the hierarchical structure of the Universe makes it possible to use models of fractal dislocations, quantum dots with variable parameters to describe individual elements of the fractal structures of the Universe [7].

When describing various nonlinear physical models, singular points (attractors), lines, surfaces, special volumetric structures (strange attractors) arise. Many physical properties near the above features are stochastic in nature, it becomes necessary to model stochastic processes [8]. In [9–12] attractors and deformation field, mutual influence of attractors and stochastic processes in the coupled fractal multilayer nanosystems were investigated. In [11] the fractal oscillator model based on the theory of fractional calculus was proposed. In [12] transient processes in a multilayer fractal nanosystem with a nonlinear fractal oscillator were investigated. In the presence of variable parameters, features arise in the behavior of such fractal nanosystems.

The aim of the paper is to describe anisotropy, transient signals from binary objects (black holes, neutron stars); modelling of the deformation field of an individual layer in a multilayer fractal nanosystem (with variable parameters), investigation of the influence of a fractal index.

2　Anisotropic Model for Binary Black Holes

For the split energy branches $2\varepsilon_{02}, 2\varepsilon_{01}$ in [7, 13, 14] there were expressions obtained, relating the rest energy of the Higgs boson $E_{H0} = 125.03238\,\text{GeV}$ and the order parameter for the Higgs field $\Delta'_0 = 21.93272771\,\text{GeV}$

$$\varepsilon_{02} = \left[E_{H0}^2 - (\Delta'_0)^2\right]^{1/2}; \ \varepsilon_{01} = \left[E_{H0}^2 + (\Delta'_0)^2\right]^{1/2};$$
$$\Delta'_0 = \psi'_0 E_{H0}; \ \psi'_0 = \delta N'_0/N'_0; \ (\xi'_0)^2 = N'_{01}/N'_{02};$$
$$N'_{02} = \left[(N'_0)^2 - (\delta N'_0)^2\right]^{1/2}; \ N'_{01} = \left[(N'_0)^2 + (\delta N'_0)^2\right]^{1/2}. \tag{1}$$

Parameters describing the presence of a Bose condensate taking into account the Higgs field are: $N'_0 = 3.7384680 \times 10^5, N'_{01} = 3.7955502 \times 10^5, N'_{02} = 3.6805005 \times 10^5, (\xi'_0)^2 = 1.031259246, \psi'_0 = 0.175416382$; energies $2\varepsilon_{01} = 253.8829698\,\text{GeV},$ $2\varepsilon_{02} = 246.1873393\,\text{GeV}$. Next, a quasi-one-dimensional lattice with two atoms in a unit cell (such as an effective atom and a Higgs boson with rest masses m_H and M_{H0}) was introduced. The basic relations between parameter $|\xi_{0H}|^2$ with the rest masses m_H and M_{H0} are following

$$|\xi_{0H}|^2 = m_H/M_{H0} = M_H/m_{H0} = E_H/E_{H0} = R_H/R_{H0}; \ M_{Ha} = N_a M_H;$$
$$M'_{H0} = N_a m_{H0}; \ R_H = 2GM_{Ha}/c_0^2; \ R_{H0} = 2GM'_{H0}/c_0^2. \tag{2}$$

Here $M_H = N_a m_H = 24.41158758\,\text{g}$ and $m_{H0} = N_a M_{H0} = 134.2770693\,\text{g}$ are molar masses of the effective atom and the Higgs boson; $E_H = 22.73090194\,\text{GeV}$ is rest energy of an effective atom; $R_H = 21.84067257\,\mu\text{m}$, $R_{H0} = 120.1356321\,\mu\text{m}$ allow us the interpretation of Schwarzschild radii of black holes with masses M_{Ha}, M'_{H0}; $G = 6.672 \times 10^{-8}\,\text{cm}^3\,\text{g}^{-1}\,\text{s}^{-2}$ is Newton's gravitational constant; Avogadro number is $N_a = 6.025438 \times 10^{23}$; c_0 is the speed of light in a vacuum. Taking into account the value $|\xi_{0H}|^2 = 0.181800122$, the main parameters of the theory $|S'_{01}| = 0.039541282$, $S'_{02} = 0.03409$, $S'_{03} = 0.460458718$, $S'_{04} = 0.53409$ were obtained. On the basis of energies $2\varepsilon_{01}$, $2\varepsilon_{02}$ from (1) and parameters S'_{0x} ($x = 1, 2, 3, 4$) we obtain energy spectra $\varepsilon_{sx} = 2\varepsilon_{01}S'_{0x}$, $\varepsilon'_{sx} = 2\varepsilon_{02}S'_{0x}$. These spectra make it possible (taking into account the Higgs field) to obtain the energy values $\varepsilon_{s3} + \varepsilon'_{s2} = 125.2951532\,\text{GeV}$ and $\varepsilon_{s3} + \varepsilon'_{s1} = 126.6371898\,\text{GeV}$ for the Higgs boson, which agree with the values of the energies 125.3 and 126.5 GeV, obtained at the LHC [15]. Based on the parameters $(\xi'_0)^2$, $|\xi_{0H}|^2$ and the molar mass M_H, we introduce the susceptibility components

$$\chi_{11} = |\xi_{0H}|^2; \; \chi_{21} = \chi_{02} = [(\xi'_0)^2 - 1]/\chi_{11}; \; \chi_{31} = -\chi_{01}; \; n_F + n'_F = 1;$$
$$n_F = M_{02}/M_H = \chi_{02}/\chi_{11}; \; n'_F = M_{01}/M_H = -\chi_{01}/\chi_{11}. \tag{3}$$

Taking into account (3), we find numerical values: $\chi_{11} = 0.181800122$, $\chi_{02} = 0.171942932$, $\chi_{01} = -0.00985719$, $n_F = 0.945780069$, $n'_F = 0.054219931$, $M_{01} = 1.323594585\,\text{g}$, $M_{02} = 23.087993\,\text{g}$.

On the basis of (3) and E_{H0} we find the characteristic energies

$$E_{1\nu} = -\chi_{01}E_{H0} = n'_F E_H; \; E_{2\nu} = \chi_{02}E_{H0} = n_F E_H; \; E_H = \chi_{11}E_{H0}.$$
$$E_{1\nu} = M_{01}c_0^2 = 1.232467935\,\text{GeV}; \; E_{2\nu} = M_{02}c_0^2 = 21.49843401\,\text{GeV}. \tag{4}$$

Next, we introduce a row-vector $\hat{\chi}_1 = (\chi_{11}, \chi_{21}, \chi_{31})$ and a column-vector $\hat{\chi}_1^+$. We find the effective susceptibility $|\chi_{ef}|$, molar mass M_{ef} from the conditions

$$\hat{\chi}_1\hat{\chi}_1^+ - |\chi_{ef}|^2 = |\chi_{01}|^2 + |\chi_{02}|^2 + |\xi_{0H}|^4 = M_{ef}^2 m_{H0}^{-2};$$
$$M_{ef}^2 = M_{01}^2 + M_{02}^2 + M_H^2. \tag{5}$$

The numerical values are $|\chi_{ef}| = 0.250425279$, $M_{ef} = 33.62637256\,\text{g}$.

To take into account the nonlinear dependences of the effective displacements $u_\mu = F(\varphi_\mu; k_\mu)$ (F is the incomplete elliptic integral of the first kind) on the angle φ_μ, the modulus k_μ of elliptic functions, we use the fractal oscillator model [11, 12]. In this model, a matrix $\hat{T}_{ef}$ with elements t_{ij} is introduced

$$t_{11} = k'_\mu; \; t_{12} = -k_\mu; \; t_{21} = k_\mu \text{cn}(u_\mu; k_\mu); \; t_{22} = k'_\mu \text{cn}(u_\mu; k_\mu);$$
$$t_{23} = -\text{sn}(u_\mu; k_\mu); \; t_{13} = 0; \; t_{31} = k_\mu \text{sn}(u_\mu; k_\mu);$$
$$t_{32} = k'_\mu \text{sn}(u_\mu; k_\mu); \; t_{33} = \text{cn}(u_\mu; k_\mu); \quad i, j = 1, 2, 3. \tag{6}$$

The action $\hat{T}_{ef}$ on $|\chi_{ef}|$ leads to a matrix $|\hat{\chi}_{ef}| = \hat{T}_{ef}|\chi_{ef}|$ with elements χ_{ij}

$$\chi_{11} = k'_\mu|\chi_{ef}|; \; \chi_{21} = k_\mu|\chi_{ef}|\mathrm{cn}(u_\mu; k_\mu); \; \chi_{31} = k_\mu|\chi_{ef}|\mathrm{sn}(u_\mu; k_\mu);$$

$$\chi_{12} = -k_\mu|\chi_{ef}|; \; \chi_{22} = \chi_{11}\cos\varphi_\mu; \; \chi_{32} = \chi_{11}\sin\varphi_\mu; \chi_{23} = -|\chi_{ef}|\sin\varphi_\mu;$$

$$\chi_{13} = 0; \; \chi_{33} = |\chi_{ef}|\cos\varphi_\mu; \; k'_\mu = \chi_{11}/|\chi_{ef}|; \; (k'_\mu)^2 + k_\mu^2 = 1. \tag{7}$$

The numerical values are $\chi_{12} = -0.172225247$, $\chi_{22} = 0.181502111$, $\chi_{32} = 0.010405201$, $\chi_{23} = -0.014332913$, $\chi_{33} = 0.250014775$, $k'_\mu = 0.725965539$, $\sin\varphi_\mu = 0.057234291$. The characteristic angles $\varphi_\mu = 3.2810763°$, $\varphi_\mu^* = \pi/2 + 2\varphi_\mu$, $\varphi'_\mu = \pi/2 - \varphi_\mu = 86.7189237°$ can be determined from the presence of peaks in X-ray structural spectra. From (1) at $\psi'_0 = 0$ we obtain the order parameter $\Delta'_0 = 0$ and the equality of the energies $2\varepsilon_{02} = 2\varepsilon_{01} = 2E_{H0}$ of the branches of the spectrum. Then from (3) follows $(\xi'_0)^2 = 1$, $\chi_{21} = 0$, and from (7) we obtain the condition $k_\mu|\chi_{ef}|\mathrm{cn}(u_\mu; k_\mu) = 0$. This condition can be fulfilled either at $k_\mu = 0$ or at $\mathrm{cn}(u_\mu; k_\mu) = 0$. Then χ_{ij} will take numerical values different from those given above. If the parameter $\Delta'_0 \neq 0$, then from (7) follows the need to analyze other row-vectors $\hat{\chi}_2 = (\chi_{12}, \chi_{22}, \chi_{32})$, $\hat{\chi}_3 = (\chi_{13}, \chi_{23}, \chi_{33})$ and the column-vectors $\hat{\chi}_2^+$, $\hat{\chi}_3^+$ the susceptibility tensor $\hat{\chi}_{ef}$.

On the basis $Q_{H6} = 1.537746366$ from [13] and the susceptibility components χ_{ij}, we write for the black hole spin tensor $\hat{n}_{hs}$ the elements in the form $n_{ij} = 2/(2Q_{H6} - z_{ij})$, where $z_{ij} = \chi_{ij}/2$. For diagonal elements we find $n_{11} = 0.6701082$, $n_{22} = 0.6700747$, $n_{33} = 0.6778548$. After the binary black holes (BBH) merger in [2] the final value of the black hole spin of 0.67 and the value of the red shift $z_s = 0.09$ are determined. Our calculated values n_{22} and $z_{22} = 0.090751056$ are close to these data. This indicates the tensor nature of the source of the black hole spin $\hat{n}_{hs}$ and redshift z_{ij}, which are related to the susceptibility χ_{ij}. The main parameter n_{A0} determines the spectrum for the occupation numbers $n_{Ax} = n_{A0}S'_{0x}$ of black holes. The number of quanta $n_{h1} = M_{h1}/M_s$, $n_{h2} = M_{h2}/M_s$ BBH before merger, and the number of black hole quanta $2n_{A4} = M_{A4}/M_s$ after merger are determined through the cosmological redshift z'_μ, the parameter Q_{H2} and n_{A0} from expressions

$$\sin^2\varphi'_{\mu\lambda} = 2z_Q/(3z_Q + 1); \; \cos^2\varphi'_{\mu\lambda} = (z_Q + 1)/(3z_Q + 1); \; z_Q = 1/(2n_{A0} - 1);$$

$$n_{A0} = (z'_{\mu\lambda})^2 - 1 = (z'_\mu + 3/2)(z'_\mu - 1/2); \; z'_{\mu\lambda} = z'_\mu + 1/2; \; 1/z'_{\mu\lambda} = \sin\varphi'_{\mu\lambda};$$

$$n'_{A0} = (z'_{\mu\lambda})^2; \; n'_{A0} - n_{A0} = 1; \; z'_Q - z_Q = 1; \; \varphi'_{\mu\lambda} = \varphi_a Q_{H2}. \tag{8}$$

Here M_{h1}, M_{h2} are masses of first, second black holes before merger; M_{A4} is black hole mass after merger; M_s is mass of the Sun. Using the values of parameters $Q_{H2} = 1/3$, $z'_\mu = 7.184181$ [13, 14], we find $\sin\varphi'_{\mu\lambda} = 0.130137486$, $n_{A0} = 58.04663887$. The angle $\varphi_a = 22.43261159°$ can be determined from the peak position on an amorphous substrate in the X-ray structural spectrum. Based on the spectrum n_{Ax}, we find the number of black hole quanta $2n_{A4} = 62.0042587$ that formed after the merger of two black holes. Number of quanta of the second black hole before merger is $n_{h2} = n_{A0}/2 = n_{A4} - n_{A2} = n_{A3} + n_{A1} = 29.02331944$. As a result of the merger of these BBH, the number of quanta $n_G = 1/Q_{H2} = 3$ is carried away by gravitational waves. The number of quanta of the first black hole before merger $n_{h1} = 35.98093926$ we obtain from equation $(n_{h1} + n_{h2}) - 2n_{A4} = n_G$.

In [13, 14], by describing the anisotropy of the CMB, connections temperatures $T_A = T_r/N_{ra} = 2.61739852$ mK, $T'_A = T_r/z'_{A2} = 2.635582153$ mK with the relict radiation temperature $T_r = 2.72548$ K were obtained, where $z'_{A2} = 1034.109294$ is the usual redshift, $N_{ra} = z'_{A2} + z'_\mu = 1041.293475$. The temperature deviation $\delta T_A = T'_A - T_A = 18.183633$ μK agrees with the experimental average value 18 μK of temperature fluctuations in the relict background in the fractal model of the Universe. On the other hand, in our model the supervoid is determined by the temperature T^*_A, the number of quanta N^*_{ra}, the parameter z^*_μ

$$T^*_A = T_r/N^*_{ra}; \quad N^*_{ra} = z'_{A2} + z^*_\mu; \quad z^*_\mu = 2n_{A4} + (n_{A1} - n_{A2}). \quad (9)$$

The parameter $z^*_\mu = 62.3206873$ allow us an interpretation as the effective cosmological shift at the early stages of the formation of the structure of the Universe after the Big Bang, and is related with numbers of black hole quanta $2n_{A4}$, n_{A1}, n_{A2}. Numerical values are $N^*_{ra} = 1096.429981$, $T^*_A = 2.4857766$ mK. The temperature deviation $\delta T^*_A = T^*_A - T'_A = -149.8055448$ μK agrees with the deviation $(-150$ μK$)$ from [5]. The sign "$-$" indicates that the area of the supervoid is colder than the neighboring areas.

3 Description of Transient Signals from Binary Objects

Busts of supernovae of type la, processes of BBH, binary neutron stars (BNS) merger can be considered as separate impulse sources in the Universe. In this case, transient gravitational-wave signals, relict radiation of photons arise. To describe the characteristic parameters and transient signals of the GW radiation from the BBH or BNS merger, we use the semiclassical superradiance model Dicke [16] and the quantum statistical theory of superradiance [17, 18]. For the radiation intensity J we have [16]

$$J = J_0[(a_0^2 - a_m^2) + (a_0 + a_m)] = J_0(a_0 + a_m)[(a_0 - a_m) + 1]. \quad (10)$$

Here J_0 is the initial radiation intensity; parameters a_0, a_m generally depend on the time, frequency and amplitude of the GW, the characteristics of the BBH or BNS. If $J = J_m$, where J_m is the maximum radiation intensity, then from (10) we obtain expressions for the critical density ρ_c ratio of the GW signal amplitude to the noise amplitude

$$\rho_c^2 = (z'_{A2} - 1)/2 + J_m/(2J_0); \quad a_0^2 = a_m^2 + z'_\mu(z'_\mu + 2)/4; \quad a_m = (z'_{A2})^{1/2}. \quad (11)$$

The numerical values of the parameters are $a_m = 32.1575698$, $a_0 = 32.4130298$, $J_m/J_0 = 81.0658042$. The value $\rho_c = 23.602701$ is close to the critical value of 23.6 for BBH from [2]. The parameter a_0 is close to the ratio of the signal amplitude to the noise amplitude of 32.4 for BNS [3].

On the basis of model I from [13, 14] we write the expressions for the Hubble constant H_0, velocity υ_0 in the model of a flat cosmology

$$H_0 = H_{01}/\Omega_{tH} = \upsilon_0/L_0; \quad \upsilon_0 = \upsilon_{01}/\Omega_{tH}; \quad \Omega_{tH} = Q_{H0} + |S'_{01}|. \tag{12}$$

Taking into account the values of Hubble constant $H_{01} = 73.2 \text{ km s}^{-1} \text{ Mpc}^{-1}$, the velocity $\upsilon_{01} = 7.32 \times 10^6 \text{ cm s}^{-1}$ (which describe the accelerated expansion of the Universe), $Q_{H0} = 1.039541282$, $L_0 = 0.30857 \times 10^{25} \text{ cm}$ [12, 13], from (12) we obtain $H_0 = 67.83540245 \text{ km s}^{-1} \text{ Mpc}^{-1}$, $\upsilon_0 = 6.783540245 \times 10^6 \text{ cm s}^{-1}$.

In [13, 14], the values for the maximum $\nu_{r1} = 160.3988698 \text{ GHz}$ and shifted $\nu_{r2} = 142.8161605 \text{ GHz}$ frequencies of relict radiation photons were obtained. According to the relict radiation, taking into account $\xi_q = \nu_{r1} - \nu_{r2}$, in [7] the value of the gap $2|\lambda|N^{1/2} = 8.396945157 \text{ GHz}$ in the spectrum and density of cold dark matter $\Omega'_{c1} = 4|\lambda|^2 N/|\xi_q|^2 = 0.228071512$ are obtained.

To calculate the wavelength $\lambda_{\gamma b}$ from the source for BNS, we use the energy spectra $\varepsilon_{\mu x} = 2\varepsilon'_{01}S'_{0x}$, $\varepsilon'_{\mu x} = 2\varepsilon'_{02}S'_{0x}$, which are written in analogy with (1) on the basis of the energy of the Higgs boson and the order parameter δ_μ

$$2\varepsilon'_{01} = 2[E^2_{H0} + (\delta_\mu)^2]^{1/2}; \quad 2\varepsilon'_{02} = 2[E^2_{H0} - (\delta_\mu)^2]^{1/2};$$
$$\delta_\mu = E_{H0} \sin\varphi_\mu/Q_{H6}; \quad \lambda_{\gamma b} = R_{H0}|\chi_{ef}|(\varepsilon_{\mu 4} + \varepsilon_{\mu 2})/(2E_{H0}). \tag{13}$$

We note that the order parameter δ_μ (describes the Bose-condensate) depends on the angle φ_μ, parameter Q_{H6}, as follows from (7) and (8). The numerical values are $\delta_\mu = 4.6536541 \text{ GeV}$, $2\varepsilon'_{01} = 250.2379072 \text{ GeV}$, $2\varepsilon'_{02} = 249.8914929 \text{ GeV}$, $\lambda_{\gamma b} = 17081.85081 \text{ nm}$. Based on the calculated wavelength $\lambda_{\gamma b}$ from the source for BNS, we find the characteristic parameters

$$\nu_{\gamma b} = \upsilon_0/\lambda_{\gamma b}; \Omega'_{c2} = \nu^2_{\gamma b}/(4|\lambda|^2 N); \upsilon^2_{\gamma b} = \upsilon^2_0/\Omega'_{c2}; \Omega_{01} = \upsilon^2_{01}/\upsilon^2_{\gamma b}. \tag{14}$$

Here $\nu_{\gamma b} = 3.974973236 \text{ GHz}$ is frequency, $\Omega'_{c2} = 0.224091707$ is density of cold dark matter, $\upsilon_{\gamma b} = 14.34353643 \times 10^6 \text{ cm s}^{-1}$ is the effective Fermi-velocity associated with neutron stars; $\Omega_{01} = 0.260441196$. Taking into account (3), relations $n'_{F\nu} = (n'_F)^2$, $n_{F\nu} = n_F(1 + n'_F)$ we find estimates of the densities of neutrinos $\Omega_{0\nu} = n'_{F\nu} = 0.0029398$, cold dark matter $\Omega_{c1} = \Omega'_{c2} + \Omega_{0\nu} = 0.2270315$ (close to the estimate of the density of cold dark matter 0.227, obtained by other authors [1]). The wavelength $\lambda_{\gamma h}$ associated with the source from black holes is determined by expressions

$$\lambda_{\gamma h} = \lambda_{\gamma b}/\eta_{bh}; \quad \eta_{bh} = (1 + \Omega_{01}/2)[1 + (n'_F)^2/2]. \tag{15}$$

Parameter is $\eta_{bh} = 1.13188191$ and wavelength is $\lambda_{\gamma h} = 15091.54856\,\text{nm}$. The calculated values $\lambda_{\gamma h}, \lambda_{\gamma b}$ are close to the values of wavelengths $15091.4, 17081.7$ nm for the sources from the BBH [2], BNS [3], respectively. From (1), (13) and (15) the relationships between Bose-condensates for black holes and neutron stars through energy E_{H0} follow. The value $z_Q = 0.008688605$ from (8) is close to the redshift of 0.009 of the source for BNS [3].

We find the spectrum for the occupation numbers n_{cx} (based on n_{ch}) and the total number of quanta n_{tot} after the BNS merger from the expressions

$$n_{cx} = n_{ch}S'_{0x} = M_{cx}/M_s; \quad n_{ch} = 1/(\psi_{ch} - 1); \quad \psi^*_{ch} = z'_Q\psi_{ch}; \quad \psi_{ch} = 1 + S'_{02}/\chi_{11};$$
$$n_{tot} = M_{tot}/M_s = 2n_{c4} - n_{1c} = 2n_{3c} + (2n_{2c} + n_{1c}); \quad n'_{tot} = 2n_{c4} - 2(n_{c1} - n_{c2}). \tag{16}$$

Numerical values are $n_{ch} = 5.332945778$; $\psi_{ch} = M_{ch}/M_s = 1.187513626$; $\psi^*_{ch} = M^*_{ch}/M_s = 1.197831463$; M_{cx} and M_{tot} are effective molar masses before and after merger. The calculated values ψ_{ch} and ψ^*_{ch} are close to 1.188 and 1.1977 for effective molar masses M_{ch} and M^*_{ch} within the GW detector from the BNS source [3]; $n_{tot} = 2.742837254$ and $n'_{tot} = 2.81920162$ are close to 2.74 and 2.82 (low-spin and high-spin approximation) [3].

For the characteristic frequency ν_{GW} of the gravitational wave, we have

$$\nu_{GW} = 4\nu_{\lambda 0} = N_{ra}\nu_{\gamma 0}; \quad \nu_{\lambda 0} = 1/(N'_0 - N'_{02})\tau_{s0}; \quad \tau_{s0} = 2(|S'_{01}| + S'_{02})/|\lambda_{\nu 0}|;$$
$$\tau_{s0} = \tau_{\lambda 2}/N'_{02} = \tau'_{\lambda 2}/N'_0; \quad \tau'_{\lambda 2} - \tau_{\lambda 2} = \tau_{\lambda 0} = 1/\nu_{\lambda 0}; \quad \nu_0 = 2z'_\mu\nu_{\lambda 0}/(2z'_\mu + 1). \tag{17}$$

The parameter $|\lambda_{\nu 0}| = 130.5593846\,\text{kHz}$ is related to parameter δ_μ from (13), describing the presence of a Bose-condensate for neutron stars. At $|\lambda_{\nu 0}| = 0$ from (17) follows, that the frequencies of the soft modes are $\nu_{\lambda 0} = 0$ and $\nu_{GW} = 0$. Further from (17), we find $\tau_{s0} = 1.127935494\,\mu\text{s}$, $\nu_{\lambda 0} = 152.9437161\,\text{Hz}$, $\nu_{GW} = 611.7748643\,\text{Hz}$, $\nu_0 = 142.9918607\,\text{Hz}$, $\tau_{\lambda 0} = 6.538352972\,\text{ms}$, $\tau_{\lambda 2} = 0.415136712\,\text{s}$, $\tau'_{\lambda 2} = 0.421675065\,\text{s}$.

The time of appearance of γ-radiation after the merger of neutron stars $\tau_{\gamma 0}$ is determined by the difference in the coalescence times τ'_{c0}, τ_{c0}

$$\tau_{\gamma 0} = \tau'_{c0} - \tau_{c0} = 1/\nu_{\gamma 0}; \quad \tau'_{c0} = \tau_{\gamma 0}n'_{A0}; \quad \tau_{c0} = 1/\nu_{c0} = \tau_{\gamma 0}n_{A0};$$
$$\tau_0 = 1/\nu_0 = \tau_{\lambda 0}(1 + 1/2z'_\mu); \quad \nu_{\gamma 0} = \nu_{c0}n_{A0} = \nu'_{c0}n'_{A0}; \quad \nu'_{c0} = 1/\tau'_{c0}. \tag{18}$$

Numerical values are $\tau_{\gamma 0} = 1.7020861\,\text{s}$, $\tau_{c0} = 98.800376\,\text{s}$, $\tau'_{c0} = 100.502462\,\text{s}$, $\nu_{\gamma 0} = 0.5875144\,\text{Hz}$, $\nu_{c0} = 0.0101214\,\text{Hz}$, $\nu'_{c0} = 0.00995\,\text{Hz}$. The delay time GW between the detectors LD and HD $\tau_0 = 6.993405\,\text{ms}$ from (18) is determined through the time $\tau_{\lambda 0}$ from (17) and the cosmological redshift z'_μ.

We write the spectra $\nu_{rx} = 2\nu_{ra}S'_{0x}$ and $\nu'_{zx} = 2\nu'_{z\mu}S'_{0x}$ on basis ν_{ra} and $\nu'_{z\mu}$

$$v_{ra} = \Omega_{ra} v_{\lambda 0} = \Omega_{ra} v_{GW}/4; \quad v'_{z\mu} = v_{ra} z'_{\mu}; \quad \Omega_{ra} = N_{ra} \Omega'_{c1}/z'_{A2};$$
$$v_{ra} = v_{r3} + v_{r1} = v_{r4} - v_{r2}; \quad v'_{z\mu} = v'_{z3} + v'_{z1} = v'_{z4} - v'_{z2};$$
$$v^*_{z\mu} = v'_{z3} + v'_{z2} = v'_{z4} - v'_{z1}. \tag{19}$$

Numerical values are $\Omega_{ra} = 0.229656$, $v_{ra} = 35.124438\,\text{Hz}$, $v'_{z\mu} = 252.340321\,\text{Hz}$, $v^*_{z\mu} = 249.589164\,\text{Hz}$. If $z'_{\mu} = 1$, then from (19) follows, that the spectrum $v'_{z\mu}$ goes to spectrum v_{ra}. When the black holes merger, the LD and HD detectors recorded the signals as a series of pulses, the frequency of which increased from 35 to 250 Hz. At the same time, the amplitude of the signals increased to the maximum value, and then dropped sharply to the noise level. Our calculated frequencies v_{ra}, $v'_{z\mu}$, $v^*_{z\mu}$ agree with the data on the detection of GW during of the BBH merger [2].

Next, we write the spectrum $v_{Wx} = v_{GW} S'_{0x}$ on the basis of frequency v_{GW}. The value of the frequency $v_{W1} = 24.190362\,\text{Hz}$ is consistent with the value of the frequency 24 Hz, that the LD detector starts detecting in the GW detection experiment during BNS merger [3].

Let h_{mL}, h_{mH} be maximum values of amplitudes of signals GW, registered by detectors LD, HD; $h_{\xi L}$, $h_{\xi H}$ are noise level after passing the GW. The GW signal on the HD detector appears later on a delay time τ_0, than on the LD detector. Taking into account (10) and (11), we write the relations for the amplitudes of the signals GW arising from the black holes merger

$$2h_{mH}/h_{mL} = (1 + a_m/a_0)[1 + (a_0 - a_m)]; \quad h_{mH} h_{\xi L} = h_{mL} h_{\xi H};$$
$$2h_{\xi H}/(2h_{mH} + h_{\xi H}) = 2h_{\xi L}/(2h_{mL} + h_{\xi L}) = |\xi_{0H}|^2. \tag{20}$$

On the basis of (20), we obtain the numerical values $h_{mL} = 0.9168936 \times 10^{-21}$, $h_{mH} = 1.146587 \times 10^{-21}$, $h_{\xi L} = 0.1833588 \times 10^{-21}$, $h_{\xi H} = 0.2292923 \times 10^{-21}$.

The maximum amplitude h'_{mL} of the GW signal, the noise level GW before and after the neutron stars merger $h^*_{\xi L}$ and $h'_{\xi L}$ (recorded by the LD detector) are determined by the formulas

$$h'_{mL} = 2a_0/n_{\Omega 2} = 2a_0 h_{mH}; \quad h^*_{\xi L} = 2a_m/n_{J0} = 2a_m h_{\xi H}; \quad 1/n_{J0} = h_{\xi H};$$
$$h'_{\xi L} = 2h'_{mL} |\xi_{0H}|^2/(2 - |\xi_{0H}|^2); \quad 1/n_{\Omega 2} = h_{mH} = |\xi_q|^2 E_G/(4|\lambda|^2 N_{ra} E_{H0} N'_{02}). \tag{21}$$

Values are $h'_{mL} = 7.4328715 \times 10^{-20}$, $h^*_{\xi L} = 1.4746969 \times 10^{-20}$, $h'_{\xi L} = 1.4864119 \times 10^{-20}$, $E_G = 12.11753067\,\mu\text{eV}$, $h'_{mL}/h^*_{\xi L} = 5.0402707$. The obtained estimates from (20) and (21) agree with the data from [2, 3].

4 Description of an Individual Layer with Variable Parameters

The hypothesis of the hierarchical structure of the Universe allows us to use models of fractal nanosystems (dislocations, quantum dots) to describe separate elements of large-scale fractal structures. When modelling the nonlinear effective displacements $u_\mu = F(\varphi_\mu; k_\mu)$ from (6), the main parameter is $b_0 = 1 - 2\text{sn}^2(u_\mu; k_\mu)$. Expressions for the four branches of effective displacements $u_{\mu i}$ ($i = 1, 2, 3, 4$) have the form [12]

$$2u_{\mu 1}(z, \alpha) = g_1 - g_2 + g_4; \quad 2u_{\mu 2}(z, \alpha) = g_1 - g_2 - g_4; \tag{22}$$

$$2u_{\mu 3}(z, \alpha) = -g_1 - g_2 + g_5; \quad 2u_{\mu 4}(z, \alpha) = -g_1 - g_2 - g_5. \tag{23}$$

The functions g_1, g_2, g_3, g_4, g_5 depend on the coordinate z, the fractal index α along the axis Oz, and are modelled by expressions

$$g_2(z, \alpha) = g_{20}|z - z_c|^{-\alpha}; \quad g_3(z, \alpha) = g_{30}|z - z_c|^{-2\alpha}; \tag{24}$$

$$g_{20}(\alpha) = 2^{-2\alpha} 3^{3\alpha - 1/2} \Gamma(\alpha + 1/3)\Gamma(\alpha + 2/3)/\sqrt{\pi}\,\Gamma(\alpha + 1/2); \tag{25}$$

$$g_{30}(\alpha) = 2 \cdot 3^{3\alpha - 1/2}\,\Gamma(\alpha + 1/3)\Gamma(\alpha + 2/3)/\pi; \quad g'_{30}(\alpha) = 1/g_{30}(\alpha); \tag{26}$$

$$g_4(z, \alpha) = \left[(g_1 + g_2)^2 - g_3\right]^{1/2}; \quad g_5(z, \alpha) = \left[(-g_1 + g_2)^2 - g_3\right]^{1/2}. \tag{27}$$

Here Γ is the gamma function; g_{20}, g_{30}, g'_{30} are nonlinear discontinuous functions of the fractal index α (Fig. 1). The functions g_2, g_3 from (24) also become nonlinear on the background of power dependencies.

The function $g_{20}(\alpha)$ has zeros at $\alpha = -(n_\alpha + 1/2)$, where $n_\alpha = 0, 1, 2 \ldots$. The function $g'_{30}(\alpha)$ has zeros at $\alpha = -(n_\alpha + 1/3)$, $\alpha = -(n_\alpha + 2/3)$. The nonlinear function g_1 depends on u_μ, α and indices n, m, j lattice nodes

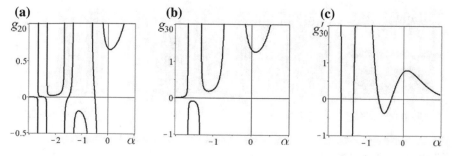

Fig. 1. Behavior of functions g_{20}, g_{30}, g'_{30} on α

$$g_1(u_\mu, \alpha; n, m, j) = k_\alpha^2(1 - 2\mathrm{sn}^2(u_\mu - u_0; k_\mu)); \quad k_\alpha^2 = (1 - \alpha)/Q; \quad (28)$$

$$Q = p_{0\alpha} - b_1(n - n_0)^2/n_c^2 - b_2(m - m_0)^2/m_c^2; \quad p_{0\alpha} = p_0 - b_3(j - j_0)^2/j_c^2. \quad (29)$$

Here u_0 is constant displacement; p_0, b_1, b_2, b_3, n_0, n_c, m_0, m_c, j_0, j_c are characteristic parameters. The function k_α depends on α, indices n, m, j lattice nodes $N_1 \times N_2 \times N_3$. In our model, this function determines the behavior of the module $k_\mu = \mathrm{sn}(u_{0\alpha}, k_\alpha)$, where $u_{0\alpha} = \mathrm{F}(\varphi_{0\alpha}, k_\alpha)$; $\varphi_{0\alpha}$ is the polar angle. As a result k_μ (implicitly depends on n, m, j) and four branches $u_{\mu i}$ from (22) and (23) become random functions. In numerical simulation, for forward $z = z_1$ and backward $z = z_2$ waves it was assumed that $z_1 = 0.053 + h_z(j_z + 33)$; $z_2 = 6.653 - h_z(j_z + 33)$, $h_z = 0.1$; $j_z = 5$; $n = \overline{1, 30}$; $m = \overline{1, 40}$; $u_0 = 29.537$; $u_{0\alpha} = \pi/5.2$; $p_0 = 1.0123$. The solution of Eqs. (22) and (23) for branches $u = u_{\mu i}$ is carried out by the iteration method on the variable m.

Consider the state of a layer without quantum dot ($b_1 = b_2 = b_3 = 0$, $Q = p_0$). The behavior branches of the displacement function of the backward wave on α is given in Fig. 2.

For the backward wave (Fig. 2a), in addition to the regular behavior of the 1, 2, 3 branches, the stochastic behavior of the 4 branch is observed, the 3 branch is characterized by the presence of the second harmonic. At $\alpha = -1.5$ (Fig. 2b) oscillations with increased amplitudes are observed for branches 1 and 4, and damped oscillations with amplitudes of order for branches 3 and 2 are observed. At $\alpha = -2.5$ (Fig. 2c), singularities appear in comparison with Fig. 2b: for branch 1 is change in shape, amplitude of oscillations; for branch 4 is doubling of the period of oscillations; for branches 3 and 2 are damped oscillations with reduced amplitudes of order $\pm 2 \times 10^{-10}$. At $\alpha = -3.5$ (Fig. 2d) branches 3 and 2 demonstrate nonlinear oscillations: on separate peaks of 3 branches there are features such as a local minimum between two humps; branches 1, 4 take zero values. With a further change α, the character of the behavior of all four branches practically does not change (similar to the behavior of Fig. 2b with the order of the branches 3, 1, 4, 2), which indicates the presence of a critical value $\alpha \approx \alpha_c = -4.5$. This is due to the nonlinear behavior of discontinuous

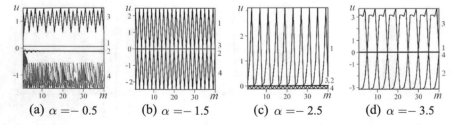

Fig. 2. Dependencies of the projections u on the plane mOu for the backward wave on m for various α: 1—$u_{\mu 1}$, 2—$u_{\mu 2}$, 3—$u_{\mu 3}$, 4—$u_{\mu 4}$

functions g_{20}, g_{30} from (25) and (26) on α. With a change α the behavior of the forward wave branches qualitatively coincides with that for the backward wave.

Next, we consider the state of the layer with a quantum dot (Fig. 3). Main parameters are $b_1 = b_2 = b_3 = 1$; $n_0 = 14.3267$; $n_c = 9.4793$; $m_0 = 19.1471$; $m_c = 14.7295$; $j_0 = 31.5279$; $j_c = 11.8247$. From (29) at $p_{0\alpha} = -3.457 \times 10^{-11}$ we find the averaged values for the layer number j: $j_1 = 19.63070035$, $j_2 = 43.42509965$. At iterations it was accepted $j = j_1$. In this case, the module k_μ implicitly depends on n, m, and becomes a random function.

The behavior of the displacement functions of all four branches for the forward (Fig. 3a–c) and the backward (Fig. 3d–f) waves is different. The 1 branch of the forward (Fig. 3a) and the backward (Fig. 3d) have peaks with large amplitudes, that confirms the state of the layer with the quantum dot. The behavior of the core of such quantum dot differs from the behavior of the cores of quantum dots from [14]. Instead of regular wave behavior [14], the cores boundaries become stochastic. Inside the cores, there are features of the type of islands, jumpers, narrowings, wells. For the 2 branch of the forward wave, the core has a convex form (Fig. 3b), and for the backward wave the core approaches to a flat form (Fig. 3e). For the 3 branch of the forward wave, the core has a concave form of the well type (Fig. 3c), and for the backward wave (Fig. 3f) the core has a flat bottom. At the boundaries of these cores there are peaks with small amplitudes (features such as additional wells, saddles,

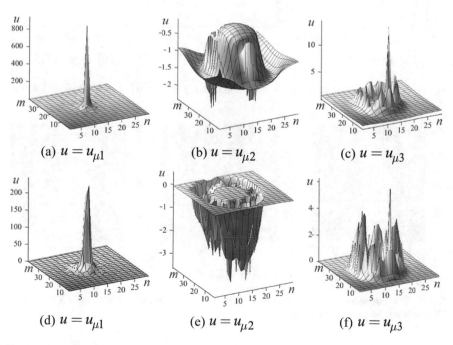

(a) $u = u_{\mu 1}$ (b) $u = u_{\mu 2}$ (c) $u = u_{\mu 3}$

(d) $u = u_{\mu 1}$ (e) $u = u_{\mu 2}$ (f) $u = u_{\mu 3}$

Fig. 3. Dependencies u on n, m at $\alpha = -0.5$ for different branches of the forward (**a, b, c**) and the backward (**d, e, f**) waves

valleys are formed). These features of the deformation field behavior indicate the appearance of an effective multi-well potential in a layer with a quantum dot.

A further change α leads to a change in the stochastic behavior of the core boundaries (Fig. 4). For the 2 branch of the backward wave, at $\alpha = -2.5$ (Fig. 4a), a stochastic peak down is observed on a stochastic background with a practically constant amplitude. At $\alpha = -4.5$ the stochastic peak disappears, and the stochastic background remains (Fig. 4b). At $\alpha = -8.5$ (Fig. 4c), in the stochastic background the formation of a failure near $m = m_0$ is observed.

When the values of the semi-axes n_c, m_c (Fig. 5) of quantum dot are changing, the behavior of the deformation field of the core and its boundary changes. For the first branch $u_{\mu 1}$ the effect of pronounced anisotropy is observed. A periodic fine structure appears on the boundaries for the forward wave (Fig. 5a), and for the backward wave there is a structure with fine wells (Fig. 5c).

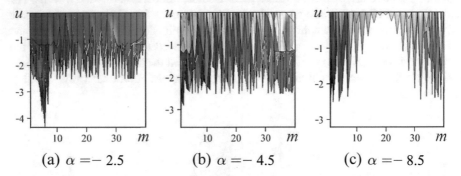

(a) $\alpha = -2.5$ (b) $\alpha = -4.5$ (c) $\alpha = -8.5$

Fig. 4. Dependencies of the projections $u = u_{\mu 2}$ on the plane mOu for backward wave on m for different α

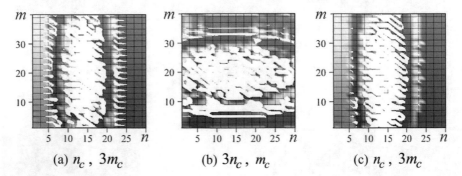

(a) n_c, $3m_c$ (b) $3n_c$, m_c (c) n_c, $3m_c$

Fig. 5. Cross-sections $u_{\mu 1} \in [0; 1.5]$ (top view) for forward (**a, b**) and backward (**c**) waves for different semi-axes, n_c, m_c at $\alpha = -0.5$

5 Conclusions

An anisotropic model is proposed for describing the main parameters of the BBH, BNS, the nature of the spin source of which has a tensor character. Taking into account the Higgs field, estimates are made of the energies of the Higgs boson, relict photons, and the temperature deviation of the relict background. It is shown that the nature of the supervoid or the "Cold relict spot" is associated with the presence of a black hole and its influence on relict photons. To describe the transition signals (gravitational waves, relict radiation), it is proposed to use the superradiance model R. H. Dicke and the quantum statistical theory of superradiance.

Based on the hypothesis of the hierarchical structure of the Universe, modelling of the deformation field of separate structures it is proposed to use of quantum model systems with variable parameters. It is shown that for a layer without a quantum dot, the presence of nonlinear oscillations is characteristic, which depend on the fractal index α. The structure of the quantum dots cores in the layer has a convex, concave, and flat form with a stochastic boundary. The formation of stochastic peak and the appearance of failure on stochastic background is possible. The change in the semi-axes of the quantum dot leads to the anisotropy effect.

References

1. M. Punturo, Opening a new window on the Universe: the future gravitational wave detectors. Europhys. News **44**(2), 17–20 (2013)
2. B.P. Abbott et al., Observation of gravitational waves from a binary black hole merger. Phys. Rev. Lett. **116**(6), 061102 (2016)
3. B.P. Abbott et al., Observation of gravitational waves from a binary neutron star inspiral. Phys. Rev. Lett. **119**(16), 161101 (2017)
4. B. Novosyadlyy, Voids—"deserts" of the Universe. Universe Space Time **6**(143), 4–11 (2016)
5. R. Mackenzie et al., Evidence against a supervoid causing the CMB Cold Spot. arXiv:1704. 03814v1 (astro-ph.CO), p 12. (Apr 2017)
6. S. Hawking, *Black Holes and Baby Universes* (Transworld Publishers, London, 1994)
7. V. Abramov, Higgs field and cosmological parameters in the fractal quantum system, in *XI International Symposium on Photon Echo and Coherent Spectroscopy (PECS-2017)*. EPJ Web Conf. 161, 02001, (2017). p 2
8. C.H. Skiadas, C. Skiadas, *Chaotic Modeling and Simulation: Analysis of Chaotic Models, Attractors and Forms* (Taylor and Francis/CRC, London, 2009)
9. O.P. Abramova, A.V. Abramov, Attractors and deformation field in the coupled fractal multi-layer nanosystem. CMSIM J. **2**, 169–179 (2017)
10. O.P. Abramova, Mutual influence of attractors and separate stochastic processes in a coupled fractal structures. Bull. Donetsk Nat. Univer. A. **1**, 50–60 (2017)
11. V.S. Abramov, Model of nonlinear fractal oscillator in nanosystem, in *Applied Non-Linear Dynamical Systems*, ed. by J. Awrejcewicz (Springer Proceedings in Mathematics & Statistics, Berlin, 2014), 93, pp. 337–350
12. V.S. Abramov, Transient processes in a model multilayer nanosystem with nonlinear fractal oscillator. CMSIM J. **1**, 3–15 (2015)
13. V.S. Abramov, Cosmological parameters and Higgs boson in a fractal quantum system. CMSIM J. **4**, 441–455 (2017)

14. V.S. Abramov, Relations of cosmological parameters and the Higgs boson in a fractal model of the Universe. Bull. Donetsk Nat. Univer. A **1**, 36–49 (2017)

15. S. Carroll, *The Particle at the End of the Universe* (Dutton, New York, 2012)

16. R.H. Dicke, Coherent in spontaneous radiation processes. Phys. Rev. **93**(1), 99–110 (1954)

17. R. Bonifacio, P. Schwendimann, F. Haake, Quantum statistical theory of superradiance. I. Phys. Rev. A. **4**(1), 302–313 (1971)

18. R. Bonifacio, P. Schwendimann, F. Haake, Quantum statistical theory of superradiance. II. Phys. Rev. A **4**(3), 854–864 (1971)

Effect of Ordering of Displacement Fields Operators of Separate Quantum Dots, Elliptical Cylinders on the Deformation Field of Coupled Fractal Structures

Olga P. Abramova and Andrii V. Abramov

Abstract By the numerical modelling method the behavior of the deformation field of the coupled fractal structure with quantum dots and elliptical cylinders was investigated. It is shown that the resulting deformation field of the structure with a number of quantum dots larger than two essentially depends on the ordering of the displacement fields operators for separate quantum dots. The coupled fractal structure with elliptical cylinders is characterized by the presence of a complex deformation field. Using zero operators for pairs of quantum dots makes it possible to obtain information about separate quantum dots. Transposition of pairs of operators allows us to transfer part of the information from one quantum dot to another.

Keywords Coupled fractal structures · Quantum dots · Elliptical cylinders · Deformation field · Ordering of operators · Numerical modelling

1 Introduction

Quantum processing of information, quantum computation requires the use of the laws of quantum mechanics [1]. In this case, there is a need for theoretical and experimental modelling of the phenomena of quantum entanglement, quantum teleportation, decoherence of quantum states of active objects [2]. Classical computers encode information in bits that can be in one of two states, 0 or 1. Quantum computers encode information in qubits, which can be in superposition states [3–5]. In this case, the phenomenon of entanglement of states arises [3–6]. The physical systems that realise qubits can be any objects having two quantum states: polarisation states of photons; electronic states of atoms or ions, separate atoms, spin states of atomic nuclei in traps; quantum dots. Separate objects placed in traps can exhibit stochastic properties, which complicates the practical realisation of qubits for quantum computers.

O. P. Abramova (✉) · A. V. Abramov
Donetsk National University, Donetsk, Ukraine
e-mail: oabramova@ua.fm

© Springer Nature Switzerland AG 2019
C. H. Skiadas and I. Lubashevsky (eds.), *11th Chaotic Modeling and Simulation International Conference*, Springer Proceedings in Complexity, https://doi.org/10.1007/978-3-030-15297-0_2

Also, nanostructures, metamaterials [7], various fractal structures can be chosen as active objects [8–11]. When creating quantum computers (which take into account the phenomena of quantum superposition and entanglement for the transmission, processing, storage of information data), it becomes necessary to theoretically and experimentally research the various physical properties of such nanostructures and metamaterials [7]. In [8, 9], the peculiarities of the stochastic state of the deformation field of coupled fractal multilayer nanosystems, nanotraps, and quantum dots taking into account the variation of the variable parameters were studied. The mutual influence of separate structures, attractors, stochastic processes on each other in the coupled fractal structure have been investigated in articles [10, 11]. From the point of view of the experimental realisation of a quantum computer, it is actual to study various fractal structures with quantum dots.

The aim of this paper is to investigate the effect of the ordering of displacement fields operators of separate fractal quantum dots and fractal elliptical cylinders onto the deformation field of a coupled fractal structure.

2 Ordering of the Displacement Fields Operators of Separate Quantum Dots

In [9] eight model nonlinear equations for the dimensionless displacement function u to describe the stochastic state of the deformation field of a multilayer nanosystem were obtained. The appearance of eight equations is due to the presence of different branches for variable modules $\pm k_u$, $\pm k'_u$ of an elliptic sine functions from an implicitly defined equation $k_u^2 + (k'_u)^2 = 1$. Variable modules k_u are functions of indices n, m, j nodes of the bulk discrete lattice with dimensions $N_1 \times N_2 \times N_3$.

As the initial equation for the dimensionless displacement function u, we consider the second branch from [9], whose equation has the form

$$u = u_2 = k_u^2\left(1 - 2sn^2(u_2 - u_0, k'_u)\right); \quad k_u^2 = (1 - \alpha)/Q, \tag{1}$$

where u_0 is the constant (critical) displacement; α is the fractal dimension of the deformation field u along the axis Oz ($\alpha \in [0, 1]$); Q determines the form of the fractal structure, the type of attractors and takes into account the interaction of the nodes of both in the main plane of the discrete rectangular lattice $N_1 \times N_2$ as well as interplane interactions. Consider a coupled fractal structure consisting of four ($i = 1, 2, 3, 4$) separate fractal quantum dots (FQD). Taking into account (1), the nonlinear equations for the dimensionless displacement function $u = u_2$ of such coupled fractal structure have the form

$$u = u_2 = \sum_{i=1}^{4} u_{Ri}; \quad u_{Ri} = R_i k_{ui}^2\left(1 - 2sn^2(u_2 - u_{0i}, k'_{ui})\right); \tag{2}$$

$$k_{ui}^2 = (1 - \alpha_i)/Q_i; \quad k_{ui}' = (1 - k_{ui}^2)^{1/2}; \tag{3}$$

$$Q_i = p_{0i} - b_{1i}(n - n_{0i})^2/n_{ci}^2 - b_{2i}(m - m_{0i})^2/m_{ci}^2 - b_{3i}(j - j_{0i})^2/j_{ci}^2. \tag{4}$$

Here parameters $p_{0i}, b_{1i}, b_{2i}, b_{3i}, n_{0i}, n_{ci}, m_{0i}, m_{ci}, j_{0i}, j_{ci}$ characterise different fractal structures; R_i determine the orientation of the deformation fields of separate structures in the coupled system. In general case these parameters may depend on the layer index j and the dimensionless time t. Nonlinear Eqs. (2)–(4) can be solved by iteration method on any of indices n, m, j. If one of these indices is considered fixed, then the result of the iteration will be the displacement function, which is a stochastic surface, depending on the other two indices. The explicit form of the resulting stochastic surface essentially depends on the order of the separate terms in the sum (2) for the deformation field u_2, which can be interpreted as separate operators describing the total stochastic process. The appearance of stochasticity is associated with the presence of variable modules k_{ui}, k_{ui}' in (2) and (3).

In this work the iterative procedure on index n simulates a stochastic process on a rectangular lattice with sizes $N_1 \times N_2$: $N_1 = 240$, $N_2 = 180$. The equations of surfaces of the considered structures do not depend on index j, thus the parameters $b_{3i} = 0$. The main parameters are chosen to be the same for all four FQD: $\alpha_i = 0.5$, $u_{0i} = 29.537$, $p_{0i} = -3.457 \times 10^{-11}$, $b_{1i} = b_{2i} = 1$, $n_{ci} = 24.4793$, $m_{ci} = 15.7295$. The parameters R_i characterise the mutual orientation of the FQD. The parameters n_{0i}, m_{0i} determine the positions of the centres of gravity FQD. Quantum dots with opposite orientation of deformation fields FQD1 ($i = 1$, $R_1 = 1$) and FQD4 ($i = 4$, $R_4 = -1$) are located at the same point with coordinates $n_{0i} = 111.1471$, $m_{0i} = 79.3267$. Quantum dots with opposite orientation of deformation fields FQD2 ($i = 2$, $R_2 = -1$) and FQD3 ($i = 3$, $R_3 = 1$) are also located at the same point, but with different coordinates $n_{0i} = 131.1471$, $m_{0i} = 99.3267$. When modelling the behavior of the total deformation field of a coupled fractal structure with quantum dots, we first consider four variants with pairs of two different quantum dots with shifted centers of gravity: (FQD1, FQD3), (FQD1, FQD2), (FQD4, FQD2) and (FQD4, FQD3). For these pairs Fig. 1 gives dependence of the total deformation field of the coupled structure on the indices (n, m) of the lattice nodes. In this case, in expressions (2) pair (FQD1, FQD3) corresponds to the ordered operator $u_2 = u_{R1} + u_{R3}$, which is a superposition of separate operators u_{R1}, u_{R3}; pair (FQD1, FQD2) corresponds to the operator $u_2 = u_{R1} + u_{R2}$; pair (FQD4, FQD2) corresponds to the operator $u_2 = u_{R4} + u_{R2}$; pair (FQD4, FQD3) corresponds to the operator $u_2 = u_{R4} + u_{R3}$. For the pairs (FQD1, FQD3) and (FQD4, FQD2) peaks down (Fig. 1a, b) and peaks up (Fig. 1g, h) are localized near FQD1, FQD3 and FQD4, FQD2 with the same orientation. In the cross-sections (Fig. 1c, i), two fractal holes are observed which are localised near the core of these quantum dots. These holes are characterised by the stochastic behavior of the deformation field. A wave behavior with a regular structure of elliptic type is observed far from the region of cores localisation. As you approach the core, this regular structure changes: features such as inflection points, breaks, narrowings, additional fine structure appear.

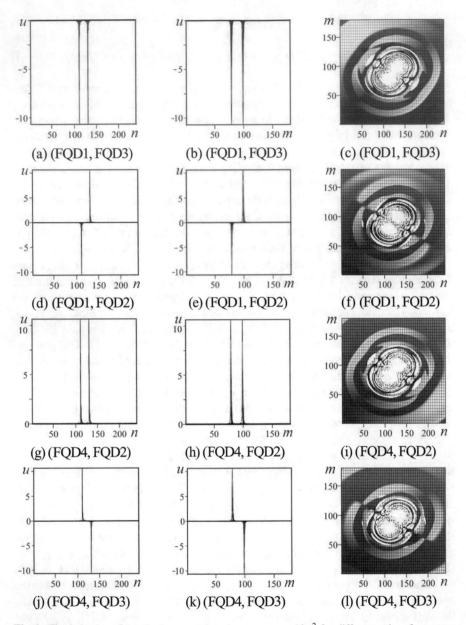

Fig. 1 The behavior of the displacement function $u = u_2 \times 10^{-2}$ for different pairs of quantum dots with shifted centres of gravity: **a, d, g, j** are the projections onto the plane nOu; **b, e, h, k** are the projections onto the plane mOu; **c, f, i, l** are cross-sections $u_2 \in [-0.1; 0.1]$ (top view)

For pairs of quantum dots with the opposite orientation (FQD1, FQD2) and (FQD4, FQD3), peaks are observed (Fig. 1d, e and Fig. 1j, k) localized near FQD1, FQD2 and FQD4, FQD3. Two fractal holes localised near cores of these quantum dots are also observed in the cross-sections (Fig. 1f and Fig. 1l). However, near cores there is a characteristic effect of the shift of wave behavior (in comparison with Fig. 1c, i), there are features such as wells, steps, saddles.

Transpositions of displacement field operators for separate quantum dots with shifted centers of gravity in pairs, for example, (FQD1, FQD3), (FQD3, FQD1) lead to deviation $\delta u_2 = (u_{R1} + u_{R3}) - (u_{R3} + u_{R1})$ of the deformation field of the coupled fractal structure within 10^{-10}.

For pairs of quantum dots with the same centres of gravity (FQD1, FQD4), (FQD2, FQD3) displacement fields u_2 are identically equal to zero, that follows from the results of modeling iterative processes.

Thus, the pair (FQD1, FQD4) corresponds to an ordered zero operator $u_2 = 0 = u_{R1} + u_{R4}$; the pair (FQD2, FQD3) corresponds to an ordered zero operator $u_2 = 0 = u_{R2} + u_{R3}$.

The possibility of using such zero operators based on pairs (FQD2, FQD3) and (FQD1, FQD4) allows us to restore information about separate FQD1, FQD4, FQD2, FQD3 on the basis of information from structures of three quantum dots (FQD2, FQD3, FQD1), (FQD2, FQD3, FQD4), (FQD1, FQD4, FQD2), (FQD1, FQD4, FQD3), respectively (Fig. 2).

In this case, in the expressions (2) the structure (FQD2, FQD3, FQD1) corresponds to the ordered operator $u_2 = u_{R2} + u_{R3} + u_{R1} = u_{R1}$; the structure (FQD2, FQD3, FQD4) corresponds to the ordered operator $u_2 = u_{R2} + u_{R3} + u_{R4} = u_{R4}$; the structure (FQD1, FQD4, FQD2) corresponds to the ordered operator $u_2 = u_{R1} + u_{R4} + u_{R2} = u_{R2}$; the structure (FQD1, FQD4, FQD3) corresponds to the ordered operator $u_2 = u_{R1} + u_{R4} + u_{R3} = u_{R3}$.

The position of the peaks and amplitudes (Fig. 2a, b), (Fig. 2d, e), (Fig. 2g, h), (Fig. 2j, k) correspond to the position of the peaks and amplitudes of separate FQD1, FQD4, FQD2, FQD3. The structure of cores (Fig. 2c), (Fig. 2f), (Fig. 2i), (Fig. 2l) completely coincides with the structure of the cores of separate FQD1, FQD4, FQD2, FQD3.

Next, consider the coupled fractal structure, which consists of four separate FQD1, FQD2, FQD3, FQD4 (Fig. 3). Thus, for example, the sequence (FQD1, FQD2, FQD3, FQD4) (Fig. 3a–d) corresponds to the ordered operator $u_2 = u_{R1} + u_{R2} + u_{R3} + u_{R4}$; the sequence (FQD1, FQD3, FQD2, FQD4) (Fig. 3e) corresponds to the ordered operator $u_2 = u_{R1} + u_{R3} + u_{R2} + u_{R4}$; the sequence (FQD3, FQD4, FQD1, FQD2) (Fig. 3f) corresponds to the ordered operator $u_2 = u_{R3} + u_{R4} + u_{R1} + u_{R2}$. For the sequence (FQD1, FQD2, FQD3, FQD4) near FQD3, a peak down an essentially small amplitude (Fig. 3a, b) is observed in comparison with the amplitude of the itself FQD3. The behavior of the deformation field near the core (Fig. 3c) of the coupled structure differs significantly from the behavior of the itself FQD3 core (Fig. 2l): stochastic wave behavior appear (instead of regular rings appear stochastic rings, which are elliptic fractal dislocations, anisotropy effect is observed).

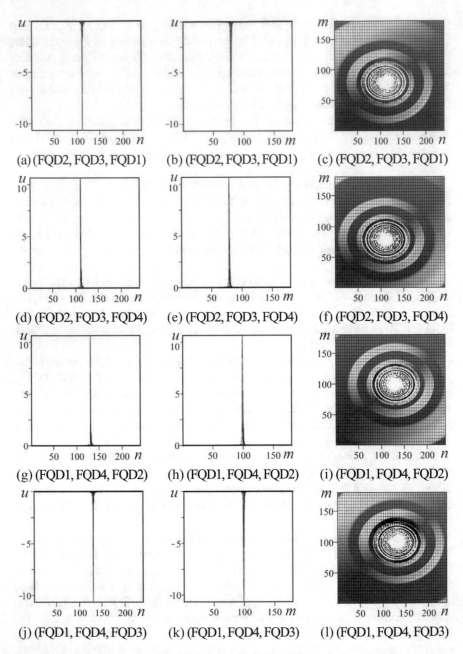

(a) (FQD2, FQD3, FQD1)   (b) (FQD2, FQD3, FQD1)   (c) (FQD2, FQD3, FQD1)

(d) (FQD2, FQD3, FQD4)   (e) (FQD2, FQD3, FQD4)   (f) (FQD2, FQD3, FQD4)

(g) (FQD1, FQD4, FQD2)   (h) (FQD1, FQD4, FQD2)   (i) (FQD1, FQD4, FQD2)

(j) (FQD1, FQD4, FQD3)   (k) (FQD1, FQD4, FQD3)   (l) (FQD1, FQD4, FQD3)

Fig. 2 The behavior of the displacement function $u = u_2 \times 10^{-2}$ for different structures from three quantum dots: **a, d, g, j** are the projections onto the plane nOu; **b, e, h, k** are the projections onto the plane mOu; **c, f, i, l** are cross-sections $u_2 \in [-0.1; 0.1]$ (top view)

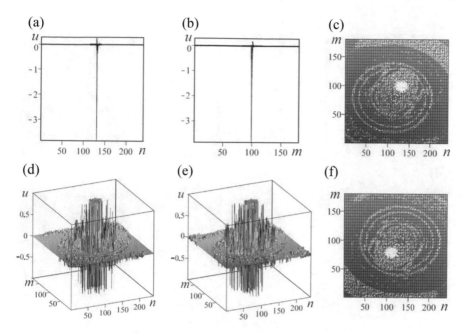

Fig. 3 The behavior of the displacement function for sequences (FQD1, FQD2, FQD3, FQD4) (**a–d**); (FQD1, FQD3, FQD2, FQD4) (**e**); (FQD3, FQD4, FQD1, FQD2) (**f**): **a, b** are projections $u = u_2 \times 10^7$ onto the planes nOu, mOu; cross-sections (**d, e**) $u = u_2 \times 10^{10} \in [-1; 1]$; (**c, f**) $u = u_2 \in [-10^{-10}; 10^{-10}]$ (top view)

Transposition of the order of the separate internal operators of the iterative process (2) u_{R2} and u_{R3} in the sequence (FQD1, FQD2, FQD3, FQD4) leads to the sequence (FQD1, FQD3, FQD2, FQD4) (Fig. 3e). In this case, all the main features of the regular behavior of the deformation field of the coupled structure are preserved, but stochastic behavior near the core changes (Fig. 3e) in comparison with Fig. 3d.

Transposition of the order of pairs of operators of the iterative process (2) $(u_{R1} + u_{R2})$ and $(u_{R3} + u_{R4})$ in the sequence (FQD1, FQD2, FQD3, FQD4) leads to the sequence (FQD3, FQD4, FQD1, FQD2) (Fig. 3f). In this case, the peak down is observed near FQD1, the anisotropy direction is reversed (Fig. 3f) in comparison with Fig. 3c. The operator pair transposition operation allows us to transfer information (for example, about the direction of the peak down) from FQD3 to FQD1.

For another coupled fractal structure consisting of four separate FQD4, FQD3, FQD2, FQD1, the behavior of the deformation field is given in Fig. 4. In this case, the sequence (FQD4, FQD3, FQD2, FQD1) (Fig. 4a–d) corresponds to the ordered operator $u_2 = u_{R4} + u_{R3} + u_{R2} + u_{R1}$; the sequences (FQD4, FQD2, FQD3, FQD1) (Fig. 4e) corresponds to the ordered operator $u_2 = u_{R4} + u_{R2} + u_{R3} + u_{R1}$; the sequence (FQD2, FQD1, FQD4, FQD3) (Fig. 4f) corresponds to the ordered operator $u_2 = u_{R2} + u_{R1} + u_{R4} + u_{R3}$. For the sequence (FQD4, FQD3, FQD2, FQD1) near FQD2, a peak of an essentially small amplitude (Fig. 4a, b) is observed in comparison

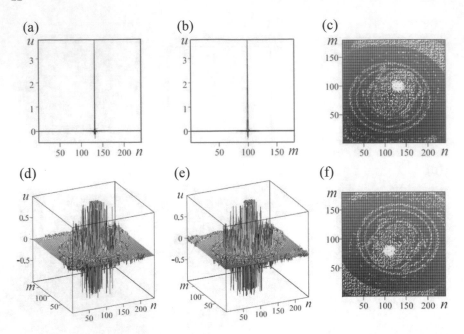

Fig. 4 The behavior of the displacement function for sequences (FQD4, FQD3, FQD2, FQD1) (**a–d**); (FQD4, FQD2, FQD3, FQD1) (**e**); (FQD2, FQD1, FQD4, FQD3) (**f**): **a, b** are projections $u = u_2 \times 10^7$ onto the planes nOu, mOu; cross-sections (**d, e**) $u = u_2 \times 10^{10} \in [-1; 1]$; (**c, f**) $u = u_2 \in \left[-10^{-10}; 10^{-10}\right]$ (top view)

with the amplitude of the itself FQD2. The behavior of the deformation field near the core (Fig. 4c) of the coupled structure differs significantly from the behavior of the itself FQD2 core (Fig. 2i): stochastic wave behavior appear (instead of regular rings appear stochastic rings, which are elliptic fractal dislocations, anisotropy effect is observed). Transposition of the order of the separate internal operators of the iterative process (2) u_{R3} and u_{R2} in the sequence (FQD4, FQD3, FQD2, FQD1) leads to the sequence (FQD4, FQD2, FQD3, FQD1) (Fig. 4e). In this case, all the main features of the regular behavior of the deformation field of the coupled structure are preserved, but stochastic behavior near the core changes (Fig. 4e) in comparison with Fig. 4d.

Transposition of the order of pairs of operators of the iterative process (2) $(u_{R4} + u_{R3})$ and $(u_{R2} + u_{R1})$ in the sequence (FQD4, FQD3, FQD2, FQD1) leads to the sequence (FQD2, FQD1, FQD4, FQD3) (Fig. 4f). In this case, the peak up is observed near FQD4, the anisotropy direction is reversed (Fig. 4f) in comparison with Fig. 4c. The operator pair transposition operation allows us to transfer information (for example, about the direction of the peak up) from FQD2 to FQD4. The mutual transposition of pairs with zero operators $(u_{R1} + u_{R4})$, $(u_{R2} + u_{R3})$, for example, in sequences (FQD1, FQD4, FQD2, FQD3), (FQD2, FQD3, FQD1, FQD4) from four quantum dots also leads to zero operators

$$u_2 = 0 = (u_{R1} + u_{R4}) + (u_{R2} + u_{R3}) = (u_{R2} + u_{R3}) + (u_{R1} + u_{R4}).$$

The deformation field for such coupled fractal structure is identically zero.

3 Coupled Fractal Structure: Quantum Dots and Elliptical Cylinders

In order to further study the effect of the ordering of displacement fields operators, we consider a coupled fractal structure consisting of two fractal quantum dots FQD1, FQD2 ($i = 1, 2$) and two fractal elliptic cylinders FEC3, FEC4 ($i = 3, 4$). Quantum dots have imaginary attractors, and elliptic cylinders have real attractors [4, 5]. The nonlinear equations for the dimensionless displacement function of such a coupled structure $u = u_2$ have the form similar to Eqs. (2)–(4), however, the parameters of the structures are different

$$u = u_2 = \sum_{i=1}^{4} u_{Ri}; \quad u_{Ri} = R_i k_{ui}^2 \big(1 - 2sn^2(u_2 - u_{0i}, k'_{ui})\big); \tag{5}$$

$$k_{ui}^2 = (1 - \alpha_i)/Q_i; \quad k'_{ui} = (1 - k_{ui}^2)^{1/2}; \tag{6}$$

$$Q_i = p_{0i} - b_{1i}(n - n_{0i})^2/n_{ci}^2 - b_{2i}(m - m_{0i})^2/m_{ci}^2 - b_{3i}(j - j_{0i})^2/j_{ci}^2. \tag{7}$$

FQD1 ($R_1 = -1$) and FQD2 ($R_2 = 1$) with opposite orientation of deformation fields have the same basic parameters: $\alpha_i = 0.5$, $p_{0i} = -3.457 \times 10^{-11}$, $u_{0i} = 29.537$, $b_{1i} = b_{2i} = 1$, $n_{ci} = 24.4793$, $m_{ci} = 15.7295$, $n_{0i} = 111.1471$, $m_{0i} = 79.3267$. FEC3 ($R_3 = 1$) and FEC4 ($R_4 = -1$) with opposite orientation of deformation fields have the same basic parameters: $\alpha_i = 0.5$, $u_{0i} = 29.537$, $p_{0i} = 1.0423$, $b_{1i} = b_{2i} = 1$, $b_{3i} = 0$, $n_{ci} = 57.4327$, $m_{ci} = 35.2153$, $n_{0i} = 121.1471$, $m_{0i} = 89.3267$. By analogy with Paragraph 2, the displacement fields for separate pairs of quantum dots (FQD1, FQD2) and (FQD2, FQD1) are identically equal to zero, that follows from the results of modelling of iterative processes. Thus, the pair (FQD1, FQD2) corresponds to an ordered zero operator $u_2 = 0 = u_{R1} + u_{R2}$; the pair (FQD2, FQD1) corresponds to an ordered zero operator $u_2 = 0 = u_{R2} + u_{R1}$. The behavior of the deformation field of an separate elliptical cylinder FEC3 ($R_3 = 1$), which is described by the operator u_{R3}, is given in Fig. 5. In this case FQD1, FQD2, FEC4 are absent, $R_1 = R_2 = R_4 = 0$. Unlike the displacement fields of FQD (Figs. 1, 2, 3 and 4), the deformation field of the FEC3 is complex. For $\text{Re}u_2 = \text{Re}u_{R3}$ the stochastic behavior of the core boundary is characteristic (Fig. 5a–c); the wave behavior inside and outside the core is observed (Fig. 5c), that is due to the presence of a variable module k_{u3} in the nonlinear Eqs. (5)–(7). The

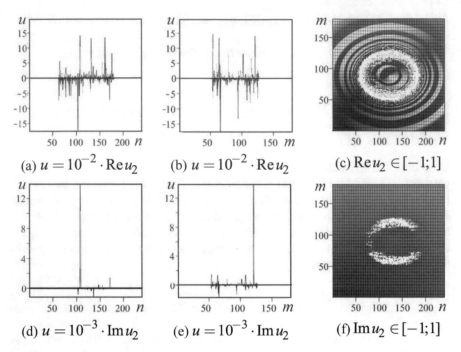

(a) $u = 10^{-2} \cdot \mathrm{Re}\, u_2$ (b) $u = 10^{-2} \cdot \mathrm{Re}\, u_2$ (c) $\mathrm{Re}\, u_2 \in [-1; 1]$

(d) $u = 10^{-3} \cdot \mathrm{Im}\, u_2$ (e) $u = 10^{-3} \cdot \mathrm{Im}\, u_2$ (f) $\mathrm{Im}\, u_2 \in [-1; 1]$

Fig. 5 The behavior of the displacement function of the FEC3: **a, b, d, e** of the projection on the planes nOu, mOu; **c, f** cross-sections (top view)

amplitudes of separate peaks on the dependences $\mathrm{Im}\, u_2 = \mathrm{Im}\, u_{R3}$ on n, m (Fig. 5d, e) are an order larger than the peaks amplitudes $\mathrm{Re}\, u_2 = \mathrm{Re}\, u_{R3}$ (Fig. 5a, b). The effective damping ($\mathrm{Im}\, u_2 = \mathrm{Im}\, u_{R3}$) is localized in the region of the stochastic core boundary (Fig. 5f).

The displacement fields for separate pairs of elliptical cylinders (FEC3, FEC4) and (FEC4, FEC3) are identically zero, which also follows from the modelling results of the iterative processes. Thus, a complex ordered zero operator $u_2 = 0 = u_{R3} + u_{R4}$ corresponds to the pair (FEC3, FEC4), and the complex zero operator $u_2 = 0 = u_{R4} + u_{R3}$ corresponds to the pair (FEC4, FEC3). In this case, for these zero operators the following conditions $\mathrm{Re}\, u_2 = 0$ and $\mathrm{Im}\, u_2 = 0$ are carried out. The structure of these complex zero operators differs from the structure of the zero operators for quantum dots. The possibility of using such zero operators based on pairs (FEC3, FEC4) and (FEC4, FEC3) allows us to restore information about separate FEC on the basis of information from structures of three FEC, for example (FEC3, FEC4, FEC3). In this case, information is extracted both on $\mathrm{Re}\, u_2$ and about $\mathrm{Im}\, u_2$ the FEC3, which is exactly coincident with Fig. 5. In this case, for the structure (FEC3, FEC4, FEC3), the ordered operator has the form $u_2 = u_{R3} + u_{R4} + u_{R3} = u_{R3}$. However, the use of zero operators based on pairs of quantum dots, for example (FQD1, FQD2), does not allow us to completely reconstruct information about a separate FEC3 on the basis of information on the structure of three elements (FQD1,

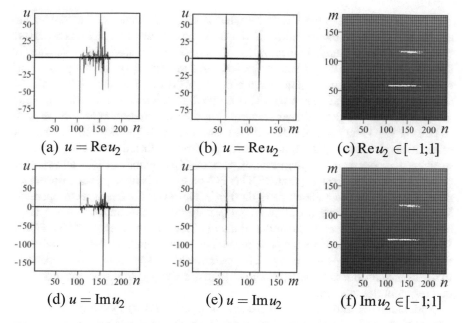

Fig. 6 The behavior of the difference of displacement function δu_2: **a, b, d, e** projections on the planes nOu, mOu; **c, f** cross-sections (top view)

FQD2, FEC3). For structure (FQD1, FQD2, FEC3) the ordered operator has the form $u_2 = u_{R1} + u_{R2} + u_{R3}$. In Fig. 6 shows the behavior of the difference of the displacement functions $\delta u_2 = (u_{R1} + u_{R2} + u_{R3}) - u_{R3}$ on the indices (n, m) of the lattice nodes. In this case, $\mathrm{Re}\delta u_2 \neq 0$ and $\mathrm{Im}\delta u_2 \neq 0$, that is explained by the phenomenon of entanglement of the eigenvalues of the operator $u_2 \neq u_{R3}$. Similar conclusions were obtained for other structures of three elements: (FQD2, FQD1, FEC3), (FEC3, FQD1, FQD2), (FEC3, FQD2, FQD1). For structures (FQD1, FEC3, FQD2) and (FQD2, FEC3, FQD1) the deformation fields also do not coincide. The type of ordering of fractal dots influences to the behavior of the displacement functions of the coupled fractal structures.

4   Conclusions

For separate FQD pairs (superpositions of two FQD with shifted centers of gravity) and a separate FEC, the behavior of the deformation field is investigated. The deformation field of FQD pairs is real. Near the FQD cores two fractal holes are observed; far from the region of cores localization wave behavior with a regular structure of elliptic type is observed. As we approach to the core, the regular structure changes: features such as inflection points, breaks, narrowings, wells, steps, saddles appear.

The deformation field of the FEC is complex: stochastic behavior of the core boundary for $\mathrm{Re}u_2$ is characteristic, inside and outside the core wave behavior is observed; effective damping ($\mathrm{Im}u_2$) is localized in the region of the stochastic boundary of the core. The stochastic surface of the deformation field u_2 of the coupled fractal structure with the number of quantum dots larger than two depends essentially on the order of the separate terms u_{Ri}. This is due to the presence of variable modules of elliptic functions in nonlinear equations for the deformation field. On the basis of pairs of FQD and pairs of FEC with the same centres of gravity and the opposite orientation of the deformation fields, zero operators are introduced. The possibility of using zero operators on the basis of pairs of FQD (pairs of FEC) allows us to restore information about separate FQD (FEC) on the basis of sequence information from structures of three elements FQD (FEC). However, the use of zero operators based on FQD pairs does not allow to completely restore information on a separate FEC based on the information on the structure of the three elements (FQD1, FQD2, FEC3). This is explained by the phenomenon of entanglement of the eigenvalues of the operator $u_{R1}+u_{R2}+u_{R3} \neq u_{R3}$. The operation of transposition pairs of operators in a sequence of four quantum dots allows us to transfer part of the information from one quantum dot to another.

The obtained results can be used in quantum algorithms for modelling the quantum superposition and entanglement phenomena for the transmission, processing, storage of information data, in quantum mechanics.

References

1. M. Nielsen, I. Chuang, *Quantum Computation and Quantum Information* (Cambridge University Press, New York, 2010)
2. D. Boumeister, A. Eckert, A. Zeilinger, *Physics of Quantum Information* (Springer, New York, 2001)
3. Y. Ozhigov, Quantum computers speed up classical with probability zero. Chaos, Solitons Fractals **10**, 1707–1714 (1999)
4. D. Castelvecchi, Quantum computers ready to leap out of the lab. Nature **541**, 9–10 (2017)
5. A.N. Omelyanchuk, E.V. Ilyichev, S.N. Shevchenko, *Quantum Coherent Phenomena in Josephson Qubits* (Naukova Dumka, Kiev, 2013)
6. M.N. Fedorov, I.A. Volkov, Yu.M Mikhailova, Coutrites and kukwarts in spontaneous parametric scattering of light, correlation and entanglement of states. JETP **142**(1(7)), 20–43 (2012)
7. Y.S. Kivshar, N.N. Rozanov (eds.), *Nonlinearities in Periodic Structures and Metamaterials* (Fizmatlit, Moscow, 2014)
8. V.S. Abramov, Quantum dots in a fractal multilayer system. Bull. Russ. Acad. Sci. Phys. **81**(5), 625–632 (2017)
9. V.S. Abramov, Alteration of the stochastic state of the deformation field in the model multilayer nanosystem. Bull. Donetsk Nat. Univ. A **2**, 81–89 (2014)
10. O.P. Abramova, A.V. Abramov, Attractors and deformation field in the coupled fractal multilayer nanosystem. CMSIM J. **2**, 169–179 (2017)
11. O.P. Abramova, Mutual influence of attractors and separate stochastic processes in a coupled fractal structures. Bull. Donetsk Nat. Univ. A **1**, 50–60 (2017)

Percolation Process in the Presence of Velocity Fluctuations: Two-Loop Approximation

Šarlota Birnšteinová, Michal Hnatič, Tomáš Lučivjanský and Lukáš Mižišin

Abstract Critical behaviour of directed bond percolation is studied in presence of turbulent mixing. The turbulent advecting velocity field is assumed to be incompressible and generated by the Kraichnan model. The model is studied by means of field-theoretic approach. The renormalization group (RG) method is used in order to analyze asymptotic large-scale behavior of the model near its critical point. The renormalization procedure is performed to the next-to-leading order of the perturbation theory. Partial results of full two-loop calculation are given.

Keywords Directed bond percolation process · Kraichnan model · Perturbative renormalization group

1 Introduction

Second order phase transitions of non-equilibrium processes represent an interesting problem not only in physics but also in other areas, such as ecology, medicine and even sociology [1]. An example that belongs to this category is a spreading process. Depending on the certain conditions spreading agent can either spread through entire population or stops after some amount of time. The first case drives a system to an active state, whereas the second one causes a system to arrive to an inactive state [2, 3]. As in the case of the static critical phenomena, non-equilibrium systems with different microscopic properties and identical macroscopic properties can be grouped

Š. Birnšteinová (✉) · M. Hnatič · T. Lučivjanský
Faculty of Sciences, P. J. Šafárik University in Košice, Košice, Slovakia
e-mail: sbirnsteinova@gmail.com

M. Hnatič · L. Mižišin
Institute of Experimental Physics, Slovak Academy of Sciences, Košice, Slovakia

M. Hnatič · L. Mižišin
Bogoliubov Laboratory of Theoretical Physics, Joint Institute for Nuclear Research, Dubna, Russia

© Springer Nature Switzerland AG 2019
C. H. Skiadas and I. Lubashevsky (eds.), *11th Chaotic Modeling and Simulation International Conference*, Springer Proceedings in Complexity, https://doi.org/10.1007/978-3-030-15297-0_3

together into the universal classes [4, 5]. Our aim here is to study the directed bond percolation model which belongs to the directed percolation universality class.

There are two approaches how to construct a mathematical model of directed percolation. A simpler case begins with the phenomenological Langevin equation and neglects terms that are irrelevant in the critical regime. The second one is to consider directed percolation as a reaction-diffusion process, for which a well known field-theoretic apparatus has been invented [6]. Both approaches turn out to yield the same field-theoretic model.

Properly constructed continuous model (usually in a form of certain stochastic differential equation) allows us to use methods of field-theoretic renormalization group (RG) for calculation of universal scaling laws. In the case of non-equilibrium phase transitions an ideal condition of a "pure" stationary critical state is hardly achievable in real systems [7]. It is believed that these inconsistencies might be caused by a strong influence of some external factor on the directed percolation processes.

In this paper, we focus on the directed bond percolation process in the presence of advective velocity fluctuations. Velocity field is assumed to be a Gaussian ensemble with prescribed statistics by the Kraichnan model [8]. Advection is assumed as passive, therefore no feedback on the velocity field is taken into consideration.

The rest of the paper is as follows. In Sect. 2 the field-theoretic action for the directed bond percolation process and advection velocity field is introduced [6]. Section 3 includes and the Feynman diagrammatic technique is applied. In Sect. 4 we treat the problem of large-scale ultraviolet (UV) divergences with renormalization group approach. In Sect. 5 main conclusions are summarized.

2 The Model

Time evolution of spreading percolating agents is effectively captured by stochastic differential equation in a following form

$$\partial_t \psi = D_0\{(-\tau_0 + \mathbf{\nabla}^2)\psi - \lambda_0\psi^2/2\} + \zeta\sqrt{\psi}, \tag{1}$$

where ψ corresponds to the density of percolating agents, $\partial_t = \partial/\partial t$ is a time derivative, $\mathbf{\nabla}^2$ is the Laplace operator, D_0 is diffusive constant, λ_0 is a coupling constant and τ_0 is the deviation from the threshold value of the infected probability. Gaussian random noise $\zeta = \zeta(t, \boldsymbol{x})$ with zero average is specified by the pair correlation function

$$\langle \zeta(t, \boldsymbol{x})\zeta(t', \boldsymbol{x}')\rangle = D_0\lambda_0\delta(t - t')\delta^d(\boldsymbol{x} - \boldsymbol{x}'). \tag{2}$$

Our aim is to investigate the spreading of the non-conserved agent in a turbulent medium and study the effect of the turbulent mixing and stirring on the critical behaviour near the phase transition between the absorbing and the active phases. Introducing the velocity field into model is realized by the replacement

$$\partial_t \rightarrow \nabla_t = \partial_t + (\boldsymbol{v} \cdot \boldsymbol{\nabla}), \tag{3}$$

where ∇_t is known as Lagrange or convective derivative in the literature. This form of replacement is due to the assumed incompressibility of the flow. The velocity field, therefore, satisfies the condition $\boldsymbol{\nabla} \cdot \boldsymbol{v} = 0$. Velocity field, modelled by the Kraichnan ensemble [2], obeys a Gaussian distribution with zero mean value and correlator

$$\langle v_i(t, \boldsymbol{x}) v_j(t', \boldsymbol{x}') \rangle = \delta(t - t') \int \frac{d^d k}{(2\pi)^d} \mathcal{D}_{ij}(k) \exp\{i\boldsymbol{k} \cdot (\boldsymbol{x} - \boldsymbol{x}')\}, \tag{4}$$

and kernel function D_{ij} is of the form

$$\mathcal{D}_{ij}(k) = P_{ij}(k) D_0 g_0 k^{-d-\xi}, \tag{5}$$

where the $P_{ij}(k) = \delta_{ij} - k_i k_j / k^2$ is transversal projector, g_0 is a coupling constant and $0 < \xi < 2$ is a free parameter with the most realistic value, Kolmogorov value $\xi = 4/3$ [2, 9]. Let us note that our problem corresponds to a passive model of advection-diffusion problem. This means that there is no feedback on the velocity field, only other fields are affected by the velocity field.

The problem of directed bond percolation process in the presence of advecting velocity field is equivalent to a field-theoretic model of the three fields $\Phi = \{\psi, \psi^\dagger, \boldsymbol{v}\}$ with the action functional consisting of the following parts:

- diffusive part

$$\mathcal{S}_{\text{diff}}(\psi, \psi^\dagger) = \int dt \int d^d x \, \psi^\dagger (-\partial_t + D_0 \partial^2 - D_0 \tau_0) \psi, \tag{6}$$

- interaction part

$$\mathcal{S}_{\text{int}}(\psi, \psi^\dagger, \boldsymbol{v}) = \int dt \int d^d x \left\{ \psi^\dagger (\boldsymbol{v} \cdot \boldsymbol{\nabla}) \psi + \frac{D_0 \lambda_0}{2} [(\psi^\dagger)^2 \psi - \psi^\dagger \psi^2] \right\}, \tag{7}$$

- quadratic term for velocity field

$$\mathcal{S}_{\text{vel}}(\boldsymbol{v}) = -\frac{1}{2} \int dt \int d^d x \int d^d x' \, v_i(t, \boldsymbol{x}) D_{ij}^{-1}(\boldsymbol{x} - \boldsymbol{x}') v_j(t, \boldsymbol{x}'), \tag{8}$$

where $D_{ij}^{-1}(\boldsymbol{x} - \boldsymbol{x}')$ is the inverse of the kernel function (5). The total action functional of this problem is given by a sum of three terms

$$\mathcal{S} = \mathcal{S}_{\text{diff}} + \mathcal{S}_{\text{int}} + \mathcal{S}_{\text{vel}}. \tag{9}$$

The partition function has the following form

$$\mathcal{Z}(A) = \mathcal{N} \int \mathcal{D}\phi \exp\{-\mathcal{S}(\phi) + A\phi\}, \tag{10}$$

where A stands for the sets of source fields, which correspond to Φ and $\int \mathcal{D}\phi$ represents integration over all possible composition of fields ϕ from set Φ. Our aim is to determine the form of connected correlation functions $G_{\phi...\phi}$. Generating functional $\mathcal{W}$ for connected correlation functions is defined as [5]

$$\mathcal{W}[A] = \ln \mathcal{Z}[A], \quad G_{\phi...\phi} = \frac{\delta\mathcal{W}}{\delta\phi...\phi}|_{A=0}. \tag{11}$$

To this end we employ the renormalization group technique in the vicinity of its upper critical dimension $d_c = 4$.

3 Feynman Diagrammatic Technique

Field-theoretic action (9) yields to the standard Feynman diagrammatic technique [5] with two propagators $\langle\psi^\dagger\psi\rangle_0$ and $\langle v_i v_j\rangle_0$ and three triple vertices $\sim (\psi^\dagger)^2\psi$, $\psi^\dagger\psi^2$ and $\psi^\dagger\psi v$. In time-momentum and frequency-momentum representation the propagators read

$$\langle\psi\psi^\dagger\rangle_0(t, k) = \theta(t) \exp\{-D_0(k^2 + \tau_0)t\},$$

$$\langle\psi\psi^\dagger\rangle_0(\omega, k) = \langle\psi^\dagger\psi\rangle_0^*(\omega, k) = \frac{1}{-i\omega + D_0(k^2 + \tau_0)}, \tag{12}$$

$$\langle v_i v_j\rangle(k) = P_{ij}(k)g_0 D_0 k^{-d-\xi}.$$

The function θ in Eq. (12) is the Heaviside step function. The propagator $\langle\psi\psi^\dagger\rangle_0$ is retarded and this fact is used in further analysis while constructing Green function. Functions built up only with the fields ψ or $\psi^\dagger$ vanish as they contain closed circuits of retarded propagators [5]. Vanishing of functions $\langle\psi...\psi\rangle$ can be viewed as a consequence of the symmetry arisen from the transformations

$$\psi(t, x) \to -\psi^\dagger(-t, x), \quad \psi^\dagger(t, x) \to -\psi(-t, x), \quad \lambda_0 \to -\lambda_0. \tag{13}$$

whereas absence of the function $\langle\psi^\dagger...\psi^\dagger\rangle$ is caused by causality, which is fulfilled for any stochastic model [5].

It can be shown, that an actual expansion parameter for this model is λ_0^2 rather than λ_0. Therefore is convenient to introduce new charge u_0 as follows

$$u_0 \equiv \lambda_0^2. \tag{14}$$

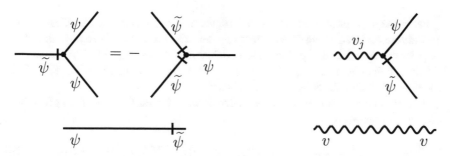

Fig. 1 Diagrammatic representation of the propagators and the percolation vertices

An interaction part of the action (7) is responsible for a description of the fluctuation effect of the percolation process and its interaction with advection field. Vertex factors have the following form [5]

$$V_m(t_1, \boldsymbol{x}_1 \ldots t_m, \boldsymbol{x}_m; \phi) = \frac{\delta^m S_{\text{int}}}{\delta \phi(t_1, \boldsymbol{x}_1) \ldots \delta \phi(t_m, \boldsymbol{x}_m)}, \tag{15}$$

where ϕ can be an arbitrary field of the model. Vertex factors of the model are

$$V_{\psi^\dagger \psi \psi} = -V_{\psi^\dagger \psi^\dagger \psi} = D_0 \lambda_0, \qquad V_{\psi^\dagger \psi v} = -i k_j, \tag{16}$$

where k_j is the momentum of the field $\psi^\dagger$. A diagrammatic form of the propagators of the model is depicted at the top part of Fig. 1. The bottom section of Fig. 1 shows the vertices in their graphical representation for the percolation process and the vertex responsible for the interaction with velocity fluctuations.

4 UV Renormalization

An analysis of ultraviolet (UV) divergences ($\boldsymbol{x} \to 0$ and $t \to 0$) is based on the analysis of the canonical dimensions [5, 10]. Dynamical models exhibit two scale dependencies, with respect to space scale, and time scale, respectively. Canonical dimensions of some quantity F are derived from the requirement that all terms in action (9) are dimensionless with respect to frequency and momentum separately and from the normalization conditions

$$d_\omega^\omega = -d_t^\omega = 1, \quad d_k^k = -d_x^k = 1, \quad d_k^\omega = d_\omega^k = 0, \tag{17}$$

A total canonical dimension of quantity F is given by $d_F = d_F^k + 2d_F^\omega$, where d_F^k is the momentum canonical dimension and d_F^ω is the frequency canonical dimension of some quantity F. Dimensions of the model (9) are given in Table 1.

Parameters without a subscript "0" represent renormalized parameters, which are introduced later. From Table 1 one can see that the model is logarithmic, both coupling constants u_0 and g_0 are dimensionless at $d = 4$ and $\xi = 0$. This means that UV divergences appear as poles in $\varepsilon = d - 4$ and ξ or their linear combinations, since the minimal subtraction scheme is used.

A total canonical dimension of an 1-irreducible Green function Γ is given by a following expression

$$d_\Gamma^k = d - N_\Phi d_\Phi^k, \quad d_\Gamma^\omega = 1 - N_\Phi d_\Phi^\omega, \quad d_\Gamma = d_\Gamma^k + 2d_\Gamma^\omega = d + 2 - N_\Phi d_\Phi, \quad (18)$$

where $N_\Phi = \{N_\psi, N_{\psi^\dagger}, N_v\}$ are numbers of fields in a given 1-irreducible Green function Γ and summation runs over all types of entering fields. The total canonical dimension δ_Γ in case of the logarithmic theory ($d = 4$ and $\xi = 0$) is used as an index of the UV divergence $\delta_\Gamma = d_\Gamma|_{\varepsilon=\xi=0}$. In order to remove UV divergences adding counterterms into action is needed for those function Γ, whose values of δ_Γ are nonnegative integer. Using relation (18) and data from Table 1 canonical dimension δ_Γ takes the form

$$\delta_\Gamma = 6 - 2N_\psi - 2N_{\psi^\dagger} - N_v. \quad (19)$$

With the restriction listed in Sect. 3 and Eq. (19), UV divergent structures are calculated and listed in Table 2. Permissible counterterms that correspond to potential UV divergent part of $\Gamma_{\psi\psi^\dagger}$ 1-irreducible Green function take the form

$$\psi^\dagger \partial_t \psi, \quad \psi^\dagger \partial^2 \psi, \quad \psi^\dagger \psi. \quad (20)$$

A possible form of counterterms for $\Gamma_{\psi^\dagger \psi v}$ 1-irreducible Green function is in the following form

$$\psi^\dagger (\boldsymbol{v} \cdot \boldsymbol{\nabla})\psi, \quad \psi(\boldsymbol{v} \cdot \boldsymbol{\nabla})\psi^\dagger, \quad (21)$$

Table 1 Canonical dimensions of the fields and parameters of the model for d dimensional space

F	ψ	$\psi^\dagger$	v	D_0	τ_0	m, μ	λ_0^2, u_0	g_0	λ, u, g
d_F^k	$d/2$	$d/2$	-1	-2	2	1	$4-d$	ξ	0
d_F^ω	0	0	1	1	0	0	0	0	0
d_F	$d/2$	$d/2$	1	0	2	1	$4-d$	ξ	0

Table 2 Canonical dimensions for the (1PI) divergent Green functions of the model

$\Gamma_{1-\text{ir}}$	$\Gamma_{\psi\psi^\dagger}$	$\Gamma_{\psi^\dagger\psi\psi}$	$\Gamma_{\psi^\dagger\psi^\dagger\psi}$	$\Gamma_{\psi^\dagger\psi v}$
d_Γ	2	$\varepsilon/2$	$\varepsilon/2$	1
δ_Γ	2	0	0	1

which are antisymmetric due to transversality of the velocity field. All of the counterterms, that are mentioned here, are present in Eq. (9), so the model described by action (9) is multiplicatively renormalizable [2].

UV divergences might also appear in in the function $\langle \psi^\dagger \psi vv \rangle$ with the canonical dimension $\delta = 0$, but the possible counterterm $\psi^\dagger \psi v^2$ is forbidden by Galilean symmetry [2]. It was shown in [11] that argument of Galilean symmetry for construction the counterterms is applicable also for synthetic velocity field modelled as Gaussian ensemble with vanishing correlation time, not only for the usual application for the velocity field governed by the Navier-Stokes equation.

Elimination of UV divergences and following addition of the counterterms leads to the renormalization of original fields

$$\psi \to Z_\psi \psi, \qquad \psi^\dagger \to Z_{\psi^\dagger} \psi^\dagger, \qquad v \to Z_v v \tag{22}$$

and parameters of the model

$$D_0 = D Z_D, \qquad g_0 = g \mu^\xi Z_g, \qquad \tau_0 = \tau Z_\tau,$$
$$\lambda_0 = \lambda Z_\lambda \mu^{\varepsilon/2}, \quad u_0 = u Z_u \mu^\varepsilon, \tag{23}$$

where μ is the renormalization mass in the minimal subtraction scheme [5], Z_e are renormalization constants and the subscript e denotes the set of the renormalized parameters $e = \{ D, \tau, g, \lambda, u \}$. In a similar manner e_0 represents the set of bare parameters $e_0 = \{ D_0, \tau_0, g_0, \lambda_0, u_0 \}$. After replacing e_0 by their renormalized equivalents in the original action (9) one get the form of the renormalization action

$$S_R = \int dt \int d^d x \left\{ \psi^\dagger [-Z_1 \nabla_t + Z_2 D \partial^2 - Z_3 D \tau] \psi \right.$$
$$\left. + Z_4 \frac{D\lambda}{2} [(\psi^\dagger)^2 \psi - \psi^\dagger \psi^2] \right\} + \int dt \int d^d x \int d^d x' \frac{v D_{ij} v}{2}. \tag{24}$$

Renormalization constants in action (24) together with Eqs. (22) and (23) yield

$$Z_1 = Z_\psi Z_{\psi^\dagger} = Z_\psi Z_{\psi^\dagger} Z_v,$$
$$Z_2 = Z_\psi Z_{\psi^\dagger} Z_D,$$
$$Z_3 = Z_\psi Z_{\psi^\dagger} Z_D Z_\tau, \tag{25}$$
$$Z_4 = Z_\psi Z_{\psi^\dagger}^2 Z_D Z_\lambda = Z_\psi^2 Z_{\psi^\dagger} Z_D Z_\lambda.$$

The renormalization constants from Eqs. (22) and (23) are easily derived in the following form

$$Z_\psi = Z_{\psi^\dagger} = Z_1^{1/2}, \qquad Z_v = 1, \qquad Z_\tau = Z_3 Z_2^{-1},$$
$$Z_D = Z_g^{-1} = Z_2 Z_1^{-1}, \quad Z_\lambda = Z_4 Z_2^{-1} Z_1^{-1/2}. \tag{26}$$

Due to the introduction of new coupling constants (14) the following relation holds

$$Z_\lambda^2 = Z_u. \tag{27}$$

The form of the renormalization constants listed in Eq. (25) can be found from the requirement that they absorb UV divergences at $\varepsilon \to 0$ and simultaneously $\xi \to 0$ in the Green functions presented in Table 2 [5, 9]. Found poles are in the following form $1/\varepsilon$, $1/\xi$ and linear combination $1/(\varepsilon + \xi)$.

In Eqs. (28) and (29) there are the examples of two of the 1-irreducible Green function from Table 2 to the next-to-leading order in perturbation theory.

$$\Gamma_{\psi^\dagger \psi} = Z_1 i\omega + Z_2 D\tau p^2 + Z_3 \tau D + \cdots \tag{28}$$

$$\Gamma_{\psi\psi^\dagger\psi^\dagger} = Z_4 D\lambda + \cdots \tag{29}$$

Symmetry factors are included in diagrams. The renormalization constants are independent on the choice of the IR regularization. From the calculation viewpoints, it is more convenient to set $\tau = 0$ in propagator (12) and cut off the momentum integrals at $k = m$ since from dimensional analysis $\tau \sim m^2$ [12].

The partial two-loop results for the renormalization constants are

$$Z_1 = 1 + \frac{u}{4\varepsilon} + \frac{u^2}{\varepsilon}\left(\frac{7}{32\varepsilon} + \frac{3}{32} - \frac{9}{64}\ln\frac{4}{3}\right) + \frac{ug}{\xi}\frac{3}{4}\left(\frac{1}{\varepsilon} + \frac{1}{4} + \frac{1}{2}c\right)$$
$$+ \frac{ug}{\varepsilon + \xi}C_1,$$

$$Z_2 = 1 + \frac{u}{8\varepsilon} - \frac{3g}{4\xi} + \frac{u^2}{8\varepsilon}\left(\frac{13}{16\varepsilon} - \frac{31}{64} + \frac{35}{32}\ln\frac{4}{3}\right) - \frac{ug}{\xi}\frac{1}{16}\left(-\frac{3}{\varepsilon} + \frac{1}{4}\right)$$
$$+ \frac{ug}{\varepsilon + \xi}C_2, \tag{30}$$

$$Z_3 = 1 + \frac{u}{2\varepsilon} + \frac{u^2}{\varepsilon}\left(\frac{1}{2\varepsilon} - \frac{5}{32}\right) + \frac{ug}{\xi}\frac{3}{2}\left(\frac{1}{\varepsilon} + \frac{1}{4} + \frac{1}{2}c\right) + \frac{ug}{\varepsilon + \xi}C_3,$$

$$Z_4 = 1 + \frac{u}{\varepsilon} + \frac{u^2}{\varepsilon}\left(\frac{7}{4\varepsilon} - \frac{7}{8}\right) + \frac{ug}{\xi}\frac{3}{4}\left(\frac{1}{\varepsilon} + \frac{1}{4} + \frac{1}{2}c\right) + \frac{ug}{\varepsilon + \xi}C_4.$$

where $c = \psi_0(3/2) - \psi_0(5/2)$, $\psi_n(z) = d^{n+1}/dz^{n+1}\ln\Gamma(z)$ and $\Gamma(z)$ is the gamma function. Finite constants C_1, C_2, C_3 are corresponding contributions from diagram

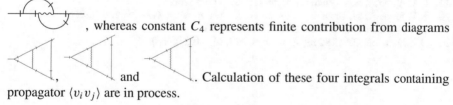

, whereas constant C_4 represents finite contribution from diagrams

, and and . Calculation of these four integrals containing propagator $\langle v_i v_j \rangle$ are in process.

5 Conclusion

Effects of the turbulent mixing and stirring on the reaction-diffusion model has been studied. This process can be viewed as an example of critical behaviour in a non-equilibrium system near transition between the active and absorbing phase. The field-theoretic model was constructed as a combination of the directed bond percolation process and advecting turbulent velocity field modelled as Kraichnan ensemble. This advecting field is considered a Gaussian variable with prescribed statistical properties. The multiplicative renormalizability of the model has been proven. Partial two-loop results for renormalization constants has been shown. We were able to compare the results for the renormalization constants in the limit of pure directed percolation process in the two-loop approximation [13, 14]. The next steps involve calculation of the Feynman diagrams containing the velocity field propagator, which has not been finished yet. It will be then possible to obtain fixed points and stable regimes to the next-to-leading order in the perturbation theory.

Acknowledgements The work was supported by VEGA grant No. 1/0345/17 of the Ministry of Education, Science, Research and Sport of the Slovak Republic.

References

1. M. Henkel, H. Hinrichsen, S. Lübeck, *Non-Equilibrium Phase Transitions: Vol. 1-Absorbing Phase Transition* (Springer, Berlin, 2008)
2. N.V. Antonov, V.I. Iglovikov, A.S. Kapustin, J. Phys. A **42**, 135001 (2008)
3. U.C. Täuber, *Critical Dynamics: A Field Theory Approach to Equilibrium and Non-Equilibrium Scaling Behavior* (Cambridge University Press, Cambridge, 2014)
4. J. Zinn-Justin, *Phase Transitions and Renormalisation Group* (Oxford University Press, Oxford, 2007)
5. A.N. Vasilev, *The Field Theoretic Renormalization Group in Critical Behaviour Theory and Stochastic Dynamics* (Chapman Hall/CRC Press, Boca Raton, FL, 2004)
6. U.C. Täuber, M. Howard, B.P. Vollmayr-Lee, J. Phys. A: Math. Gen. **38**, R79–R131 (2005)
7. N.V. Antonov, A.S. Kapustin, A.V. Malyshev, Theor. Math. Phys. **169**, 1470 (2011)
8. G. Falkovich, K. Gawedzki, M. Vergassola, Rev. Mod. Phys. **73**, 913 (2001)
9. N.V. Antonov, Phys. Rev. E. **60**, 6691 (1999)
10. D.J. Amit, *Field Theory, the Renormalization Group, and Critical Phenomena* (World Scientific, Singapore, 1984)
11. N.V. Antonov, A.A. Ignatieva, J. Phys. A **39**, 13593 (2006)
12. N.V. Antonov, M. Hnatich, J. Honkonen, J. Phys. A: Math. Gen. **39**, 7867 (2006)
13. H.-K. Janssen, J. Stat. Phys. **103**, 801–839 (2001)
14. H.-K. Janssen, U.C. Täuber, Ann. Phys. **315**, 147–192 (2005)

Phase Transition in Incompressible Active Fluid: Effect of Long-Range Interactions

Šarlota Birnšteinova, Juha Honkonen, Tomáš Lučivjanský
and Viktor Škultéty

Abstract Phase transitions in active fluids attracted a significant attention in the last two decades. In the recent work (Chen et al in New J Phys 17:042002, 2015 [1]) authors showed that an order-disorder phase transition in incompressible active fluids belongs to a new universality class and suggested a potential experimental realization to be systems with long-range (LR) repulsive interactions. In this work, we study the effect of LR interactions by introducing non-local shear stress into the microscopic description of the model. Using methods of field-theoretic renormalization group we investigate the large-scale properties in critical region. We have found that the effect of LR interactions can change the universality class to the Model A class with LR interactions or destroy the relevance of the non-linearities completely which leads to the mean-field values for critical exponents.

Keywords Active fluid · Phase transitions · Renormalization group · Long-range interactions

Š. Birnšteinova · T. Lučivjanský
Faculty of Sciences, Šafarik University, Moyzesova 16, 040 01 Košice, Slovakia
e-mail: sbirnsteinova@gmail.com

T. Lučivjanský
e-mail: tomas.lucivjansky@upjs.sk

J. Honkonen
National Defence University, 00861 Helsinki, Finland
e-mail: juha.honkonen@helsinki.fi

V. Škultéty (✉)
School of Physics and Astronomy, The University of Edinburgh,
Peter Guthrie Tait Road, Edinburgh EH9 3FD, UK
e-mail: viktoroslavs@gmail.com

© Springer Nature Switzerland AG 2019
C. H. Skiadas and I. Lubashevsky (eds.), *11th Chaotic Modeling
and Simulation International Conference*, Springer Proceedings
in Complexity, https://doi.org/10.1007/978-3-030-15297-0_4

1 Introduction

Nonequilibrium physics constitutes fascinating research topic to which considerable effort has been devoted in last decades. Typical problems are notoriously known for their complexity and unreliability of perturbation approaches. However, a great simplification is permissible in a vicinity of critical region, where new symmetry related to scale invariance emerges. An immediate hallmark of it is divergence of the correlation length, which results in strong fluctuations on all length and times scales. The system then effectively forgets about microscopic details and can be effectively described by a few collective coarse-grained quantities.

Investigation of the phase transition in the active fluid began more than two decades ago with Viscek's numerical work on the order-disorder phase transition in the collective movement of the bird-like objects [2]. About the same time, a continuum Toner-Tu model [3] was derived for the theoretical investigation of this type of transition. Their model represents a very elegant hydrodynamic description for compressible active fluid with violated Galilean invariance, which is a crucial property for the existence of the phase transition.

Since then most of the attention has been attracted by the phase transition in the compressible active fluids [3–6]. In the recent work, the existence of the order-phase transition in the incompressible active fluid was shown using perturbative renormalization group [1], where authors considered the incompressible version of the previously proposed Toner-Tu model. One of the proposed experimental realizations was the model with long-range repulsive interactions. In such scenario for example, active particles (say birds) can perceive through the entire flock, hence the repulsive energy per particle scales like the system volume, which force the compressibility vanish in the thermodynamic limit.

In this work, we study the effect of the long-range interactions on the incompressible active fluid phase transition from a slightly different point of view. Our aim is to take into account the effect of the non-local stress, which is responsible for the non-local energy dissipation. This adjustment will lead to a modification of the propagators of the model and the large scale behaviour will then be investigated using well known methods of renormalization group.

As in the case of previous works of field-theories with LR interactions, such as directed percolation [7] or fully developed turbulence [8], the non-locality is described by means of an analytic regulator. This allow us to calculate the fixed point values and the corresponding stability in a combined double expansion scheme of the analytic and dimensional regularization [9].

2 Microscopic Formulation

Starting from the "Navier-Stokes" equation [2, 10]

$$\partial_t v_i + v_j \partial_j v_i = \partial_j \mathcal{E}_{ji} - \partial_i p + \mathcal{F}_i, \tag{1}$$

where $\partial_t \equiv \partial/\partial t$ is time derivative, $\partial_j = \partial/\partial x_j$, v_i is the velocity field, $\mathcal{E}_{ij}$ is the strain rate tensor, and p is the pressure field. The force $\mathcal{F}_i$ is derived from the Landau potential

$$\mathcal{F}_i = \frac{\delta U[v]}{\delta v_i}, \quad U[v] = \frac{\tau_0}{2}|v|^2 + \frac{g_{10}}{4!}|v|^4, \tag{2}$$

where τ_0 and g_0 are the microscopic deviation from the criticality and coupling constant, respectively. Following earlier works [11, 12] we postulate components of the modified strain rate tensor to be

$$\mathcal{E}_{ij} = \nu_0 \varepsilon_{ij} + S_0 Q_{ij}, \tag{3}$$

where $\varepsilon_{ij} = (\partial_i v_j + \partial_j v_i)$ is the classical strain rate tensor and $Q_{ij} = v_i v_j - \frac{\delta_{ij}}{d}|v|^2$ is the active nematic stress tensor [12] and S_0 is the microscopic amplitude. General hydrodynamic arguments [13] imply, that $S_0 < 0$ for push-swimmers like *E. coli* or *Bacillus subtilis*, whereas $S_0 > 0$ for puller type microswimmers such as *Chlamydomonas* algae.

The resulting Toner-Tu model is [1]

$$\partial_t v_i + \lambda_0 v_j \partial_j v_i = \nu_0 \partial^2 v_i - \partial_i (p + \lambda_0'|v|^2) - (\tau_0 + g_{10}|v|^2/3!)v_j + f_i, \tag{4}$$

where $\lambda_0 = 1 - S_0$, $\lambda_0' = S_0/d$. The second term on the right hand side also represents the modified pressure term. Note that in [1] the velocity modification of the pressure term was not present since in the case of incompressible fluid the pressure term does not influence the critical properties of the system. In order to mimic the random movement of active particles we also introduce the random force f_i obeying Gaussian statistics with the two point correlator

$$\langle f_i(x, t) f_j(x', t') \rangle = \delta(t - t') \int_{k \geq \sqrt{\tau}} \mathrm{d}^d k \, D_{ij}^v(k) e^{ik \cdot r}, \quad r = x - x', \tag{5}$$

$$D_{ij}^v(k) = \nu_0 P_{ij}(k), \quad P_{ij}(k) = \delta_{ij} - k_i k_j/k^2. \tag{6}$$

Since we work with an incompressible active fluid, random force was chosen to be purely transversal in order to avoid generation of sound modes in a system. Note that the model (4) is not Galilean invariant and the violation of this symmetry is crucial for the existence of the phase transition.

2.1 Effect of Long-Range Interactions

Here we assume that the stress in the active fluid is non-local, i.e. it depends not only on the nearest neighbouring points, but the effects of the long range correlations play important role. Generalizing the approach of [14] to the case of an incompressible active fluid, we further modify the expression for the classical strain rate tensor as follows

$$\varepsilon_{ij}(x) \rightarrow \varepsilon_{ij}(x) + \int d^d y \, \varepsilon_{ij}(y) \kappa(|y - x|). \tag{7}$$

The first term in the above equation is the classical local term and the second represents the non-local contributions. The kernel function κ weights the contribution from long-distance points y on the point x and by assuming isotropicity of the system, we expect its decay to be an ordinary power law $\kappa(|r|) \sim 1/|r|^{d-2\alpha}$. Using the definition of the Riesz fractional integro-differentiation formula (see for example [15], Chap. 25)

$$[I^{2\alpha}\varepsilon_{ij}](x) \equiv \frac{1}{C_{2\alpha}} \int d^d y \, \frac{\varepsilon_{ij}(y)}{|x - y|^{d-2\alpha}} \rightarrow (-\partial_x^2)^{-\alpha}, \tag{8}$$

$$C_\alpha = 2^\alpha \pi^{d/2} \frac{\Gamma(\alpha/2)}{\Gamma((d - \alpha)/2)}, \tag{9}$$

where $\Gamma(x)$ is the Gamma function, the Toner-Tu model with long-range interactions will then be

$$\partial_t v_i + \lambda_0 v_j \partial_j v_i = \nu_0 \partial^2 v_i - w_0(-\partial^2)^{1-\alpha} v_i - \partial_i(p + \lambda_0'|v|^2)$$
$$- (\tau_0 + g_{10}|v|^2/3!)v_j + f_i, \tag{10}$$

where w_0 is some microscopic amplitude.

3 Field-Theoretic Approach

Following general procedures [16, 17], De Dominicis-Janssen response functional for the active fluid model with LR interactions is

$$S[v', v] = - v_i' D_{ij}^v v_j'/2 + v_i'\{\partial_t + \nu_0(-\partial^2 + w_0(-\partial^2)^{1-\alpha} + \tau_0)\}v_i$$
$$+ \nu_0 v_i'(\lambda_0 v_j \partial_j + g_{10}|v|^2/3!)v_i, \tag{11}$$
$$D_{ij}^v(k) = \nu_0 P_{ij}(k), \tag{12}$$

where we have rescaled all parameters with the viscosity due to the dimensional reasons $\tau_0 \to \nu_0 \tau_0$, $g_{10} \to \nu_0 g_{10}$, $\lambda_0 \to \nu_0 \lambda_0$, $w_0 \to \nu_0 w_0$. Let us note that the modified pressure term has disappeared due to the assumed transversality of the response field v'_i and therefore does not affect the critical properties of the model as has mentioned before.

The field-theoretic formulation implies, that all the correlation and response functions can be calculated from the following generating functional

$$\mathcal{Z}[A] = \int \mathcal{D}v' \mathcal{D}v \, \exp\{-\mathcal{S}[v', v] + \varphi A\}, \tag{13}$$

$$\varphi \equiv \{v', v\}, \tag{14}$$

$$A \equiv \{A_{v'}, A_v\}, \tag{15}$$

by taking the variational derivative with respect to the corresponding source field A. For example, the linear response function is obtained as

$$\langle v_i(\boldsymbol{x}) v'_j(\boldsymbol{x}', t') \rangle = \frac{\delta^2 \mathcal{Z}[A]}{\delta A_v(\boldsymbol{x}, t) \delta A_{v'}(\boldsymbol{x}, t')}. \tag{16}$$

Note that last term in exponent of Eq. (13) should be interpreted as scalar product between corresponding terms, i.e.

$$\varphi A \equiv \boldsymbol{v} \cdot \boldsymbol{A}_v + \boldsymbol{v}' \cdot \boldsymbol{A}_{v'}.$$

In general, interacting field-theoretic models such as (13) are not exactly solvable and one may treat them only within some perturbation method. In order to simplify the calculation, it is convenient to work with the effective potential Γ, which is defined as a Legendre transformation of the generating functional $\mathcal{Z}$

$$\Gamma[\Phi] = \ln \mathcal{Z}[A] - A\Phi, \quad \Phi(\boldsymbol{x}, t) = \frac{\delta \ln \mathcal{Z}[A]}{\delta A(\boldsymbol{x}, t)}. \tag{17}$$

The following relation can be shown

$$\Gamma[\Phi] = -\mathcal{S}[\Phi] + \text{(loop corrections)} \tag{18}$$

to hold [16–18]. The effective potential is also a generating functional of vertex functions that can be calculated as

$$\Gamma^{\Phi \cdots \Phi'}(\{\boldsymbol{x}_i, t_i\}) = -\frac{\delta \mathcal{S}[\Phi]}{\delta \Phi(\boldsymbol{x}, t) \cdots \delta \Phi'(\boldsymbol{x}', t')} + \begin{pmatrix} \text{corrections from} \\ \text{amputated diagrams} \end{pmatrix} \tag{19}$$

The above corrections from amputated diagrams are represented in terms of Feynman diagrams which are constructed from the Feynman rules of the corresponding model.

Fig. 1 Feynman
correspondence rules for the
incompressible active fluid
model

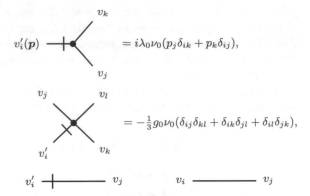

$$v_i'(\boldsymbol{p}) \quad = i\lambda_0\nu_0(p_j\delta_{ik} + p_k\delta_{ij}),$$

$$= -\tfrac{1}{3}g_0\nu_0(\delta_{ij}\delta_{kl} + \delta_{ik}\delta_{jl} + \delta_{il}\delta_{jk}),$$

The propagators are identified from the free (quadratic) part and the vertex factors
from the interaction part. The exact form of propagators and interaction vertices is
the following

$$\langle v_i v_j'\rangle_0(\boldsymbol{k}, \omega) = \frac{P_{ij}(\boldsymbol{k})}{-i\omega + \nu_0(k^2 + w_0 k^{2(1-\alpha)} + \tau_0)}, \tag{20}$$

$$\langle v_i v_j\rangle_0(\boldsymbol{k}, \omega) = \frac{\nu_0 P_{ij}(\boldsymbol{k})}{\omega^2 + \nu_0^2(k^2 + w_0 k^{2(1-\alpha)} + \tau_0)^2}, \tag{21}$$

$$V^{(3)}_{v_i'(p)v_j v_k} = i\lambda_0\nu_0(p_j\delta_{ik} + p_k\delta_{ij}), \tag{22}$$

$$V^{(4)}_{v_i'v_j v_k v_l} = -\tfrac{1}{3}g_{10}\nu_0(\delta_{ij}\delta_{kl} + \delta_{ik}\delta_{jl} + \delta_{il}\delta_{jk}), \tag{23}$$

and their graphical representation can be seen in Fig. 1. It can be easily noticed that
from a formal point of view this model represents a combination of ϕ^4 and $\partial\phi^3$
theory.

4   Renormalization Group Analysis

It is well known that the perturbative RG analysis is based on the analysis of the
canonical dimensions of the model [16, 17]. In order to obtain the renormalize the
model, we must eliminate divergences arising in the vertex functions (25) with non-
negative value of UV exponent

$$\delta_\Gamma = d + 2 - \sum_\Phi d_\Phi n_\Phi, \tag{24}$$

where d_Φ and n_Φ are the canonical dimension and the total number of the field Φ.
The table of canonical dimensions for the active fluid model is shown in Table 1 and
the degree of divergence for this model is then

Table 1 Canonical dimensions of the bare fields and bare parameters for the active fluid model

Q	v	v'	$\sqrt{\tau}, \mu, \Lambda$	ν_0, ν	$g_{10}, g_{20} = \lambda_0^2$	w_0	w, λ, g_1, g_2
d_Q^k	$\frac{d}{2} - 1$	$\frac{d}{2} + 1$	1	-2	ε	2α	0
d_Q^ω	0	0	0	1	0	0	0
d_Q	$\frac{d}{2} - 1$	$\frac{d}{2} + 1$	1	0	ε	2α	0

$$d_\Gamma = d + 2 - N_v \left(\frac{d}{2} - 1\right) - N_{v'} \left(\frac{d}{2} + 1\right), \tag{25}$$

Divergent vertex functions are the following

$$\Gamma_{v'v} : \quad \text{with counterterms} \quad v'\partial_t v, \ v'\partial^2 v, \ \tau v' v, \tag{26}$$
$$\Gamma_{v'vv} : \quad \text{with counterterm} \quad v'(v\partial)v, \tag{27}$$
$$\Gamma_{v'vvv} : \quad \text{with counterterm} \quad v'vvv. \tag{28}$$

In order to eliminate these divergences one must renormalize parameters of the model in the following way

$$g_1 = \mu^\varepsilon g_{10} Z_{g_1}, \quad g_{20} = \mu^\varepsilon g_2 Z_{g_2} \quad \lambda = \mu^{\varepsilon/2} \lambda_0 Z_\lambda, \tag{29}$$
$$w = \mu^{2\alpha} Z_w w_0, \quad \tau = Z_\tau \tau_0, \quad \nu = \nu_0 Z_\nu. \tag{30}$$

with Z_x being the renormalization constants. The renormalized response functional reads

$$\mathcal{S}_R[v', v] = + v'_i \{Z_1 \partial_t + \nu(Z_2(-\partial^2) + w\mu^{2\alpha}(-\partial^2)^{(1-\alpha)} + Z_3 \tau)\} v_i$$
$$+ \nu v'_i (Z_4 \lambda \mu^{\varepsilon/2} v_j \partial_j + Z_5 g_1 \mu^\varepsilon |v|^2/3!) v_i - v'_i D^v_{ij} v'_j / 2 \tag{31}$$
$$D^v_{ij}(k) = Z_6 \nu P_{ij}(k), \tag{32}$$

where

$$Z_1 = Z_\nu Z_{v'}, \quad Z_2 = Z_\nu Z_v Z_{v'}, \quad Z_3 = Z_\nu Z_\tau Z_v Z_{v'} \quad Z_4 = Z_\lambda Z_\nu Z_v^2 Z_{v'}, \tag{33}$$
$$Z_5 = Z_{g_1} Z_\nu Z_v^3 Z_{v'}, \quad Z_6 = Z_\nu Z_{v'}^2, \quad Z_w Z_\nu Z_v Z_{v'} = 1. \tag{34}$$

From above, one can obtain inverse relations between the renormalization constants

$$Z_v = (Z_1 Z_2 Z_6^{-1})^{\frac{1}{2}}, \ Z_{v'} = (Z_1 Z_2^{-1} Z_6)^{\frac{1}{2}}, \ Z_\nu = Z_2 Z_1^{-1}, \ Z_\tau = Z_3 Z_2^{-1}, \tag{35}$$
$$Z_\lambda = (Z_1^{-1} Z_2^{-3} Z_4^2 Z_6^{-1})^{\frac{1}{2}}, \ Z_{g_1} = Z_1^{-1} Z_2^{-2} Z_5 Z_6, \ Z_w = Z_2^{-1}. \tag{36}$$

Now one must calculate the loop corrections to the divergent vertex functions, which in the present case are

$$\Gamma_{v_i' v_j} = i\Omega Z_1 - \nu k^2 Z_2 - \nu w k^{2(1-\alpha)} - \nu\tau Z_3 + \;\text{⟨diagram⟩}\; +$$
$$+ \frac{1}{2} \;\text{⟨diagram⟩}\; + \dots \tag{37}$$

$$\Gamma_{v_i'(p) v_j v_k} = i\lambda\nu Z_4 (p_j \delta_{ik} + p_k \delta_{ij}) + \;\text{⟨diagram⟩}\; + \;\text{⟨diagram⟩}\; +$$
$$+ \frac{1}{2} \;\text{⟨diagram⟩}\; + \;\text{⟨diagram⟩}\; + \dots \tag{38}$$

$$\Gamma_{v_i' v_j v_k v_l} = -\frac{g_1 \nu}{3} Z_5 (\delta_{ij}\delta_{kl} + \delta_{ik}\delta_{jl} + \delta_{il}\delta_{jk}) + 3 \;\text{⟨diagram⟩}\; +$$
$$+ \frac{3}{2} \;\text{⟨diagram⟩}\; + 3 \;\text{⟨diagram⟩}\; + \dots$$
$$\Gamma_{v_i' v_j'} = \nu Z_6 P_{ij}(\boldsymbol{k}) + \dots \tag{39}$$

with possible higher order corrections. The corresponding renormalization constants to one-loop order are

$$Z_1 = 1 \tag{40}$$

$$Z_2 = 1 - \frac{(d-1)}{4(d+2)} \sum_{k=0}^{\infty} \binom{-2}{k} \frac{g_2 w^k}{\varepsilon + 2\alpha k}, \tag{41}$$

$$Z_3 = 1 + \frac{(d-1)(d+2)}{12d} \sum_{k=0}^{\infty} \binom{-1}{k} \frac{g_1 w^k}{\varepsilon + 2\alpha k}, \tag{42}$$

$$Z_4 = 1 + \left(\frac{(d^2-2)}{6d(d+2)} + \frac{(d-1)}{12(d+2)} \right) \sum_{k=0}^{\infty} \binom{-2}{k} \frac{g_1 w^k}{\varepsilon + 2\alpha k}, \tag{43}$$

$$Z_5 = 1 + \frac{(d^3 + 9d^2 + 2d - 12)}{12d(d+2)} \sum_{k=0}^{\infty} \binom{-2}{k} \frac{g_1 w^k}{\varepsilon + 2\alpha k}, \tag{44}$$

$$Z_6 = 1, \tag{45}$$

where $g_2 = \lambda^2$, and we have performed a substitution $g_i \bar{S}_d \rightarrow g_i$, with $\bar{S}_d = S_d/(2\pi)^d$ and S_d being the surface of d dimensional sphere.

5 Critical Properties

The investigation of large-scale requires analysis of the Green's functions at different scales. The relation between renormalized and bare green functions is

$$G_0(\{k_i\}, e_0) = Z_{v'}^{N_{v'}}(g) Z_v^{N_v}(g) G(\{k_i\}, e, \mu) \tag{46}$$

where $e_0 \equiv \{g_{10}, g_{20}, \nu_0, \tau_0\}$ is the set of all bare parameter, $e = e(\mu)$ are their renormalized counter parts at the scale μ and $g = g(\mu) \equiv \{g_1, g_2\}$ is the set of all renormalized charges.

We now define operators $\mathcal{D}_x = x\partial_x|_e$ and $\tilde{\mathcal{D}}_x = x\partial_x|_{e_0}$ to be logarithmic differential operators with respect to the renormalized parameters and bare parameters fixed respectively. The investigation at different scales requires performing the logarithmic partial derivative with respect to the μ while holding bare parameters e_0 fixed

$$\{\mathcal{D}_\mu + \beta_g \partial_g - \gamma_\nu \mathcal{D}_\nu - \gamma_\tau \mathcal{D}_\tau + N_v \gamma_v + N_{v'} \gamma_{v'}\} G(\{k_i\}, e, \mu) = 0 \tag{47}$$

with

$$\beta_g = \tilde{\mathcal{D}}_\mu g, \quad \gamma_x = \tilde{\mathcal{D}}_\mu \ln Z_x \tag{48}$$

being corresponding beta and gamma function. The first one describes the running coupling constant while the latter represents the anomalous scaling dimension observable after the coarse-graining process. Gamma functions are calculated from normalization constants providing the fact, that they must be UV finite. The relation between anomalous dimensions can be then found from the relation between the normalization constants and the results are

$$\gamma_v = (\gamma_1 + \gamma_2 - \gamma_6)/2, \quad \gamma_{v'} = (\gamma_1 - \gamma_2 + \gamma_6)/2, \quad \gamma_\nu = \gamma_2 - \gamma_1, \tag{49}$$

$$\gamma_\lambda = (-\gamma_1 - 3\gamma_2 + 2\gamma_4 - \gamma_6)/2, \quad \gamma_{\bar{g}_1} = (-\gamma_1 - 2\gamma_2 + \gamma_5 + \gamma_6), \tag{50}$$

$$\gamma_\tau = \gamma_3 - \gamma_2, \quad \gamma_{\bar{g}_2} = 2\gamma_\lambda, \quad \gamma_w = -\gamma_2. \tag{51}$$

Beta functions are then found

$$\beta_{g_1'} = -g_1 \left(\varepsilon - \frac{17g_1}{24(1+w)^2} - \frac{g_2}{4(1+w)^2} \right), \tag{52}$$

$$\beta_{g_2'} = -g_2 \left(\varepsilon - \frac{5g_1}{18(1+w)^2} - \frac{3g_2}{8(1+w)^2} \right), \tag{53}$$

$$\beta_w = -w \left(2\alpha - \frac{g_2}{8(1+w)^2} \right), \tag{54}$$

where $\beta_g = 0$ at a fixed point. The stability is then determined by the eigenvalues λ of the matrix

$$\Omega_{ij} = \partial_{g_i} \beta_{g_j}, \tag{55}$$

where $\mathrm{Re}[\lambda] > 0$ for a stable fixed point.

5.1 Fixed Points for Finite w

It seems that the actual expansion parameters are $g_i' = g_i/(1+w)^2$ but the transformation into these new variables is not necessary in order to obtain results for finite w. The above set of equations, however does not have a fixed point with nonzero value of w^*. Therefore, all fixed points belong to the universality classes already found in the model with SR interactions - SR Gaussian, SR Navier-Stokes, SR Model A and SR Active fluid models.

5.2 Fixed Points for $w \to \infty$

In order to study the limit $w \to \infty$, i.e. to study the true LR limit (TLR), we perform the following substitution

$$\tilde{v}' = v'/w_0^{1/2}, \quad \tilde{v} = vw_0^{1/2}, \quad \tilde{\nu}_0 = \nu_0 w_0, \quad \tilde{\tau}_0 = \tau_0/w_0, \tag{56}$$

$$\tilde{g}_{10} = g_{10}/w_0^2, \quad \tilde{\lambda}_0 = \lambda_0/w_0^{3/2}, \quad \tilde{g}_{20} = g_{20}/w_0^3, \tag{57}$$

where the bare response functional will attain the following form

$$S[v', v] = -\tilde{v}_i' D_{ij}^v \tilde{v}_j'/2 + \tilde{v}_i' \{ \partial_t + \tilde{\nu}_0 (-\partial^2/w_0 + (-\partial^2)^{1-\alpha} + \tilde{\tau}_0) \} \tilde{v}_i$$
$$+ \tilde{\nu}_0 \tilde{v}_i' (\tilde{\lambda}_0 \tilde{v}_j \partial_j + \tilde{g}_{10} |\tilde{v}|^2/3!) \tilde{v}_i, \tag{58}$$

$$D_{ij}^v(k) = \tilde{\nu}_0 P_{ij}(k). \tag{59}$$

Table 2 Canonical dimensions of the bare fields and bare parameters for AF model. Note that in this case $z = 2(1 - \alpha)$

Q	$\tilde{v}$	$\tilde{v}'$	μ, Λ	$\tilde{\tau}_0, \tilde{\tau}$	$\tilde{\nu}_0, \tilde{\nu}$	$\tilde{g}_{10}$	$\tilde{g}_{20} = \tilde{\lambda}_0^2 / \lambda_0^2$	w	$w_0, \tilde{\lambda},$ $\tilde{g}_1, \tilde{g}_2$
d_Q^k	$\frac{d}{2} - (1-\alpha)$	$\frac{d}{2} + (1-\alpha)$	1	$2(1-\alpha)$	$-2(1-\alpha)$	$\varepsilon - 4\alpha$	$\varepsilon - 6\alpha$	2α	0
d_Q^ω	0	0	0	0	1	0	0	0	0
d_Q	$\frac{d}{2} - (1-\alpha)$	$\frac{d}{2} + (1-\alpha)$	1	0	0	$\varepsilon - 4\alpha$	$\varepsilon - 6\alpha$	2α	0

An important point to stress out here is that since the original LR coupling constant scales as $w_0 \sim \Lambda^{2\alpha}$, the parameters and fields in the above rescaled response functional will have different canonical dimensions - see Table 2. Note, that for the rescaled response functional $z = 2(1 - \alpha)$. Another interesting fact is, that parameters $\tilde{g}_{10}$ and $\tilde{g}_{20}$ have no longer the same canonical dimensions. Since for $\alpha > 0$ the coupling constants have $d_{\tilde{g}_{10}} > d_{\tilde{g}_{20}}$ one can in principle expect the absence of a NS fixed point with long-range interactions. The results below demonstrate these expectations.

Beta functions for this model can be calculated from the already known gamma functions in the following way

$$\beta_{\tilde{g}_1} = \mathcal{D}_\mu \left(\frac{g_1}{w^2} \right) = -\tilde{g}_1 \left(\varepsilon - 4\alpha + \gamma_{g_1} - 2\gamma_w \right) \big|_{\{g_1, g_2\} \to \{w^2 \tilde{g}_1, w^3 \tilde{g}_1\}}, \tag{60}$$

$$\beta_{\tilde{g}_2} = \mathcal{D}_\mu \left(\frac{g_2}{w^3} \right) = -\tilde{g}_1 \left(\varepsilon - 6\alpha + \gamma_{g_2} - 3\gamma_w \right) \big|_{\{g_1, g_2\} \to \{w^2 \tilde{g}_1, w^3 \tilde{g}_1\}}, \tag{61}$$

$$\beta_w = -w(2\alpha + \gamma_w) \big|_{\{g_1, g_2\} \to \{w^2 \tilde{g}_1, w^3 \tilde{g}_1\}}. \tag{62}$$

The last step is to perform the transformation $w = 1/f$ and study β_f in the limit $f \to 0$ instead the limit $w \to \infty$

$$\beta_{\tilde{g}_1} = -\tilde{g}_1 \left(\varepsilon - 4\alpha - \frac{17}{24} \frac{\tilde{g}_1}{(1+f)^2} \right), \tag{63}$$

$$\beta_{\tilde{g}_2} = -\tilde{g}_2 \left(\varepsilon - 6\alpha - \frac{8}{15} \frac{\tilde{g}_1}{(1+f)^2} \right), \tag{64}$$

$$\beta_f = f \left(2\alpha - \frac{3}{8} \frac{\tilde{g}_2}{f(1+f)^2} \right). \tag{65}$$

The list of all fixed points is shown in Table 3. In this case, we have found only two new fixed points - TLR Gaussian and TLR Model A fixed point. It is interesting that in the one loop approximation there is no TLR Active fluid nor TLR Navier-Stokes fixed point. The physically most interesting is the case $\varepsilon = 1(d = 3)$, where for the small α, the model belongs to the universality class of SR Active fluid. However,

Table 3 Fixed points with their stability. SR - short-range, TLR - true long-range

FP/g_i^*|λ_i	g_1^*|$\bar{g}_1^*$	g_2^*|$\bar{g}_2^*$	w^*	λ_1	λ_2	λ_3
FP0 - SR Gaussian	0	0	0	$-\varepsilon$	$-\varepsilon$	-2α
FPI - SR Navier-Stokes	0	$\frac{8}{3}\varepsilon$	0	$-\frac{1}{3}\varepsilon$	ε	$\frac{1}{3}(\varepsilon - 6\alpha)$
FPII - SR Model A	$\frac{24}{17}\varepsilon$	0	0	ε	$-\frac{31}{51}\varepsilon$	-2α
FPIII - SR Active fluid	$\frac{72}{113}\varepsilon$	$\frac{248}{113}\varepsilon$	0	ε	$\frac{31}{113}\varepsilon$	$\frac{1}{133}(31\varepsilon - 226\alpha)$
FPIV - TLR Gaussian	0	0	∞	$4\alpha - \varepsilon$	$6\alpha - \varepsilon$	2α
FPVI - TLR Model A	$\frac{24}{17}(\varepsilon - 4\alpha)$	0	∞	$\varepsilon - 4\alpha$	$\frac{1}{51}(226\alpha - 31\varepsilon)$	2α

at $\alpha > 31/266 \approx 0.137$ a crossover to the TLR Model A universality occurs and for $\alpha > 0.25$ the degree of non-locality is so strong that the relevance of all non-linearities is destroyed completely and the mean-field approximation is valid.

6 Conclusion

In this work, we have been studying the effect of non-local interactions on the order-disorder phase transition in the incompressible active fluid. Starting from the Toner-Tu theory, we have constructed the model by including non-local shear stress to the hydrodynamic description of the system. Using standard approach, we have obtained the De Dominicis-Janssen response functional. The renormalization group procedure was then based on the analysis of the UV divergences of the corresponding formulation. In total, nine Feynman diagrams were calculated and the model was successfully renormalized. Anomalous dimensions and beta functions were obtained which allowed us to find six fixed points corresponding to six different universality classes and investigate their stability. In the case of $d = 3$, we have found that although for small values of α the system belongs to the universality class of the incompressible active fluid, for $\alpha \gtrsim 1.37$ there is a crossover to the universality class of the well known Model A with long range interactions. In addition, for $\alpha > 0.25$ the magnitude of the LR interactions destroys the relevance of the non-linearities and the mean-field approximation becomes valid.

Acknowledgements The work was supported by VEGA grant No. 1/0345/17 of the Ministry of Education, Science, Research and Sport of the Slovak Republic.

References

1. L. Chen et al., New J. Phys. **17**, 042002 (2015)
2. T. Vicsek, Phys. Rev. Lett. **75**, 1226 (1995)
3. J. Toner, Y. Tu, Phys. Rev. Lett. **75**, 4326 (1995)
4. Y.-H. Tu, J. Toner, M. Ulm, Phys. Rev. Lett. **80**, 4819 (1998)
5. J. Toner, Y.-H. Tu, Phys. Rev. E **58**, 4828 (1998)
6. J. Toner, Y.-H. Tu, S. Ramaswamy, Ann. Phys. **318**, 170 (2005)
7. H.-K. Janssen, U.C. Täuber, Ann. Phys. **315**, 147 (2005)
8. L.T. Adzhemyan, N.V. Antonov, A.N. Vasiliev, *The Field Theoretic Renormalization Group in Fully Developed Turbulence* (CRC Press, Boca Raton, FL, 1999), p. 208. ISBN: 9789056991456
9. M. Hnatič, J. Honkonen, T. Lučivjanský, Acta Phys. Slovaca **66**, 69 (2016)
10. L.D. Landau, E.M. Lifshitz, *Fluid Mechanics*, 2nd edn. (Pergamon Press, Oxford) [Russian eds. (1944, 1953, 1986, 1987)]
11. J. Dunkel et al., New J. Phys. **15**, 045016 (2013)
12. J. Swift, P.C. Hohenberg, Phys. Rev. A **15**, 319–28 (1977)
13. T.J. Pedley, Exp. Mech. **50**, 1293 (2010)
14. M. Di Paola, M. Zingales, Int. J. Solids Struct. **45**(21), 5642–5659 (2008)
15. G.S. Samko, A.A. Kilbas, O.I. Marichev, *Fractional Integrals and Derivatives: Theory and Applications* (Gordon and Breach, New York, NY, 1993)
16. A.N. Vasil'ev, *The Field Theoretic Renormalization Group in Critical Behavior Theory and Stochastic Dynamics* (Chapman and Hall/CRC Press, New York, 2004)
17. U.C. Täuber, *Critical Dynamics: A Field Theory Approach to Equilibrium and Non-Equilibrium Scaling Behavior* (Cambridge University Press, Cambridge, 2014)
18. J. Zinn-Justin, *Quantum Field Theory and Critical Phenomena*, 4th edn. (Oxford University Press, Oxford, 2002)

On the Exponential Decay of Solutions in Dual-Phase-Lag Porous Thermoelasticity

José R. Fernández, Antonio Magaña and Ramón Quintanilla

Abstract In the last years, a big interest has been developed to understand the time decay of solutions for the porous thermoelasticity with different thermal mechanisms. We here want to consider the problem of the one-dimensional porous thermoelasticity when the heat conduction is given by means of the dual-phase-lag theory. We want to give suitable conditions in order to guarantee that the decay of solutions is controlled by a negative exponential. We also want to provide conditions for the slow decay of the solutions.

Keywords One-dimensional porous thermoelasticity · Time decay · Dual-phase-lag · Semi-group arguments · Spectral arguments

1 Introduction and Basic Equations

Experimental observation shows that the classical heat continuum theory cannot be used to describe satisfactorily some thermal phenomena. At the same time, the behavior of the thermal waves obtained from the combination of the Fourier law with the equation

$$c\dot{\theta} = -\nabla \mathbf{q} \tag{1.1}$$

where θ denotes the temperature, $\mathbf{q}$ is the heat flux vector and c is the thermal capacity ($c > 0$), violates the principle of causality. In order to overcome these difficulties, several alternative proposals have been stated. In this sense, it is suitable to recall the damped hyperbolic equation proposed by Cattaneo and Maxwell [2]. This heat

J. R. Fernández
Departamento de Matemática Aplicada I, Universidade de Vigo,
ETSI Telecomunicación, Campus As Lagoas Marcosende s/n, 36310 Vigo, Spain

A. Magaña (✉) · R. Quintanilla
Departament de Matemàtiques, Universitat Politècnica de Catalunya,
C. Colom 11, 08222 Terrassa, Barcelona, Spain
e-mail: antonio.magana@upc.edu

© Springer Nature Switzerland AG 2019
C. H. Skiadas and I. Lubashevsky (eds.), *11th Chaotic Modeling and Simulation International Conference*, Springer Proceedings in Complexity, https://doi.org/10.1007/978-3-030-15297-0_6

51

conduction model has been extended to a couple of thermoelastic theories: one proposed by Lord and Shulman [14] and the other by Green and Lindsay [7]. We can also recall the theories of Green and Naghdi [8–10] which determine three possible models named as type I, II and III, respectively. The easier one, the type I, is similar to the classical theory. The type II is also known as the *thermoelasticity without energy dissipation* because in this theory no dissipation for the energy is assumed. In this setting, a system of conservative hyperbolic equations is proposed. The most general is the type III, which contains the two previous models as limit cases.

Following this line of alternative proposals, we can recall that in 1995 Tzou [25] proposed a theory in which the heat flux and the gradient of the temperature have a delay in the constitutive equations. When this consideration is taken into account, it is usual to speak about phase-lag theories. In this case, the constitutive equations are given by

$$q_i(\mathbf{x}, t + \tau_1) = -k\theta_{,i}(\mathbf{x}, t + \tau_2) \tag{1.2}$$

where τ_1 and τ_2 are the delay parameters, which are assumed to be positive and k is the thermal conductivity ($k > 0$).

Unfortunately, if we combine Eqs. (1.1) and (1.2), the problem that arises is ill posed in the sense of Hadamard [5]. At the same time, as it is pointed out in [6], this model disagrees with the Second Law of Thermodynamics. In fact, the solutions have a very explosive behavior and, therefore, the proposed model does not seem to be a good candidate to describe the heat conduction phenomenon, nor from a mathematical point of view neither from a thermomechanical perspective. However, many scientists have been attracted by the theories obtained when some Taylor approximations are considered in the model. This consideration has deserved a lot of research [19–24].

In this paper we propose a second-order Taylor approximation for the heat flux vector and a first-order approximation for the temperature. That is, we take

$$\begin{aligned}
\mathbf{q}(\mathbf{x}, t + \tau_1) &\approx \mathbf{q}(\mathbf{x}) + \tau_1 \dot{\mathbf{q}}(\mathbf{x}) + \frac{\tau_1^2}{2}\ddot{\mathbf{q}}(\mathbf{x}), \\
\theta(\mathbf{x}, t + \tau_2) &\approx \theta(\mathbf{x}) + \tau_2 \dot{\theta}(\mathbf{x}).
\end{aligned} \tag{1.3}$$

If we combine these two approximations with the heat equation (1.2) we find

$$q_i(t) + \tau_1 \dot{q}_i(t) + \frac{\tau_1^2}{2}\ddot{q}(t) = -k\left(\theta_{,i}(t) + \tau_2\dot{\theta}_{,i}(t)\right). \tag{1.4}$$

On the other hand, Nunziato and Cowin [17] established a theory for the behavior of porous solids in which the skeletal or matrix material is elastic and the interstices are voids of the material. The intended applications of this theory are to geological materials such as rocks and soils or to manufactured porous materials as ceramics or pressed powders. It is worth recalling that the linear theory of elastic materials with voids has been established by Cowin and Nunziato [4]. Later, thermal effects were incorporated to this theory [12].

Many contributions have been written to analyze the asymptotic behavior (with respect to the time) of the solutions to the thermo-porous-elasticity problem [1, 15, 16]. This behavior strongly depends on the model of heat conduction that is assumed. In this work we also analyze the asymptotic behavior of the solutions for the one-dimensional model of the dual-phase-lag thermo-porous-elasticity that comes from Eq. (1.4).

The plan of the paper is the following. In the next section we recall the basic equations and conditions to the problem. Existence and uniqueness are obtained in Sect. 3. Sections 4 and 5 are devoted to study the asymptotic behavior of the solutions for two different situations. First we prove that if thermal and porous dissipation are present in the system, then the solutions are exponentially stable. Later we show that the thermal dissipation is not strong enough to lead the system to the equilibrium in an exponential way. We finish the paper by given some conclusions.

2 Preliminaries

In this section we consider the one-dimensional model of the dual-phase-lag thermo-porous-elasticity. It is worth noting that we can obtain the system in a similar way as Chandrasekharaiah proposed the system of equations [3]. We have

$$
\begin{cases}
\rho \ddot{u} = \mu u_{xx} + a\theta_x + b\varphi_x \\
I\ddot{\varphi} = \alpha\varphi_{xx} - bu_x - \xi\varphi + m\theta - \epsilon\dot{\varphi} \\
c(\dot{\theta} + \tau_1\ddot{\theta} + \frac{\tau_1^2}{2}\dddot{\theta}) = k\theta_{xx} + k\tau_2\dot{\theta}_{xx} + a(\dot{u}_x + \tau_1\ddot{u}_x + \frac{\tau_1^2}{2}\dddot{u}_x) - m(\dot{\varphi} + \tau_1\ddot{\varphi} + \frac{\tau_1^2}{2}\dddot{\varphi})
\end{cases}
\tag{2.1}
$$

Here, u is the displacement, θ is the temperature and φ is the volume fraction. As usual, ρ denotes the mass density and c the thermal capacity, μ is the shear elastic modulus, a is the coupling tensor between the displacement and the temperature, m is the coupling between the porosity and the temperature, the meaning of I is also well known, τ_1 and τ_2 are the relaxation parameters and k is the thermal conductivity.

We want to study our system in $[0, \pi] \times [0, \infty)$. To have a well posed problem we need to impose initial and boundary conditions. As initial conditions we assume

$$
\begin{aligned}
&u(\mathbf{x}, 0) = u^0(\mathbf{x}), \dot{u}(\mathbf{x}, 0) = v^0(\mathbf{x}), \varphi(\mathbf{x}, 0) = \varphi^0(\mathbf{x}), \dot{\varphi}(\mathbf{x}, 0) = \phi^0(\mathbf{x}) \\
&\theta(\mathbf{x}, 0) = \theta^0(\mathbf{x}), \dot{\theta}(\mathbf{x}, 0) = \zeta^0(\mathbf{x}), \ddot{\theta}(\mathbf{x}, 0) = \psi^0(\mathbf{x}) \quad \text{for } \mathbf{x} \in [0, \pi].
\end{aligned}
\tag{2.2}
$$

And we impose homogeneous boundary conditions:

$$
u(0, t) = u(\pi, t) = \varphi_x(0, t) = \varphi_x(\pi, t) = \theta_x(0, t) = \theta_x(\pi, t) = 0 \quad \text{for } t \geq 0.
\tag{2.3}
$$

In this paper we assume that:

$$
\rho > 0, \ \mu > 0, \ I > 0, \ \mu\xi > b^2, \ c > 0, \ k > 0, \ 2\tau_2 > \tau_1, \ \epsilon \geq 0.
\tag{2.4}
$$

It is well known that the axioms of thermomechanics imply that the thermal conductivity, k, cannot have negative sign. The meaning of the positivity of ρ, I and c is clear. The assumptions on μ, ξ, α and b are related to the stability of the system and the assumption on the delays is usual in the dual-phase-lag heat conduction which is related to the stability of the heat conduction equation. We do not restrict the sign of a, m and b, but it will be relevant to assume that a must be different from zero. Concerning the parameter ϵ we will assume that it is non-negative to guarantee that the dissipation is positive. We will prove the exponential stability of our system when ϵ is positive and the slow decay of solutions when ϵ vanishes.

3 Existence of Solutions

In this section we give an existence result for the problem determined by system (2.1) with initial conditions (2.2) and boundary conditions (2.3) whenever conditions (2.4) hold. We will use the notation

$$\hat{f} = f + \tau_1 \dot{f} + \frac{\tau_1^2}{2} \ddot{f}. \tag{3.1}$$

System (2.1) can be expressed as

$$
\begin{cases}
\dot{\hat{u}} = \hat{v} \\
\dot{\hat{v}} = \rho^{-1}(\mu \hat{u}_{xx} + b\hat{\varphi}_x + a(\theta_x + \tau_1 \zeta_x + \frac{\tau_1^2}{2}\psi_x)) \\
\dot{\hat{\varphi}} = \hat{\phi} \\
\dot{\hat{\phi}} = I^{-1}(\alpha\hat{\varphi}_{xx} - b\hat{u}_x - \xi\varphi - \epsilon\hat{\phi} + m(\theta + \tau_1\zeta + \frac{\tau_1^2}{2}\psi)) \\
\dot{\theta} = \zeta \\
\dot{\zeta} = \psi \\
\dot{\psi} = \frac{2}{c\tau_1^2}(k\theta_{xx} + k\tau_2\zeta_{xx}) + \frac{2a}{c\tau_1^2}\hat{v}_x - \frac{2m}{c\tau_1^2}\hat{\phi} - \frac{2}{\tau_1^2}\zeta - \frac{2}{\tau_1}\psi
\end{cases}
\tag{3.2}
$$

Notice that, if we find $\hat{u}$, therefore we can also find u by solving a second-order ordinary differential equation. Hence, we omit the hat in our equations to simplify the notation. The system becomes

$$
\begin{cases}
\dot{u} = v \\
\dot{v} = \rho^{-1}(\mu u_{xx} + b\varphi_x + a(\theta_x + \tau_1\zeta_x + \frac{\tau_1^2}{2}\psi_x)) \\
\dot{\varphi} = \phi \\
\dot{\phi} = I^{-1}(\alpha\varphi_{xx} - bu_x - \xi\varphi - \epsilon\phi + m(\theta + \tau_1\zeta + \frac{\tau_1^2}{2}\psi)) \\
\dot{\theta} = \zeta \\
\dot{\zeta} = \psi \\
\dot{\psi} = \frac{2}{c\tau_1^2}(k\theta_{xx} + k\tau_2\zeta_{xx}) + \frac{2a}{c\tau_1^2}v_x - \frac{2m}{c\tau_1^2}\phi - \frac{2}{\tau_1^2}\zeta - \frac{2}{\tau_1}\psi
\end{cases}
\tag{3.3}
$$

To prove the existence and uniqueness of solutions to the problem determined by this system with the initial and boundary conditions that we have imposed we consider the Hilbert space

$$\mathcal{H} = \{(u, v, \varphi, \phi, \theta, \zeta, \psi), u \in W_0^{1,2}, \ v \in L^2, \ \varphi, \theta, \zeta \in W_*^{1,2}, \phi, \psi \in L_*^2\}$$

where $W_0^{1,2}, L^2$ are the usual Hilbert spaces and

$$L_*^2 = \{f \in L^2, \int_0^\pi f\,dx = 0\}, \ \ W_*^{1,2} = \{f \in W^{1,2}, \int_0^\pi f\,dx = 0\}.$$

In this Hilbert space, we define the inner product

$$< (u, v, \varphi, \phi, \theta, \zeta, \psi), (u^*, v^*, \varphi^*, \phi^*, \theta^*, \zeta^*, \psi^*) = \frac{1}{2}\int_0^\pi W\,dx, \qquad (3.4)$$

where

$$W = \rho v \bar{v}^* + \mu u_x \bar{u}_x^* + I \phi \bar{\phi}^* + \alpha \varphi_x \bar{\varphi}_x^* + b(u_x \bar{\varphi}^* + \bar{u}_x^* \varphi)$$

$$+ c(\theta + \tau_1 \zeta + \frac{\tau_1^2}{2}\psi)(\bar{\theta}^* + \tau_1 \bar{\zeta}^* + \frac{\tau_1^2}{2}\bar{\psi}^*) \qquad (3.5)$$

$$+ k(\tau_1 + \tau_2)\theta_x \bar{\theta}_x^* + \frac{k\tau_1^2 \tau_2}{2}\zeta \bar{\zeta}_x^* + \frac{k\tau_1^2}{2}(\theta_x \bar{\zeta}_x^* + \zeta_x \bar{\theta}_x^*),$$

where the bar means the complex conjugate.

It is worth noting that the norm induced by this inner product is equivalent to the usual norm in $\mathcal{H}$.

We define the matrix operator

$$\mathcal{A} = \begin{pmatrix} 0 & Id & 0 & 0 & 0 & 0 & 0 \\ \frac{\mu}{\rho}D^2 & 0 & \frac{b}{\rho}D & 0 & \frac{a}{\rho}D & \frac{a\tau_1}{\rho}D & \frac{a\tau_1^2}{\rho}D \\ 0 & 0 & 0 & Id & 0 & 0 & 0 \\ -\frac{b}{I}D & 0 & \frac{\alpha D^2 - \xi}{I} & -\frac{\epsilon}{I} & \frac{m}{I} & \frac{m\tau_1}{I} & \frac{m\tau_1^2}{I} \\ 0 & 0 & 0 & 0 & 0 & Id & 0 \\ 0 & 0 & 0 & 0 & 0 & 0 & Id \\ 0 & \frac{2a}{c\tau_1^2}D & 0 & -\frac{2m}{c\tau_1^2} & \frac{2k}{c\tau_1^2}D^2 & \frac{2k\tau_2}{c\tau_1^2}D^2 - \frac{2}{\tau_1^2} & -\frac{2}{\tau_1} \end{pmatrix}, \qquad (3.6)$$

where Id is the identity operator and D denotes the derivative with respect to x.

We can write our problem as a Cauchy problem in the Hilbert space $\mathcal{H}$ as

$$\frac{dU}{dt} = \mathcal{A}U, \quad U_0 = (u_0, v_0, \varphi_0, \phi_0, \theta_0, \zeta_0, \psi_0), \qquad (3.7)$$

where $U = (u, v, \varphi, \phi, \theta, \zeta, \psi)$. In this case the domain of the operator $\mathcal{A}, \mathcal{D}(\mathcal{A})$, is the set of $U \in \mathcal{H}$ such that $\mathcal{A}U \in \mathcal{H}$. Therefore the domain is a dense subspace of $\mathcal{H}$

Lemma 1 *The operator $\mathcal{A}$ is dissipative. That is*

$$\Re\langle \mathcal{A}U, U \rangle \le 0, \tag{3.8}$$

for every $U \in \mathcal{D}(\mathcal{A})$.

Proof If we take into account the boundary conditions and the evolution equation we see that

$$2\Re\langle \mathcal{A}U, U \rangle = -\epsilon \int_0^\pi |\phi|^2 dx - k \int_0^\pi |\theta_x|^2 dx - \frac{k\tau_1}{2}(2\tau_2 - \tau_1) \int_0^\pi |\zeta_x|^2 dx. \tag{3.9}$$

Since we assume that $\epsilon \ge 0, k > 0$ and that $2\tau_2 > \tau_1$, the lemma is proved. $\square$

Lemma 2 *The operator $\mathcal{A}$ satisfies that Range $(\mathcal{A}) = \mathcal{H}$.*

Proof For $F = (f_1, f_2, f_3, f_4, f_5, f_6, f_7) \in \mathcal{H}$, the equation $\mathcal{A}U = F$ can be seen as

$$
\begin{cases}
v = f_1 \\
\mu u_{xx} + b\varphi_x + a(\theta_x + \tau_1\zeta_x + \frac{\tau_1^2}{2}\psi_x) = \rho f_2 \\
\phi = f_3 \\
\alpha\varphi_{xx} - bu_x - \xi\varphi - \epsilon\phi + m(\theta + \tau_1\zeta + \frac{\tau_1^2}{2}\psi) = I f_4 \\
\zeta = f_5 \\
\psi = f_6 \\
k\theta_{xx} + k\tau_2\zeta_{xx} + av_x - m\phi - c\zeta - c\tau_1\psi = \frac{c\tau_1^2}{2} f_7
\end{cases}
\tag{3.10}
$$

We obtain straightforwardly the values of v, ϕ, ζ and ψ. Substituting them into the other equations we obtain the system

$$
\begin{cases}
\mu u_{xx} + b\varphi_x = -a\theta_x - a\tau_1 f_{5,x} - \frac{a\tau_1^2}{2} f_{6,x} + \rho f_2 \\
\alpha\varphi_{xx} - bu_x - \xi\varphi = \epsilon f_3 - m\theta - m\tau_1 f_5 - \frac{m\tau_1^2}{2} f_6 + I f_4 \\
k\theta_{xx} = -k\tau_2 f_{5,xx} - af_{1,x} + mf_5 + cf_3 + c\tau_1 f_6 + \frac{c\tau_1^2}{2} f_7
\end{cases}
\tag{3.11}
$$

Now, we consider

$$f_1 = \sum_{n\ge 1} f_1^n \sin nx, \quad f_2 = \sum_{n\ge 1} f_2^n \sin nx, \quad f_3 = \sum_{n\ge 1} f_3^n \cos nx, \quad f_4 = \sum_{n\ge 1} f_4^n \cos nx,$$

$$f_5 = \sum_{n\ge 1} f_5^n \cos nx, \quad f_6 = \sum_{n\ge 1} f_6^n \cos nx, \quad f_7 = \sum_{n\ge 1} f_7^n \cos nx.$$

As $F \in \mathcal{H}$, we know that the following conditions are satisfied:

$$\sum_{n=1}^\infty n^2 |f_i^n|^2 < \infty, \text{ for } i = 1, 3, 5, 6 \text{ and } \sum_{n=1}^\infty |f_j^n|^2 < \infty, \text{ for } i = 2, 4, 7.$$

We can find solutions for u, φ and θ of the form

$$u = \sum_{n\geq 1} u^n \sin nx, \quad \varphi = \sum_{n\geq 1} \varphi^n \cos nx, \quad \theta = \sum_{n\geq 1} \theta^n \cos nx,$$

where u^n, φ^n and θ^n must satisfy

$$\sum_{n=1}^{\infty} n^2 |u^n|^2 < \infty, \quad \sum_{n=1}^{\infty} n^2 |\varphi^n|^2 < \infty \text{ and } \sum_{n=1}^{\infty} |\theta^n|^2 < \infty. \tag{3.12}$$

From the third equation of system (3.11) we can find θ. To be precise, we have found

$$\theta^n = -\frac{-anf_1^n + \frac{1}{2}c\tau_1^2 f_7^n + c\tau_1 f_6^n + cf_3^n + (kn^2\tau_2 + m) f_5^n}{kn^2}.$$

Therefore we can obtain u and φ (we have used Mathematica to find them):

$$u^n = -\frac{f_1^n \left(-\frac{a^2(\alpha n^2 + \xi)}{k} - \frac{abm}{k}\right)}{n^2 \left(b^2 - \mu\left(\alpha n^2 + \xi\right)\right)}$$

$$-\frac{f_6^n \left(\frac{ac\tau_1(\alpha n^2 + \xi)}{kn} - \frac{1}{2}an\tau_1^2\left(\alpha n^2 + \xi\right) + \frac{bcm\tau_1}{kn} - \frac{1}{2}bmn\tau_1^2\right)}{n^2 \left(b^2 - \mu\left(\alpha n^2 + \xi\right)\right)}$$

$$-\frac{f_7^n \left(\frac{ac\tau_1^2(\alpha n^2 + \xi)}{2kn} + \frac{bcm\tau_1^2}{2kn}\right)}{n^2 \left(b^2 - \mu\left(\alpha n^2 + \xi\right)\right)} - \frac{f_3^n \left(\frac{ac(\alpha n^2 + \xi)}{kn} + \frac{bcm}{kn} + bn\epsilon\right)}{n^2 \left(b^2 - \mu\left(\alpha n^2 + \xi\right)\right)}$$

$$-\frac{f_5^n \left(\frac{a(\alpha n^2 + \xi)(kn^2(\tau_2 - \tau_1) + m)}{kn} + \frac{bm(kn^2\tau_2 + m)}{kn} - bmn\tau_1\right)}{n^2 \left(b^2 - \mu\left(\alpha n^2 + \xi\right)\right)}$$

$$-\frac{bf_4^n J}{n\left(b^2 - \mu\left(\alpha n^2 + \xi\right)\right)} - \frac{f_2^n \rho\left(-\alpha n^2 - \xi\right)}{n^2 \left(b^2 - \mu\left(\alpha n^2 + \xi\right)\right)}$$

$$\varphi^n = -\frac{f_7^n \left(abc\tau_1^2 + c\mu m\tau_1^2\right)}{2kn^2 \left(\mu\left(\alpha n^2 + \xi\right) - b^2\right)} - \frac{f_6^n \left(2abc\tau_1 - abkn^2\tau_1^2 + 2c\mu m\tau_1 - k\mu mn^2\tau_1^2\right)}{2kn^2 \left(\mu\left(\alpha n^2 + \xi\right) - b^2\right)}$$

$$-\frac{f_3^n \left(abc + c\mu m + k\mu n^2\epsilon\right)}{kn^2 \left(\mu\left(\alpha n^2 + \xi\right) - b^2\right)}$$

$$-\frac{f_5^n \left(2kn^2\tau_2(ab + \mu m) - 2abkn^2\tau_1 + 2abm - 2k\mu mn^2\tau_1 + 2\mu m^2\right)}{2kn^2 \left(\mu\left(\alpha n^2 + \xi\right) - b^2\right)}$$

$$+\frac{af_1^n(ab + \mu m)}{kn \left(\mu\left(\alpha n^2 + \xi\right) - b^2\right)} - \frac{f_4^n J\mu}{\mu\left(\alpha n^2 + \xi\right) - b^2} + \frac{bf_2^n \rho}{n \left(\mu\left(\alpha n^2 + \xi\right) - b^2\right)}$$

It is clear that u, φ and θ satisfy conditions (3.12). Furthermore: we can see that $\|U\| \leq K\|F\|$ for some K independent of F.

If we recall the Lumer-Phillips corollary to the Hille-Yosida theorem we obtain the following result.

Theorem 3 *The operator $\mathcal{A}$ generates a contractive semigroup, $S(t) = \{e^{t\mathcal{A}}\}$, in $\mathcal{H}$, and for each $U_0 \in \mathcal{D}$ there exists a unique solution $U(t) \in C^1([0, \infty), \mathcal{H}) \cap C^0([0, \infty), \mathcal{D})$ to the problem determined by system (2.1) with initial conditions (2.2) and boundary conditions (2.3).*

4 Case $\epsilon > 0$. Exponential Decay of Solutions

The aim of this section is to prove that when ϵ is strictly positive the solutions of our problem are exponentially stable. Therefore, in this section, we assume that $\epsilon > 0$. We will use the following characterization, stated in the book of Liu and Zheng, that ensures the exponential decay (see [11, 13] or [18]).

Theorem 4 *Let $S(t) = \{e^{\mathcal{A}t}\}_{t\geq 0}$ be a C_0-semigroup of contractions on a Hilbert space. Then $S(t)$ is exponentially stable if and only if the following two conditions are satisfied:*

(i) $i\mathbb{R} \subset \rho(\mathcal{A})$, (here $\rho(\mathcal{A})$ means the resolvent of $\mathcal{A}$).
(ii) $\overline{\lim}_{|\lambda|\to\infty} \|(i\lambda\mathcal{I} - \mathcal{A})^{-1}\|_{\mathcal{L}(\mathcal{H})} < \infty$.

The main theorem of this section is:

Theorem 5 *The operator $\mathcal{A}$ defined in (3.6) generates a semigroup which is exponentially stable.*

Proof First, we prove the first condition of Theorem 4. Following the arguments given by Liu and Zheng ([13], page 25), the proof consists of the following steps:

(i) Since 0 is in the resolvent of $\mathcal{A}$, using the contraction mapping theorem, we have that, for any real λ such that $|\lambda| < \|\mathcal{A}^{-1}\|^{-1}$, the operator $i\lambda\mathcal{I} - \mathcal{A} = \mathcal{A}(i\lambda\mathcal{A}^{-1} - \mathcal{I})$ is invertible. Moreover, $\|(i\lambda\mathcal{I} - \mathcal{A})^{-1}\|$ is a continuous function of λ in the interval $(-\|\mathcal{A}^{-1}\|^{-1}, \|\mathcal{A}^{-1}\|^{-1})$.

(ii) If $\sup\{\|(i\lambda\mathcal{I} - \mathcal{A})^{-1}\|, |\lambda| < \|\mathcal{A}^{-1}\|^{-1}\} = M < \infty$, then by the contraction theorem, the operator

$$i\lambda\mathcal{I} - \mathcal{A} = (i\lambda_0\mathcal{I} - \mathcal{A})\left(\mathcal{I} + i(\lambda - \lambda_0)(i\lambda_0\mathcal{I} - \mathcal{A})^{-1}\right)$$

is invertible for $|\lambda - \lambda_0| < M^{-1}$. It turns out that, by choosing λ_0 as close to $\|\mathcal{A}^{-1}\|^{-1}$ as we can, the set $\{\lambda, |\lambda| < \|\mathcal{A}^{-1}\|^{-1} + M^{-1}\}$ is contained in the resolvent of $\mathcal{A}$ and $\|(i\lambda\mathcal{I} - \mathcal{A})^{-1}\|$ is a continuous function of λ in the interval $(-\|\mathcal{A}^{-1}\|^{-1} - M^{-1}, \|\mathcal{A}^{-1}\|^{-1} + M^{-1})$.

(iii) Let us assume that the intersection of the imaginary axis and the spectrum is not empty, then there exists a real number ϖ with $||\mathcal{A}^{-1}||^{-1} \leq |\varpi| < \infty$ such that the set $\{i\lambda, |\lambda| < |\varpi|\}$ is in the resolvent of $\mathcal{A}$ and $\sup\{||(i\lambda\mathcal{I} - \mathcal{A})^{-1}||, |\lambda| < |\varpi|\} = \infty$. Therefore, there exist a sequence of real numbers λ_n with $\lambda_n \to \varpi$, $|\lambda_n| < |\varpi|$ and a sequence of vectors $U_n = (u_n, v_n, \varphi_n, \phi_n, \theta_n, \zeta_n, \psi_n)$ in the domain of the operator $\mathcal{A}$ and with unit norm such that

$$\|(i\lambda_n\mathcal{I} - \mathcal{A})U_n\| \to 0. \tag{4.1}$$

If we write (4.1) in components, we obtain the following conditions:

$$i\lambda_n u_n - v_n \to 0, \text{ in } W^{1,2} \tag{4.2}$$

$$i\lambda_n v_n - \frac{1}{\rho}\left(\mu D^2 u_n + bD\varphi_n + a(D\theta + \tau_1 D\zeta_n + \frac{\tau_1^2}{2}D\psi_n)\right) \to 0, \text{ in } L^2 \tag{4.3}$$

$$i\lambda_n \varphi_n - \phi_n \to 0, \text{ in } W^{1,2} \tag{4.4}$$

$$i\lambda_n \phi_n - \frac{1}{I}\left(\alpha D^2\varphi - bDu_n - \xi\varphi_n - \epsilon\phi_n + m(\theta_n + \tau_1\zeta_n + \frac{\tau_1^2}{2}\psi_n)\right) \to 0, \text{ in } L^2 \tag{4.5}$$

$$i\lambda_n \theta_n - \zeta_n \to 0, \text{ in } W^{1,2} \tag{4.6}$$

$$i\lambda_n \zeta_n - \psi_n \to 0, \text{ in } W^{1,2} \tag{4.7}$$

$$i\lambda_n \psi_n - \frac{2}{c\tau_1^2}\left(kD^2\theta_n + k\tau_2 D^2\zeta_n + aDv_n - m\phi_n - c\zeta_n - \tau_1\psi_n\right) \to 0, \text{ in } L^2 \tag{4.8}$$

If we consider $\Re\langle((i\lambda_n\mathcal{I} - \mathcal{A})U_n), U_n\rangle$ we see that ϕ_n tends to zero in L^2 and θ_n, ζ_n tend also to zero in $W^{1,2}$. It also implies that $\lambda_n^{-1}\psi_n$ tends to zero in $W^{1,2}$ and $\lambda_n\varphi_n$ tends to zero in L^2. From (4.5) we obtain that $\lambda_n^{-1}(\alpha D^2\varphi_n - bDu_n) \to 0$ is bounded in L^2. If we multiply by $\lambda_n\varphi_n$ in L^2 we see that $D\varphi_n$ tends to zero.

We now multiply (4.8) by $\lambda_n^{-1}\psi_n$ to obtain that

$$i\|\psi_n\|^2 + \frac{2k}{\lambda_n c\tau_1^2} < D\theta_n, D\psi_n > + \frac{2k\tau_2}{\lambda_n c\tau_1^2} < D\zeta_n, D\psi_n >$$

$$- \frac{2a}{\lambda_n c\tau_1^2} < Dv_n, \psi_n > + \frac{2m}{\lambda_n c\tau_1^2} < \phi_n, \psi_n >$$

$$+ \frac{2}{\lambda_n \tau_1^2} < \zeta_n, \psi_n > + \frac{2}{\lambda_n \tau_1}\|\psi_n\|^2 \to 0.$$

But

$$\lambda_n^{-1} < Dv_n, \psi_n > \sim -i < v_n, D\zeta_n > \to 0.$$

Therefore we also obtain that ψ_n tends to zero in L^2. Again, from (4.8) we see that

$$\lambda_n^{-1}(kD^2\theta_n + k\tau_2 D^2\zeta_n + aDv_n) \to 0 \text{ in } L^2.$$

We then obtain that

$$\lambda_n^{-1}(kD^2\theta_n + k\tau_2 D^2\zeta_n) + aDu_n \to 0 \text{ in } L^2,$$

and taking the inner product by Du_n we see

$$\langle \lambda_n^{-1}(kD^2\theta_n + k\tau_2 D^2\zeta_n), Du_n \rangle + a\|Du_n\|^2 \to 0.$$

In view of the boundary conditions this is equivalent to

$$-\langle (kD\theta_n + k\tau_2 D\zeta_n), \lambda_n^{-1} D^2 u_n \rangle + a\|Du_n\|^2 \to 0.$$

In view of (4.3) we see that $\lambda_n^{-1} D^2 u_n$ is bounded and then we conclude that Du_n tends to zero. Now, if we consider again (4.3) we arrive to

$$i\rho\|v_n\|^2 + \mu\langle Du_n, \lambda_n^{-1} Dv_n \rangle + \frac{a\tau_1^2}{a}\langle \lambda_n^{-1} D\psi_n, v_n \rangle \to 0,$$

and then v_n tends to zero in L^2, which contradicts the assumption that the sequence has unit norm and, in consequence, the first condition of Theorem 4 holds.

Now, we will prove the second condition of Theorem 4. Again we use a contradiction argument. If this condition is not true, then there exist a sequence λ_n such that $|\lambda_n| \to \infty$ and a sequence of unit norm vectors in the domain of the operator such that (4.1) holds. But for this point we can use the same argument we have used to prove the first condition because the only relevant fact is that the sequence of the λ_n does not tend to zero. We see that both conditions of Theorem 4 hold and then Theorem 5 is proved. □

5 Case $\epsilon = 0$. Slow Decay of Solutions

The aim of this section is to prove that when the parameter ϵ vanishes the decay of the solutions of our system cannot be controlled by an exponential. To this end we are going to use the spectral study of the problem with the help of the Routh-Hurwitz lemma.

Theorem 6 *Let (u, φ, θ) be a solution to the problem determined by (2.1), (2.2) and (2.3). Then (u, φ, θ) decays in a slow way.*

Proof We will prove that there exists a solution to system (2.1) of the form

$$u = K_1 e^{\omega t} \sin(nx), \quad \varphi = K e^{\omega t} \cos(nx), \quad \theta = K_3 e^{\omega t} \cos(nx)$$

such that $\Re(\omega) > -\varepsilon$ for all positive ε. Hence, a solution ω as close as desired to the imaginary axis can be found. Imposing that u, φ and θ are as above and replacing them in (2.1) the following homogeneous system in the unknowns K_1, K_2 and K_3 is obtained:

$$B \begin{pmatrix} K_1 \\ K_2 \\ K_3 \end{pmatrix} = \begin{pmatrix} 0 \\ 0 \\ 0 \end{pmatrix}$$

where

$$B = \begin{pmatrix} \mu n^2 + \rho \omega^2 & bn & an \\ -bn & -\alpha n^2 - I\omega^2 - \xi & m \\ -an\omega \left(\omega^2 \tau_1^2 + 2\omega\tau_1 + 2 \right) & m\omega \left(\omega^2 \tau_1^2 + 2\omega\tau_1 + 2 \right) & 2k \left(\omega\tau_2 + 1 \right) n^2 + c\omega \left(\omega^2 \tau_1^2 + 2\omega\tau_1 + 2 \right) \end{pmatrix}.$$

This system will have non trivial solutions if and only if the determinant of the coefficients matrix is equal to zero. We denote by $p(x)$ the determinant once ω is replaced by x. Straight calculations (made using Mathematica) show that $p(x)$ is a seventh degree polynomial.

To prove that $p(x)$ has roots as close as we want to the complex axis, we will show that for any $\varepsilon > 0$ there are roots of $p(x)$ located on the right hand side of the vertical line $\Re(z) = -\varepsilon$. This fact will be shown if the polynomial $p(x - \varepsilon)$ has a root with positive real part. To prove that, we use the Routh–Hurwitz theorem. It assesses that, if $b_0 > 0$, then all the roots of polynomial

$$b_0 x^7 + b_1 x^6 + b_2 x^5 + b_3 x^4 + b_4 x^3 + b_5 x^2 + b_6 x + b_7$$

have negative real part if and only if all the leading minors of matrix

$$\begin{pmatrix} b_1 & b_0 & 0 & 0 & 0 & 0 & 0 \\ b_3 & b_2 & b_1 & b_0 & 0 & 0 & 0 \\ b_5 & b_4 & b_3 & b_2 & b_1 & b_0 & 0 \\ b_7 & b_6 & b_5 & b_4 & b_3 & b_2 & b_1 \\ 0 & 0 & b_7 & b_6 & b_5 & b_4 & b_3 \\ 0 & 0 & 0 & 0 & b_7 & b_6 & b_5 \\ 0 & 0 & 0 & 0 & 0 & 0 & b_7 \end{pmatrix}$$

are positive. We denote by L_i, for $i = 1, 2, 3, 4, 5, 6, 7$, the leading minors of this matrix.

Direct computations show that the sixth leading minor, L_6, is a eighteenth degree polynomial on n:

$$L_6 = \left(A + B\varepsilon + C\varepsilon^2 \right) \varepsilon n^{18} + R(n),$$

where

$$A = -8a^2 \alpha c I k^2 \mu \tau_1^2 (\tau_1 - 2\tau_2)^2 \rho^2 \left(\alpha \tau_1^2 \left(a^2 J - \alpha c \rho + c I \mu \right) + 2I k \tau_2 (\alpha \rho - I \mu) \right)^2$$

and $R(n)$ is a polynomial on n of degree 16 and A, B and C are constants related to the constitutive coefficients of the system. For ε small enough, but positive, we see that $A + B\varepsilon + C\varepsilon^2$ is negative. Thus, for n large enough, L_6 becomes negative. Therefore

$p(x - \varepsilon)$ has infinite roots with positive real part. (We have used Mathematica to compute L_6.)

This argument shows that the solutions to system (2.1) decay in a slow way, or, in other words, that a uniform rate of decay of exponential type for all the solutions cannot be obtained. Therefore, Theorem 6 is proved. $\square$

6 Conclusions

In this paper we have considered the problem determined by the one dimensional dual-phase-lag porous thermoelasticity. Two cases has been considered: when the porous dissipation is present and when it is not. We have proved that the decay of the solutions can be controlled for an exponential in the first case, but this is not possible in the second case.

Acknowledgements The work of J. R. Fernández has been supported by the Ministerio de Economía y Competitividad under the research project MTM2015-66640-P (with the participation of FEDER). The work of A. Magaña and R. Quintanilla has been supported by the Ministerio de Economía y Competitividad under the research project "Análisis Matemático de Problemas de la Termomecánica", MTM2016-74934-P, (AEI/FEDER, UE).

References

1. P. Casas, R. Quintanilla, Exponential stability in thermoelasticity with microtemperatures. Int. J. Eng. Sci. **43**, 33–47 (2005)
2. C. Cattaneo, On a form of heat equation which eliminates the paradox of instantaneous propagation. C. R. Acad. Sci. Paris **247**, 431–433 (1958)
3. D.S. Chandrasekharaiah, Hyperbolic thermoelasticity: a review of recent literature. Appl. Mech. Rev. **51**, 705–729 (1998)
4. S.C. Cowin, J.W. Nunziato, Linear elastic materials with voids. J. Elast. **13**, 125–147 (1983)
5. M. Dreher, R. Quintanilla, R. Racke, Ill-posed problems in thermomechanics. Appl. Math. Lett. **22**, 1374–1379 (2009)
6. M. Fabrizio, F. Franchi, Delayed thermal models: stability and thermodynamics. J. Therm. Stresses **37**, 160–173 (2014)
7. A.E. Green, K.A. Lindsay, Thermoelasticity. J. Elast. **2**, 1–7 (1972)
8. A.E. Green, P.M. Naghdi, On undamped heat waves in an elastic solid. J. Therm. Stresses **15**, 253–264 (1992)
9. A.E. Green, P.M. Naghdi, Thermoelasticity without energy dissipation. J. Elast. **31**, 189–208 (1993)
10. A.E. Green, P.M. Naghdi, A unified procedure for contruction of theories of deformable media. I. Classical continuum physics, II. Generalized continua, III. Mixtures of interacting continua. Proc. R. Soc. Lond. A **448**, 335–356, 357–377, 379–388 (1995)
11. F.L. Huang, Strong asymptotic stability of linear dynamical systems in Banach spaces. J. Differ. Equ. **104**, 307–324 (1993)
12. D. Iesan, *Thermoelastic Models of Continua* (Kluwer Academic Publishers, Dordrecht, 2004)
13. Z. Liu, S. Zheng, *Semigroups Associated with Dissipative Systems*, vol. 398, Chapman & Hall/CRC Research Notes in Mathematics (Chapman & Hall/CRC, Boca Raton, FL, 1999)

14. H.W. Lord, Y. Shulman, A generalized dynamical theory of thermoelasticity. J. Mech. Phys. Solids **15**, 299–309 (1967)
15. A. Magaña, R. Quintanilla, On the time decay of solutions in one-dimensional theories of porous materials. Int. J. Solids Struct. **43**, 3414–3427 (2006)
16. A. Magaña, R. Quintanilla, On the time decay of solutions in porous-elasticity with quasi-static microvoids. J. Math. Anal. Appl. **331**, 617–630 (2007)
17. J.W. Nunziato, S.C. Cowin, A nonlinear theory of elastic materials with voids. Arch. Ration. Mech. Anal. **72**, 175–201 (1979)
18. J. Prüss, On the spectrum of C_0-semigroups. Trans. Am. Math. Soc. **284**(2), 847–857 (1984)
19. R. Quintanilla, Exponential stability in the dual-phase-lag heat conduction theory. J. Non Equilibr. Thermodyn. **27**, 217–227 (2002)
20. R. Quintanilla, R. Racke, Qualitative aspects in dual-phase-lag thermoelasticity. SIAM J. Appl. Math. **66**, 977–1001 (2006)
21. R. Quintanilla, R. Racke, A note on stability of dual-phase-lag heat conduction. Int. J. Heat Mass Transfer **49**, 1209–1213 (2006)
22. R. Quintanilla, R. Racke, Qualitative aspects in dual-phase-lag heat conduction. Proc. R. Soc. Lond. A **463**, 659–674 (2007)
23. R. Quintanilla, R. Racke, Spatial behavior in phase-lag heat conduction. Differ. Integr. Equ. **28**, 291–308 (2015)
24. S.A. Rukolaine, Unphysical effects of the dual-phase-lag model of heat conduction. Int. J. Heat Mass Transfer **78**, 58–63 (2014)
25. D.Y. Tzou, A unified approach for heat conduction from macro to micro-scales. ASME J. Heat Transfer **117**, 8–16 (1995)

Dynamical Invariant Calculations Involving Evolution Equations with Discontinuities

Avadis Hacınlıyan and Engin Kandıran

Abstract Many models of physical systems involving electronic circuit elements [6], population dynamics [5] involve evolution equations with discontinuities. The key to understand such systems is to hope that the discontinuity does not adversely affect the integration process. There are also three variable chaotic dynamical system examples, such as the Sprott systems for deriving jerky dynamics that have also become of interest [10]. In order to calculate dynamical invariants in chaotic systems such as characteristic exponents and fractal dimensions we often need to find the Jacobian; this often requires attempting to differentiate discontinuous functions. Therefore finding a suitable continuous approximation to the discontinuities becomes important. In previous communications, two example systems had been used with two parametrizations for approximating discontinuous functions with continuous ones, one of which is the same as that used in the literature. In this work, we will use further examples to optimize the parameters of the continuous approximation to discontinuities using different examples in order to test the degree of applicability of this approach. Where possible, the invariants calculated by this method will be compared to the corresponding invariants calculated from its time series.

Keywords Chaotic systems · Sprott systems · Fractal dimension · Lyapunov exponents · Simulation · Chaotic simulation

A. Hacınlıyan
Department of Physics, Yeditepe University, Ataşehir, Istanbul, Turkey
e-mail: avadis@yeditepe.edu.tr

A. Hacınlıyan · E. Kandıran (✉)
Department of Information Systems and Technologies, Yeditepe University,
Ataşehir, Istanbul, Turkey
e-mail: engin.kandiran@yeditepe.edu.tr

E. Kandıran
The Institute for Graduate Studies in Sciences and Engineering,
Yeditepe University, Istanbul, Turkey

© Springer Nature Switzerland AG 2019
C. H. Skiadas and I. Lubashevsky (eds.), *11th Chaotic Modeling and Simulation International Conference*, Springer Proceedings in Complexity, https://doi.org/10.1007/978-3-030-15297-0_7

1 Introduction

Jerky dynamics is an interesting research area for understanding the dynamic behavior of 3-D autonomous system with nonlinearity in dynamical system and chaos theory. It is shown that chaotic dynamical behavior can be observed with three or more dimensions with nonlinearities [1, 4]. In Newtonian mechanics $\ddot{x}$ and $\dot{x}$ are called acceleration and velocity respectively and $\dddot{x}$ is called jerk. A jerk system is described by a third order differential equation which represents the time evolution of a single scalar variable x according to the dynamics:

$$\dddot{x} = J(x, \dot{x}, \ddot{x}) \tag{1}$$

Here $J(x, \dot{x}, \ddot{x})$ is called the jerk function. This representation of function is used to find the answer to the question, which simplest dynamical systems with smooth continuous functions result in chaos. One of the simplest examples of jerk systems is the Coullet system which can be written as:

$$\dddot{x} + a\ddot{x} + \dot{x} = g(x) \tag{2}$$

Here $g(x)$ is nonlinear function such as $g(x) = b(x^2 - 1)$; a and b are constants. For the values $a = 0.6$ and $b = 0.58$, the Coullet system for the given $g(x)$ displays chaotic behavior. In addition to this, many well-known dynamical systems such as Lorenz and Rössler systems, can be written in jerk form [3]. In [8, 9], C. J. Sprott gives 7 jerk system cases which display chaotic behavior with one positive, one negative and one zero Lyapunov exponents. In addition this, Eichorn et al. [2] shows that 14 of the Sprott Systems [7] can be written as jerk system of a single variable x. JD0 and JD1 systems [8] are studied in great detail in [2]. Recently, [11], 3-D jerk chaotic systems with two cubic nonlinearities was studied and it has been shown that the system displays chaotic behavior. In addition to jerk systems, another class of dynamical systems are called hyperjerk systems which are in the form $\frac{d^x}{dt^4} = J(\dddot{x}, \ddot{x}, \dot{x}, x)$ [9].

Jerk systems can be written as 3-D autonomous systems as in the following form:

$$\begin{aligned} \dot{x} &= y \\ \dot{y} &= z \\ \dot{z} &= J(x, y, z) \end{aligned} \tag{3}$$

It is important to note that for all jerk systems only the third equation of the system determines the different behavior of the dynamics.

In this study, we try to analyze the effect of nonlinear terms that are not differentiable in jerk systems (such as $\dot{x}sqn(x)$), but for this study we change the not differentiable function with continuous one with comparable behavior: for instance, instead of $sgn(x)$, we use $tanh(x)$ by using the result of our previous study. (Reference verelim) In Sect. 2, we analyze the JD3 and JD4 systems. Finally, we apply adaptive control method for demonstrating a possible stabilization alternative for the JD3 system.

2 Jerk Systems

2.1 JD3 System

The JD3 system in general form is:

$$\dddot{x} = a\ddot{x} - \dot{x} + bx^2 + x\dot{x} + c \tag{4}$$

where a, b, c are constants Equations of JD3 system in 3-D form is:

$$
\begin{aligned}
\dot{x} &= y \\
\dot{y} &= z \\
\dot{z} &= az - y + bx^2 + xy - c
\end{aligned} \tag{5}
$$

There are two fixed points of the systems: $(-\sqrt{\frac{c}{b}}, 0, 0)$ and $(\sqrt{\frac{c}{b}}, 0, 0)$. For practical advantages it is necessary to transform the fixed point to the origin by letting:

$$\bar{x} = x \pm \sqrt{\frac{c}{b}} \quad \bar{y} = y \quad \text{and} \quad \bar{z} = z \tag{6}$$

Then the system takes the following form and note we drop the overbar in notation:

$$
\begin{aligned}
\dot{x} &= y \\
\dot{y} &= z \\
\dot{z} &= az - y + bx^2 + xy + 2c \pm 2b\sqrt{\frac{c}{b}}x \pm \sqrt{\frac{c}{b}}y
\end{aligned} \tag{7}
$$

To study the stability of the system the Jacobian at the origin is:

$$
J_{(0,0,0)} = \begin{bmatrix} 0 & 1 & 0 \\ 0 & 0 & 1 \\ 2b \pm \sqrt{\frac{c}{b}} & -1 + \sqrt{\frac{c}{b}} & a \end{bmatrix}
$$

The characteristic equation of the Jacobian matrix is also calculated:

$$\lambda^3 - a\lambda^2 + \sqrt{\frac{c}{b}}\lambda + 2b\sqrt{\frac{c}{b}} = 0 \tag{8}$$

According to criterion of Routh-Hurwitz, the fixed point is stable if the following conditions are satisfied:

$$a > 0, \quad \sqrt{\frac{c}{b}} > 0, \quad 2b\sqrt{\frac{c}{b}} > 0 \quad \text{and} \quad a\sqrt{\frac{c}{b}} + 2b\sqrt{\frac{c}{b}} < 0 \tag{9}$$

where the first two given conditions and the last expression vanishes (equals to zero) in Eq. 9 is satisfied, **a limit cycle arises via Hopf-bifurcation**. For a $= -0.6$, b $= 0.9$, c $= 1.0$, the fixed point $(0, 0, 0)$ is stable and the system dynamics displays chaotic behavior as described in [9] and the trajectory of the system is given in Fig. 1. The time evolution Lyapunov spectrum of the system is given in Fig. 2. The Lyapunov exponents of the system are $(0.0809, 0, -0.6810)$.

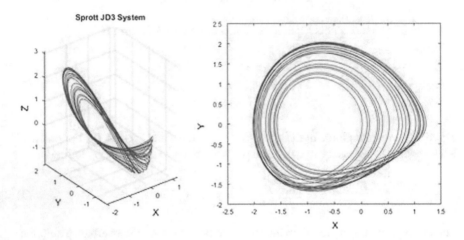

Fig. 1 Trajectory of JD3 system

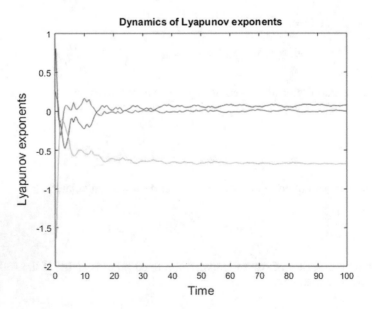

Fig. 2 Lyapunov exponents spectrum of JD3 system

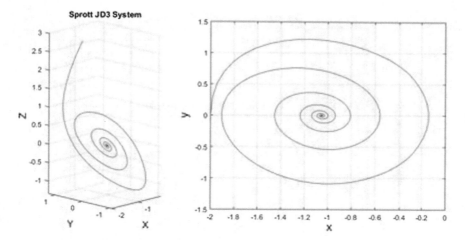

Fig. 3 Trajectory of modified JD3 system

After the analysis of original JD_3 system, we have added a nonlinear term $\dot{x}\tanh x$ to the original jerk function, we have the following jerk form:

$$\dddot{x} = a\ddot{x} - \dot{x} + bx^2 + x\dot{x} + c + \dot{x}\tanh x \tag{10}$$

Then, the new system is studied numerically. With the same set of initial conditions and parameter values we simulate the new system (see Fig. 3). The maximal Lyapunov exponents of the system are $(0.047, -0.1214, 0.4833)$ and the Lyapunov exponent spectrum of modified system is given in Fig. 4.

2.2 JD4 System

The JD4 system in general form is:

$$\dddot{x} = a\ddot{x} + b\dot{x} + cx^2 + x\dot{x} + d \tag{11}$$

where a, b, c, d are constants. This system can be written in 3-D form as:

$$\begin{aligned}\dot{x} &= y \\ \dot{y} &= z \\ \dot{z} &= az + by + cx^2 + xz + d\end{aligned} \tag{12}$$

There are two fixed point of the system: $(\pm\frac{\sqrt{d}}{c}, 0, 0)$ which means only possible choice for the system to have a real fixed point is $d \leq 0$. To study the stability of the

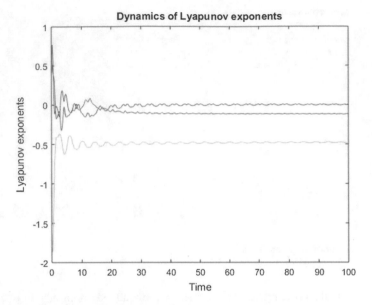

Fig. 4 Lyapunov exponents spectrum of modified JD3 system

system the Jacobian is calculated:

$$J_{(0,0,0)} = \begin{bmatrix} 0 & 1 & 0 \\ 0 & 0 & 1 \\ 2cx + z & y & a + x \end{bmatrix}$$

For $a = -1$, $b = -0.7$, $c = 1$ and $d = -1$, in [7], it is shown that the system displays chaotic behavior with Lyapunov exponents $(0.0734, 0, -1.7769)$. The trajectory of the system with the given parameter values is given in Fig. 5. The dynamics of Lyapunov exponents is given in Fig. 6.

For $d = 0$, the only fixed point of the system is the origin. In that case, Jacobian J has the following characteristic equation.

$$\lambda(\lambda^2 - a\lambda - b) = 0 \tag{13}$$

and the eigenvalues of the Jacobian J are $\lambda_1 = 0$ and $\lambda_{2,3} = \frac{\pm\sqrt{a^2+4b}+a}{2}$. So by looking at the eigenvalues it is not possible to reach a conclusion about the stability of the fixed point.

For $d < 0$ case, the characteristic equation is:

$$\lambda^3 - a\lambda^2 - \sqrt{\frac{d}{c}}\lambda - 2c\sqrt{\frac{d}{c}} = 0 \tag{14}$$

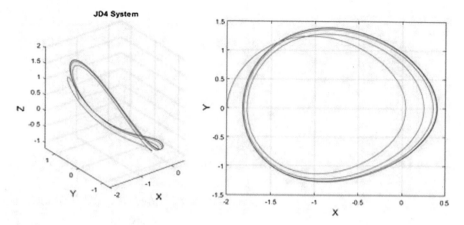

Fig. 5 Trajectory of JD4 system

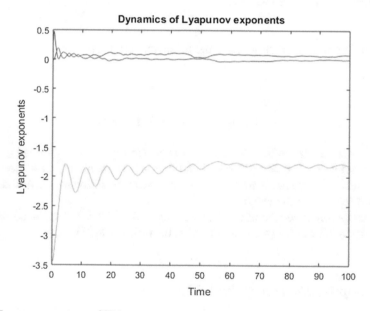

Fig. 6 Lyapunov spectrum of JD4

For the following condition:

$$a < 0, \quad \sqrt{\frac{d}{c}} < 0, \quad 2c\sqrt{\frac{d}{c}} < 0 \quad \text{and} \quad (a - 2c)\sqrt{\frac{d}{c}} > 0$$

the fixed point can be stable. However, for $d < 0$ the condition given above cannot be satisfied since the given quantities which involve d are not real numbers. After the analysis of of original JD4 system, we have added the same nonlinear term $\dot{x} \tanh x$

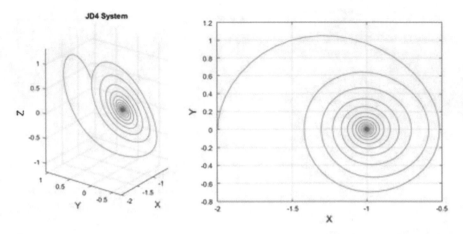

Fig. 7 Trajectory of modified JD4 system

to the original jerk function we have the following jerk form:

$$\dot{x} = y$$
$$\dot{y} = z \qquad \qquad (15)$$
$$\dot{z} = az + by + cx^2 + xz + d + y \tanh x$$

The new system has the following phase portrait in Fig. 7 for the following values of parameter a $= -1$, b $= -0.7$, c $= 1$ and d $= -1$. These parameter values are the same as the Sprott use in his study and we see that new additional term turn the limit cycle like structure into the attractor.

The Lyapunov spectrum of the system is plotted in Fig. 8 and the maximal Lyapunov exponents of the system are $(1.2961, 0.4166, -0.4627)$.

2.3 Adaptive Control of the JD3 System

In this section, we design an adaptive control law for globally stabilizing the chaotic JD3 system when the parameter values are unknown. Thus, we consider the controlled JD3 system given by the dynamics:

$$\dot{x} = y + u_1$$
$$\dot{y} = z + u_2 \qquad \qquad (16)$$
$$\dot{z} = az - y + bx^2 + xy - c + u_2$$

where u_1, u_2 and u_3 are feedback controllers to be designed using the states and estimates of the unknown parameters of the system.

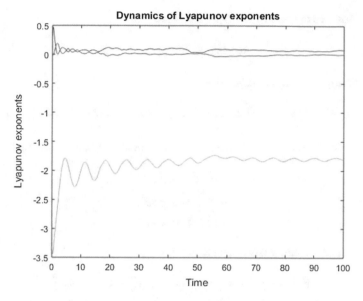

Fig. 8 Dynamics of Lyapunov exponents of modified JD4 system

In order to ensure that the controlled system globally converges to the fixed point, we consider the following adaptive control functions:

$$
\begin{aligned}
u_1 &= -y - k_1 x \\
u_2 &= -z - k_2 y \\
u_3 &= -\bar{a}z + y - \bar{b}x^2 - xy + \bar{c} - k_3 z
\end{aligned}
\tag{17}
$$

where $\bar{a}, \bar{b}, \bar{c}$ are estimates of the parameters a, b and c, respectively, and k_i's, (i = 1, 2, 3) are positive constants. Substituting the control law into the JD3 system

$$
\begin{aligned}
\dot{x} &= -k_1 x \\
\dot{y} &= -k_2 y \\
\dot{z} &= e_a z + e_b x^2 - e_c - k_3 z
\end{aligned}
\tag{18}
$$

where

$$
e_a = (a - \bar{a}) \quad e_b = (b - \bar{b}) \quad e_c = (c - \bar{c})
\tag{19}
$$

defines the error. For the derivation of the update law for adjusting the parameter estimates $\bar{a}, \bar{b}, \bar{c}$, the Lyapunov function V:

$$
V(x, y, z, e_a, e_b, e_c) = \frac{1}{2}(x^2 + y^2 + z^2 + e_a^2 + e_b^2 + e_c^2)
\tag{20}
$$

which is positive definite in $\mathbb{R}^5$. Note also that:

$$\dot{e}_a = -\dot{\bar{a}} \quad \dot{e}_b = -\dot{\bar{b}} \quad \dot{e}_c = -\dot{\bar{c}} \tag{21}$$

Differentiating V along the trajectories:

$$\dot{V} = -k_1 x^2 - k_2 y^2 - k_3 z^2 + e_a(-z^2 - \dot{\bar{a}}) + e_c(-z - \dot{\bar{c}}) + e_b(-x^2 z - \dot{\bar{b}}) \tag{22}$$

In view of Eq. 21, the estimated parameters are updated by the following law:

$$\begin{aligned} \dot{\bar{a}} &= z^2 + e_a k_4 \\ \dot{\bar{b}} &= -x^2 z + e_b k_5 \\ \dot{\bar{z}} &= -z + e_c k_6 \end{aligned} \tag{23}$$

where k_4, k_5 and k_6 are positive constants. So:

$$\dot{V} = -k_1 x^2 - k_2 y^2 - k_3 z^2 - k_4 e_a^2 - k_6 e_c^2 - k_5 e_b^2 \tag{24}$$

is negative definite. According to this calculation, we obtain the following results about Lyapunov stability: JD3 system with unknown parameters is globally and exponentially stabilized for all initial conditions $(x_0, y_0, z_0) \in \mathbb{R}^3$ by the adaptive control law, where the update law for the parameters is given by Eq. 21 and, k_i, $k = 1 \ldots 6$, are positive constants.

3 Conclusion

In this study, we try to analyze JD3 and JD4 system systematically and we try to estimate effect of new nonlinear term on the chaotic dynamics of the systems. We have demonstrated that both system has an attractor with the addition of new term. Then we theoretically show that the JD3 system can be controlled so as to lead to convergence to the fixed points by using an adaptive control method and Lyapunov stability.

References

1. L.O. Chua, Global unfolding of Chua's circuits. Technical Report UCB/ERL M93/7, EECS Department, University of California, Berkeley, 1993
2. R. Eichhorn, S.J. Linz, H. Peter, Simple polynomial classes of chaotic jerky dynamics (2002)
3. S.J. Linz, Nonlinear dynamical models and jerky motion. Am. J. Phys. **65**(6), 523–526 (1997)
4. E.N. Lorenz, Deterministic nonperiodic flow. J. Atmos. Sci. **20**, 130–148 (1963)

5. S. Motesharrei, J. Rivas, E. Kalnay, Human and nature dynamics (HANDY): modeling inequality and use of resources in the collapse or sustainability of societies. Ecol. Econ. **101**, 90–102 (2014)
6. L. Pecora, T. Carroll, Synchronization in chaotic systems. Phys. Rev. Lett. **64**, 821–824 (1990)
7. J.C. Sprott, Some simple chaotic flows. Phys. Rev. E **50**, R647–R650 (1994)
8. J.C. Sprott, Some simple chaotic jerk functions. Am. J. Phys. **65**(6), 537–543 (1997)
9. J.C. Sprott, *Elegant Chaos: Algebraically Simple Chaotic Flows* (World Scientific Publishing Company Pte Limited, Singapore, 2010)
10. K. Sun, J.C. Sprott, A simple jerk system with piecewise exponential nonlinearity. Int. J. Nonlinear Sci. Numer. Simul. **10**(11), 1443–1450 (2009)
11. S. Vaidyanathan, A new 3-D jerk chaotic system with two cubic nonlinearities and its adaptive backstepping control. **27**, 1 (2017)

Gröbner Basis Method in FitzHugh-Nagumo Model

Veronika Hajnová

Abstract The FitzHugh-Nagumo model is a two dimensional system of differential equations with polynomial right-hand sides. The model describes an excitable system and explains basic phenomena in dynamics of neuron activity, for example spike generations in a neuron after stimulation by external current input. The system is slow-fast, meaning system with different time scales for each state variable. We analyse bifurcation manifolds of the FitzHugh-Nagumo system in whole parameter space using algebraic approach based on Gröbner basis.

Keywords FitzHugh-Nagumo model · Gröbner basis · slow-fast system · Hopf bifurcation · Fold bifurcation

1 Introduction

Whole paper is focused on the well-known two dimensional FitzHugh-Nagumo model of neuron in form

$$V' = V - \frac{1}{3}V^3 - W + i$$
$$W' = a\,(Vb - Wc + d)\,, \tag{1}$$

where state variable V is the membrane potential and W is a recovery variable. Parameter i is the magnitude of stimulus current. Other parameters a, b, c, d are set to be constant for specific type of neuron. Bifurcation analysis of this model was done by Rocsoreanu et al. [5]. In Fig. 1 there is a bifurcation diagram for two parameters a, i. Model exhibits complicated phenomena near the point where the Hopf bifurcation curve intersect itself. For example in the neighbourhood of the point

V. Hajnová (✉)
Department of Mathematics and Statistics, Faculty of Science,
Masaryk University, Brno, Czech Republic
e-mail: xhajnovav@math.muni.cz

© Springer Nature Switzerland AG 2019
C. H. Skiadas and I. Lubashevsky (eds.), *11th Chaotic Modeling and Simulation International Conference*, Springer Proceedings in Complexity, https://doi.org/10.1007/978-3-030-15297-0_8

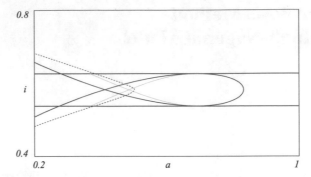

Fig. 1 Bifurcation diagram of the FitzHugh-Nagumo model for $b = 1$, $c = 1.2$, $d = 0.7$. Bifurcation manifolds: Hopf (red), fold (black), LPC (black dashed), separatrix-saddle loop (grey)

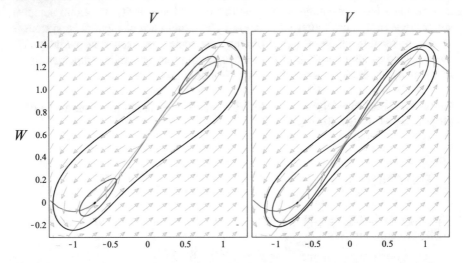

Fig. 2 Phase portraits for $b = 1$, $c = 1.2$, $d = 0.7$, $i = \frac{c}{d}$ and $a = 0.44$ (left) or $a = 0.49$ (right). Stable (black) and unstable (dark blue) invariant sets, direction field (grey), nulclines (red, light blue)

where the separatrix saddle loop curve intersect itself two topologically different phase portraits occurs, as you can see in Fig. 2.

The aim of this paper is to describe a different method of finding bifurcation manifolds in (1) without computing equilibria using the Gröbner basis. To provide description of the method let us assume a general system of polynomial differential equations in following form

$$x' = \frac{dx}{dt} = f(x, \varepsilon),$$
(2)

Table 1 One-parameter bifurcation manifolds. J denotes Jacobi matrix of system (1), $I_n \in \mathbb{R}^{n \times n}$ denotes identity matrix. Symbol $\odot$ denotes bi-alternate matrix product [3]. For $n = 2$ Eq.(6) can be rewritten in form trace $J(x, \varepsilon) = 0$

Bifurcation	System
Fold	$f(x, \varepsilon) = 0$ (3) $\det(J(x, \varepsilon)) = 0.$ (4)
Hopf	$f(x, \varepsilon) = 0$ (5) $\det(2J(x, \varepsilon) \odot I_n) = 0.$ (6)

where $x \in \mathbb{R}^n$ is a vector of state variables, $\varepsilon \in \mathbb{R}^k$ is a vector of parameters and f is a set of polynomials.

One-parameter bifurcation points lies on manifolds described, in space of both state variable and parameters, by equations listed in Table 1. Besides one-parameter bifurcation points, degenerated bifurcation points, e.g. pitchfork or transcritical bifurcation points, neutral-saddles or two-parameter bifurcation generally lies on those manifolds [3].

Because f is a set of polynomials systems of Eqs. (3)–(6) are also polynomial, it is possible to compute a Gröbner basis of set of polynomials given by right-hand sides of Eqs. (3), (4) or (5), (6) with lexicographic order $x_1 > \cdots > x_n > \varepsilon_1 > \cdots > \varepsilon_k$. Polynomials in the basis which contains only parameters $\varepsilon_1, \ldots, \varepsilon_k$, state variables $x_1 \ldots x_n$ are eliminated, gives an implicit description of the bifurcation manifolds in space of parameters.

Similar approach can be use for two-parameter bifurcation manifolds. There are two ways how two-parameter bifurcation can arise:

- violation of non-degeneracy conditions for one-parameter bifurcation, e.g. cusp bifurcation
- intersection of one-parameter bifurcation manifolds, e.g. Bogdanov-Takens bifurcation.

Therefore it is possible to describe multi-parameter bifurcation, in space of both state variable and parameters, adding additional conditions to systems listed in Table 1.

Niu and Wang [4] already used similar approach to compute Hopf bifurcation manifold. It is also possible to find bifurcation manifolds for systems of difference equations using this approach. Similar technique was already use by Kotsireas and Karamanos [2] to exact computation of the bifurcation point B4 of the logistic map.

2 Results

Using approach described in previous section it was possible to compute bifurcation manifolds in the model (1). Results are listed in following Tables 2, 4 and 5. In top right cell of Tables 2, 4 and 5 implicit description of the bifurcation manifold in

Table 2 Fold bifurcation manifold

Fold	$g_1 = V - 1/3\,V^3 - W + i$
	$g_2 = Vab - Wac + ad$
	$g_3 = V^2ac + ab - ac$

Gröbner basis

$$p_1 = 9ac^3i^2 - 18c^2di + 4ab^3 - 12cb^2a + 12abc^2 - 4ac^3 + 9acd^2$$
$$p_2 = 6Wac^2i - 9ac^2i^2 - 6Wacd + 12acdi + 4b^2a - 8cba + 4ac^2 - 3d^2a$$
$$p_3 = 2cbWa - 2Wac^2 - 3icba + abd + 2acd$$
$$p_4 = 6W^2acd - 15Wacdi + 9acdi^2 - 4Wab^2 + 4Wac^2 + 3d^2Wa$$
$$\qquad +6ib^2a + 6icba - 3d^2ia - 4abd - 8acd$$
$$p_5 = 4W^2ac^2 - 9ac^2$$
$$p_6 = i^2 - 8Wacd + 18acdi + 8b^2a - 12cba + 4ac^2 - 5d^2a$$
$$p_7 = 4W^2ac - 12Waci + 9aci^2 + 2daV + 2dWa - 3dia + 4ab - 4ac$$
$$p_8 = 2caV - 2Wac + 3ica - ad, Vab - Wac + ad$$
$$p_9 = V^3 - 3V + 3W - 3i$$

Table 3 Cusp bifurcation manifold

Cusp bifurcation manifold

Implicit description	Parametrization
$p_1\,(a, b, c, d, i) = 0$	$a = t$
$\frac{dp_1(a,b,c,d,i)}{di} = 18ac^2(ci - d) = 0$	$i = \frac{d}{b}$
	$c = b,\ t \in \mathbb{R}$

space of both state variables V, W and parameters a, b, c, d, i is shown. Bifurcation manifolds in parameter space are given by the first polynomial, or the first and the second polynomial in case of Bogdanov-Takens bifurcation, of the Gröbner basis. In Table 3 cusp bifurcation manifold is derived. Figures 3, 4 and 5 shows bifurcation manifolds in space of parameters a, i, c and $b = 1, d = 0.7$. Figures 6 and 7 depict two dimensional sections of those three dimensional manifolds.

3 Discussion

Our paper focuses on usage Gröbner basis to find bifurcation manifolds in the FitzHugh-Nagumo model.

Problem of finding bifurcation manifolds of system of differential equations generally leads to a system of algebraic equations. Usage of Gröbner basis is possible for a system of polynomial equations. This requirement is fulfilled, for instance, for systems of differential equations with polynomial or rational right-hand side. Approach

Table 4 Hopf bifurcation manifold

Hopf	$g_1 = V - 1/3\,V^3 - W + i$
	$g_2 = Vab - Wac + ad$
	$g_4 = -V^2 - ac + 1$

Gröbner basis

$$q_1 = a^4c^5 - 6a^3c^3b + 3a^3c^4 + 9a^2b^2c - 6a^2c^2b + 9ac^2i^2 - 18acdi$$
$$\quad -9b^2a + 12cba - 4ac^2 + 9d^2a$$
$$q_2 = -a^4c^4 + 3a^3bc^2 - 3a^3c^3 + 9Waci + 3a^2bc - 9aci^2 - 9dWa + 9dia$$
$$\quad -6ab + 4ac$$
$$q_3 = -a^2c^2W + a^2cd + 3bWa - 2Wac - 3iba + 2ad$$
$$q_4 = a^3c^3 + 3a^2c^2 + 9W^2 - 18Wi + 9i^2 - 4$$
$$q_5 = caV + 2V - 3W + 3i$$
$$q_6 = a^3c^3 - 3a^2bc + a^2c^2 + 3daV + 6Vi - 9Wi + 3ab - 2ac + 9i^2$$
$$q_7 = Wac^2 - acd + 2bV - 3bW + 3bi$$
$$q_8 = a^3c^3 - 3a^2bc + 3a^2c^2 + 3daV + 6VW - 9Wi + 3ab + 9i^2 - 4$$
$$q_9 = V^2 + ac - 1$$

Table 5 Bogdanov-Takens bifurcation manifold

Bogdanov-Takens	$g_1 = V - 1/3\,V^3 - W + i$
	$g_2 = Vab - Wac + ad$
	$g_3 = V^2ac + ab - ac$
	$g_4 = -V^2 - ac + 1$

Gröbner basis

$$r_1 = 4a^4c^5 - 12a^3c^4 + 12c^3a^2 + 9ac^2i^2 - 18acdi - 4c^2a + 9d^2a$$
$$r_2 = -a^2c^2 + ab$$
$$r_3 = -8a^4c^4 + 24a^3c^3 + 18Wa^2cd - 27a^2cdi - 24a^2c^2 + 9a^2d^2 - 18aci^2$$
$$\quad -18dWa + 36dia + 8ac$$
$$r_4 = 2a^2c^2W - 3a^2c^2i + a^2cd - 2Wac + 2ad$$
$$r_5 = 4a^5c^4d + 8a^4c^4i - 24ic^3a^3 - 12a^3c^2d + 9a^2cdi^2 - 18Wa^2d^2$$
$$\quad +24a^2c^2i + 9a^2d^2i + 18$$
$$r_6 = aci^3 + 18Wadi + 8a^2cd - 36adi^2 - 8ica$$
$$r_7 = 2a^4c^4 + 9Waci - 6a^2c^2 - 9aci^2 - 9dWa + 9dia + 4ac$$
$$r_8 = a^3c^3 + 3a^2c^2 + 9W^2 - 18Wi + 9i^2 - 4$$
$$r_9 = 2Wac - 3ica + ad + 4V - 6W + 6i$$

Fig. 3 Fold bifurcation
manifold (blue),
two-parameter bifurcation
sub-manifolds (red): cusp,
Bogdanov-Takens in space
of parameters a, i, c and
$b = 1$, $d = 0.7$

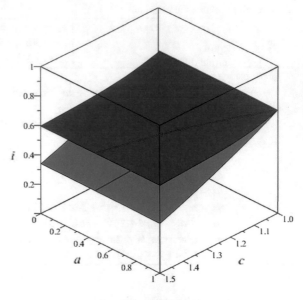

Fig. 4 Hopf bifurcation
manifold (grey),
two-parameter bifurcation
sub-manifolds:
Bogdanov-Takens (red) in
space of parameters a, i, c
and $b = 1$, $d = 0.7$

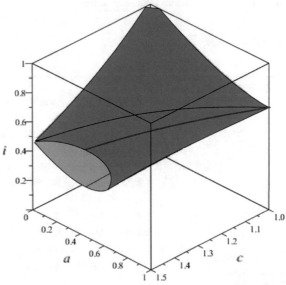

Fig. 5 Complete bifurcation diagram in space of parameters a, i, c and $b = 1, d = 0.7$. Fold bifurcation manifold (blue), Hopf bifurcation manifold (grey), two-parameter bifurcation sub-manifolds: Bogdanov-Takens (red)

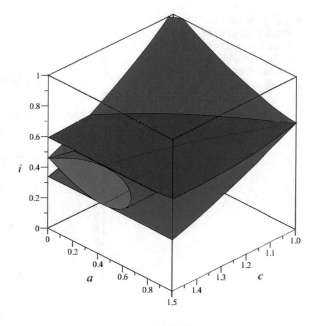

Fig. 6 Bifurcation diagram in space of parameters a, i and $b = 1, d = 0.7$, $c = 1.225$. Fold bifurcation manifold (black), Hopf bifurcation manifold (red)

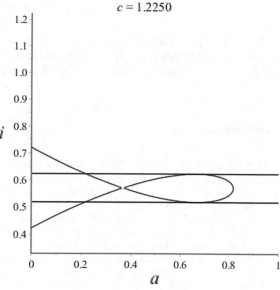

Fig. 7 Bifurcation diagram
in space of parameters
c, i and $b = 1, d = 0.7$,
$a = 0.25$. Fold bifurcation
manifold (black), Hopf
bifurcation manifold (red)

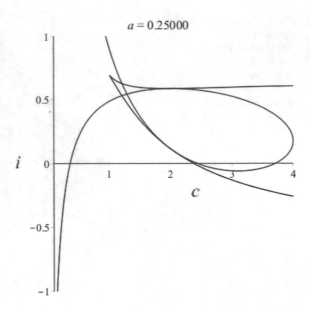

through Gröbner basis allows to compute bifurcation manifolds without computing equilibria.

As an example of system polynomial differential equations FitzHugh-Nagumo model was used in this paper. In comparison to original bifurcation analysis, which was done by Rocsoreanu et al. [5], computations were simplified.

References

1. D.H. Bailey, J.M. Borwein, V. Kapoor, E.W. Weisstein, Ten problems in experimental mathematics. Am. Math. Monthly **113**(6), 481–509 (2006)
2. I.S. Kotsireas, K. Karamanos, Exact computation of the bifurcation point B4 of the logistic map and the Bailey-Broadhurst conjectures. Int. J. Bifurcat. Chaos **14**(07), 2417–2423 (2004)
3. Y.A. Kuznetsov, *Elements of Applied Bifurcation Theory*, 2nd edn. (Springer, New York, 1998)
4. W. Niu, D. Wang, Algebraic analysis of bifurcation and limit cycles for biological systems, in *Algebraic Biology: Third International Conference, AB 2008, Castle of Hagenberg, Austria, July 31–August 2, 2008 Proceedings*, vol. 5147 (2008), pp. 156–171
5. C. Rocsoreanu, A. Georgescu, N. Giurgiteanu, *The FitzHugh-Nagumo Model: Bifurcation and Dynamics*, 1st edn. (Kluwer Academic Publishers, Boston, 2000)
6. Gröebner Basis: compute a Gröebner basis. 2016. *Maplesoft* (online). Waterloo: Maplesoft, a division of Waterloo Maple Inc (Accessed 29 July 2016). Available at https://www.maplesoft.com/support/help/Maple/view.aspx?path=Groebner

Digital Signature: Quantum Chaos Approach and Bell States

Nafiseh Hematpour, Sodeif Ahadpour and Sohrab Behnia

Abstract The quantum mechanics applied by quantum digital signatures (QDSs) is applied to guarantee the nonrepudiation, unforgeability, and transferability of a signature. Previously, the security of QDSs schemes only depended on the length of the signature. Also, they required unreliable security channels and have been written for one-bit messages. In this paper, quantum digital signature schemes based on bell states sequence and synchronization of coupled chaotic map is introduced. In our schemes, using the synchronization of coupled chaotic map when signing up, security increases against repudiation and forgery. Furthermore, to sign a regular message, repeating the signature steps is half as much as the previous ones.

Keywords Quantum digital signature · Synchronization of coupled chaotic map · Bell states · Repudiation · Forgery

1 Introduction

Cryptography is essential for the security of bankers, shoppers and other Internet users because attackers are recording, and forging, vast volumes of human communication. A mathematical scheme for demonstrating the authenticity of digital messages or documents is called a digital signature. First time, Whitfield Diffie and Martin Hellman described a digital signature scheme in 1976 [1]. And so far used in software distribution, financial transactions, contract management software, and so on [2, 3]. The security of such classical digital signature (CDS) schemes, with

N. Hematpour (✉) · S. Ahadpour
University of Mohaghegh Ardabili, Ardabil, Iran
e-mail: n_hematpour@uma.ac.ir

S. Ahadpour
e-mail: ahadpour@uma.ac.ir

S. Behnia
Department of Physics, Urmia University of Technology, Urmia, Iran
e-mail: s.behnia@sci.uut.ac.ir

© Springer Nature Switzerland AG 2019
C. H. Skiadas and I. Lubashevsky (eds.), *11th Chaotic Modeling and Simulation International Conference*, Springer Proceedings in Complexity, https://doi.org/10.1007/978-3-030-15297-0_9

the rapid development of quantum computing, is seriously challenged. Therefore, Gottesman and Chuang proposed the first Quantum digital signature protocol [4]. Quantum digital signatures (QDSs) apply quantum mechanics, with information-theoretic security, to the problem of forging and repudiation which types of these schemes have been developed [5, 6]. These QDS proposals deal with the problem of sending single bit messages while no-forging and non-repudiation are guaranteed. These plans need to be further developed to enhance security against attackers and speed up regular message signatures.

In order to increase the security, the nonlinear dynamical systems generate a kind of deterministic random-like process, which is called chaos. Chaotic systems have some the properties such as sensitivity to initial condition [7], system parameter, ergodicity [8] and mixing [9], random-like behavior, unstable periodic orbits with long periods and desired diffusion and confusion properties, etc. Chaos bring much promise application in the cryptography field. Given the distinctions of quantum and classical spaces, we can use the realities of quantum spaces to increase the key space and the capabilities of the dynamic systems in cryptography. One of the popular fields of quantum chaos addresses quantum maps. Furthermore, the iteration process is one-way. Various chaotic maps were used in Cryptography, Watermarking [10], Random number generators [11], Complex system [12], Image encryption, Quantum blind signature [13] and lots of other fields. On the other hand, a big number of researches have been dedicated to quantum maps as paradigms of quantum chaos [14].

A good encryption scheme should be sensitive to all the secret keys, and the key space should be large enough to make brute force attacks infeasible. If the key is small enough, the cryptosystem will be broken and no matter how strong and well designed the algorithm might be. Key space size is the total number of different keys that can be used in an encryption algorithm. From the cryptographical point of view, the key space should not be smaller than 2^{100} in order to provide a high level of security [15].

We consider quantum digital signature based on bell states that capability of the bell states increases the length of the signature than the same ones. Also, this signature can be used in quantum computer. The dynamically designed system based on synchronization of coupled chaotic map and bell states under current attacks called denial and forgery designed for the safety of the protocol.

The rest of the paper is organized as follows. The model explained in Sect. 2 and also the bell states presented in Sect. 3. The QDS protocol has been proposed in Sect. 4, with the results in Sect. 5 and the security analysed in Sect. 6. Finally, the Conclusion is presented.

2 The Model

The pair-coupled chaotic map can be considered as a two-dimensional dynamical map defined as:

$$\phi(x, y) = \begin{cases} X = F(x, y) \\ Y = F(y, x) \end{cases}$$

Complete synchronization in the coupled chaotic map, means the existence of an invariant sub-manifold ($x = y \Leftrightarrow X = Y$), synchronization is one of the invariant manifold of the dynamical systems. The corresponding invariant measure is(a similar calculation has been presented [16])

$$\mu(X) = \sum_{xk \in \phi^{-1}(x, y)} \frac{\mu(x_k)}{\left| \frac{\partial F(x, y)}{\partial x} \right|_{x=y^+} \frac{\partial F(x, y)}{\partial y} \bigg|_{x=y}}$$

$$= \sum_{xk \in \phi^{-1}(x, x)} \left| \frac{\partial F(x_k, x_k)}{\partial x_k} \right|^{-1} \mu(x_k) \tag{2}$$

where $x_k \in \phi^{-1}(x, x)$, i.e., x_k is one of the roots of the map.

One can calculate the ks-entropy by considering the invariant measure, which leads to

$$h_{ks}(\phi_{syn}) = \lim_{n \to \infty} \frac{1}{n} \ln \left| \frac{\partial F(x, y)}{\partial x} \right|_{x=y} + \frac{\partial F(x, y)}{\partial y} \bigg|_{x=y} \right|$$

$$+ \lim_{n \to \infty} \frac{1}{n} \ln \left| \frac{\partial F(y, x)}{\partial x} \right|_{x=y} + \frac{\partial F(y, x)}{\partial y} \bigg|_{x=y} \right| \tag{3}$$

where "n" is iteration of synchronization of coupled chaotic map.

In the proposed algorithm, synchronization of coupled maps are employed to achieve the goal of steganography in image. As an example, we may consider the following a generic symmetric non-linearly coupled chaotic map

$$\phi_{coupled(x, y)} = \begin{cases} X = \left[(1 - \epsilon)(f_1(x))^p + \epsilon(f_2(y))^p \right]^{\frac{1}{v}} \\ Y = \left[(1 - \epsilon)(f_1(y))^p + \epsilon(f_2(x))^p \right]^{\frac{2}{v}} \end{cases} \tag{4}$$

where in general, "p" is an arbitrary parameter, "ϵ" the strength of the coupling. Obviously, by choosing p = 1, we get ordinary linearly coupled chaotic maps. The functions $f_1(x)$ and $f_2(x)$ are two arbitrary one-dimensional maps

$$f_1(x) = \frac{\alpha^2 4x(1 - x)}{1 + (\alpha^2 - 1)4x(1 - x)},$$

$$f_2(x) = \frac{\beta^2(2x - 1)^2}{4x(1 - x) + \beta^2(2x - 1)^2}. \tag{5}$$

In synchronization state of coupled chaotic map

Fig. 1 Bifurcation diagram
of dynamical system
[Eq. (4)] for $x(1) = 0.0002$,
$y(1) = 0.0002$, $\alpha = 3.95$, $p = 0.98$, $\epsilon = 0.678$

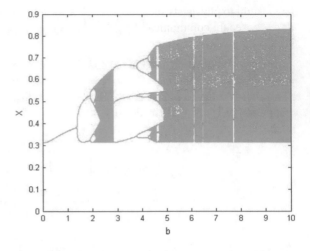

Fig. 2 Curve shows the
variation of Lyapunov
exponent of dynamical
system [Eq. (4)], in terms of
the parameter α

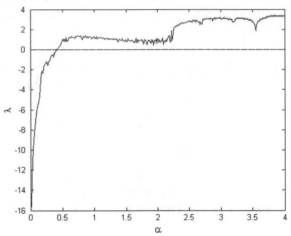

$$\phi_{coupled}^{(x,x)} = X = [(1 - \epsilon)(f_1(x)) + \epsilon(f_2(x))] \tag{6}$$

The corresponding bifurcation diagram of dynamical system [Eq. (4)] is shown
in Fig. 1. A possible way to describe the key space might be in terms of positive
Lyapunov exponents (see Fig. 2).

3   Bell States

In 1964, Bell presented a theory to resolve quantum mechanics issues. According to
this theory, the two particles can communicate with one another and find each other
without regard to the distance between them. This theory has the potential to give a

new insight to physicists such as Dr. Eckert, the Bell theory will be able to guarantee perfectly secure communications even when you have purchased your communications equipment from your enemies. This paper presents a quantum digital signature scheme based on bell states.

The following states are called bell states [17, 18]:

$$|\phi^+\rangle = \frac{1}{\sqrt{2}}(|00\rangle + |11\rangle)$$

$$|\phi^-\rangle = \frac{1}{\sqrt{2}}(|00\rangle - |11\rangle)$$

$$|\psi^+\rangle = \frac{1}{\sqrt{2}}(|01\rangle + |10\rangle)$$

$$|\psi^-\rangle = \frac{1}{\sqrt{2}}(|01\rangle - |10\rangle)$$

All of these states are perpendicular to each other and provide a basis for the two qubit space $C^2 \otimes C^2$. These states can be used in encryption.

4 QDS Scheme

One of the important subjects in quantum cryptography is QDS schemes. In this section, by considering one signer and two participants, a protocol for QDS scheme is introduced. Alice signs the message. Given that the roles of Bob and Charlie are arbitrary, Bob is taken as the first authenticator. He authenticates the message received from Alice and then forwards it to Charlie. Charlie verifies that the initial source was Alice. QDS can be explained based on the distribution and the messaging stage.

Distribution stage

Alice chooses the initial value and control parameter. These are according to the Lyapunov's curve and in chaotic interval of dynamical system [Eq. (4)]. Then the initial keys has been sent to Bob and Charlie. Bob and Charlie create the signature and exchange the signature elements (Fig. 3). Finally, Bob and Charlie record an eliminated signature.

Messaging stage

Alice converts the message M into a binary and then encodes into the bell states. Alice sends the elements of message to Bob together with her corresponding private key. He controls private key against the key she sent for the message. By considering the signature length L, Bob checks its mismatches if it is less than a threshold value $s_a L$, He accepts the message.

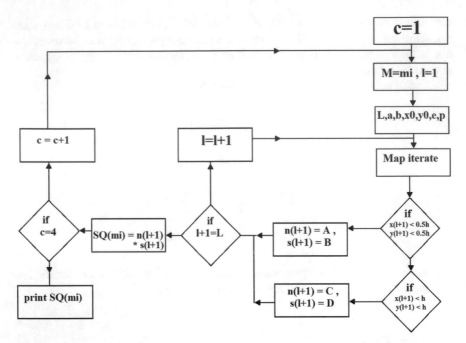

Fig. 3 Creating a signature algorithm for $\{A = |\phi^+\rangle, B = |\phi^-\rangle, C = |\psi^+\rangle, D = |\psi^-\rangle\}$, $\{a = \alpha, b = \beta, e = \varepsilon\}$

Bob forwards the obtained two-bit message m(A, B, C or D) and private key to Charlie. At the same check by considering the mismatches threshold $s_b L$, Charlie accepts the forwarded message. The message could be repudiated with a high probability if the threshold for accepting directly from Alice is similar to the threshold for accepting a forwarded message. The authentication security threshold a and b can be determined $(0 \leq s_a < s_b < 1)$ [19].

5 Results

Finding the dynamical system's critical points, dividing the dynamical system respect to critical points and choosing the alphabet ($|A\rangle$, $|B\rangle$, $|C\rangle$ and $|D\rangle$) for each domain (see Table 1), are necessary for generating the signature.

Initial key is consist of y(0), x(0), the control parameter of dynamical system in chaotic domain. Alice selects them for one of the element of signature (see Fig. 3).

To utilize the QDS protocol, Alice selects a series of the bell states by iterating of the dynamical system to the length of the signature L. Each element of message is signed using a quantum signature

Table 1 Divided domain for chaotic regen of Eq. (4)

Bell states	Domain
A	0–0.25
B	0.25–0.5
C	0.5–0.75
D	0.75–1

$$QS_m = \overset{L}{\underset{l=1}{\otimes}} |\psi_l^m\rangle, \ |\psi_l^m\rangle \in \{A = |\phi^+\rangle, B = |\phi^-\rangle, C = |\psi^+\rangle, D = |\psi^-\rangle\}.$$

She sends a copy of the chosen value to Bob and Charlie. After creating the digital signature, they exchange the signature elements randomly. To authenticate the signature, Bob generates an eliminated signature by measuring the states. We must ensure that Bob expects on average L/2 signature elements from Charlie.

To send a signed message, Alice sends the two-bit message and, y(0), x(0), a, b, e and p to Bob. He accept message after checking mismatches. Then expressed values has been forwarded to Charlie. He accepts the message coming from Alice if mismatches is below his threshold.

6 The Security Analyze

We require security both against message forging by Bob and against repudiation by Alice. Unforgeability means that a given piece of message indeed comes from the signer and remains intact during transmission, namely, no one can forge a valid signature that can be accepted by other honest recipients. Nonrepudiation means that once the signer signs a message, he or she cannot deny having signed it.

6.1 Security Against Repudiation

In the protocol, by using a chaotic map in the signature process and dependence on initial value and control parameter, Alice cannot deny declared values. If the initial values are expressed by a difference of 10^{-14}, the results of the chaotic map vary considerably. In addition, Table 1 is used to select values of A, B, C and D. Changing these values also creates another signature and Bob and Charles cannot obtain a joint sign. It itself increases security against repudiation.

6.2   Security Against Forging

In the protocol, because Bob sends the initial values and control parameters related to chaotic map, it cannot change these values. If the initial values are expressed by a difference of 10^{-14}, the results of the chaotic map vary considerably. This small change creates a different signature with its signature. That is, Bob and Charlie do not get the same signature.

7   Conclusions

In this work, by considering the model and probing the chaotic region for selecting the keys, a QDS scheme is introduced. Proposed QDS scheme based on the bell states can be used for protection of copy-right. In the protocol, using the chaotic map makes it impossible to deny and forge. This signature can be used in quantum computer. Quantum computing are developing, and much work and research is needed to reach the implementation stage.

References

1. W. Diffie, M.E. Hellman, New directions in cryptography. IEEE Trans. Inf. Theory **22**, 644–654 (1976)
2. R.L. Rivest et al., A method for obtaining digital signatures and public-key cryptosystems. Commun. ACM **21**, 120–126 (1978)
3. S. Vaudenay, *A Classical Introduction to Cryptography Applications for Communications Security* (Springer Science & Business Media Inc, New York, 2006), pp. 260–261
4. D. Gottesman, I. Chuang, Quantum digital signatures (2001). arXiv:quant-ph/0105032v2
5. E. Andersson et al., Experimentally realizable quantum comparison of coherent states and its applications. Phys. Rev. A **74**, 022304 (2006)
6. P.J. Clarke et al., Experimental demonstration of quantum digital signatures using phase–encoded coherent states of light. Nat. Commun. **3**, 1174 (2012)
7. M.W. Lee et al., Transmission system using chaotic delays between light waves. IEEE J. Quantum Electron. **39**, 931–935 (2003)
8. R. Brown, L.O. Chua, Clarifying chaos: examples and counterexamples. Int. J. Bifurcat. Chaos **6**, 219–242 (1996)
9. M.A. Jafarizadeh, Hierarchy of chaotic maps with an invariant measure. J. Math. Phys. **104**, 1013–1028 (2001)
10. S. Behnia et al., Design and implementation of coupled chaotic maps in watermarking. J. Appl. Soft Comput. **21**, 481–490 (2014)
11. M.A. Jafarizadeh et al., Hierarchy of rational order families of chaotic maps with an invariant measure. J. Pramana **67**, 1073–1086 (2006)
12. S. Ahadpour et al., Markov-binary visibility graph: a new method for analyzing complex systems. J. Inf. Sci. **274**, 286–302 (2014)
13. X. Lou et al., A weak quantum blind signature with entanglement permutation. Int. J. Theor. Phys. **54**, 2605–2612 (2015)

14. A. Lakshminarayan, N.L. Balazs, On the quantization of linear maps. Ann. Phys. **212**, 220–234 (1991)
15. B. Schneier, *Applied Cryptography: Protocols, Algorithms, and Source Code in C* (Wiley, 1996)
16. S. Ahadpour et al., Synchronization in pair coupled maps with invariant measure. Commun. Nonlinear Sci. and Numer. Simul. **14**, 2916–2922 (2009)
17. J.J. Sakurai, *Modern Quantum Mechanics* (Addison-Wesley Publication Company, Inc., 1985)
18. A. Michael et al., *Quantum Computation and Quantum Information* (Cambridge University Press, New York, 2000)
19. H.L. Yin et al., Experimental quantum digital signature over 102 km. Phys. Rev. A **95**, 0323341–0323349 (2017)

Large Scale Behavior of Generalized Stochastic Magnetohydrodynamic Turbulence with Mirror Symmetry Breaking

Michal Hnatič, Georgii Kalagov, Tomáš Lučivjanský and Peter Zalom

Abstract The field theoretic renormalization group techniques is applied to an inherently classical problem of a general A model of active scalar advection with A representing a continuous real parameter that governs the interaction structure of the system. The model encompasses the important magnetohydrodynamic scenario as well as the important $A = 0$ model and the model of linearized Navier-Stokes equations. The turbulent environment is modeled in the limit of infinitely large Reynolds number, i.e., in the regime of fully developed turbulence. Additionally, spatial parity breaking is incorporated via the continuous parameter ρ.

Keywords General A model · Passive and active admixtures · Navier-Stokes equation · Renormalization-group theory

M. Hnatič (✉) · G. Kalagov · T. Lučivjanský
Faculty of Sciences, Šafarik University, Moyzesova 16, 040 01 Košice, Slovakia
e-mail: hnatic@saske.sk

M. Hnatič
Bogoliubov Laboratory of Theoretical Physics, Joint Institute for Nuclear Research,
Dubna, Moscow Region 141 980, Russian Federation

M. Hnatič
Institute of Experimental Physics, SAS, Watsonova 47, 040 01 Košice, Slovakia

G. Kalagov
Department of Physics, Saint-Petersburg State University,
7/9 Universitetskaya nab., St. Petersburg 199034, Russian Federation

P. Zalom
Institute of Physics, The Czech Academy of Sciences,
Na Slovance 2, 18221 Prague 8, Czech Republic

RUDN University, 6 Miklukho-Maklaya St, Moscow 117198, Russian Federation

1   Introduction

Systematic study of matter may roughly be divided into the physics of fundamental
interactions among few particles (as performed in the scope of high energy physics)
and to the study of their macroscopic manifestations in many-body problems. The
advances on the field of fundamental microscopic behavior are connected with the
theoretical development of the Standard model of particle physics and its experimental verification that proceeds up to nowadays [1]. Problems encompassed within the
many-body physics include a broad variety of systems like, for example, classical
turbulent flows, systems at phase transitions or quantum condensed matter systems
[2]. Seemingly unrelated, a versatile theoretical tool emerging in both microscopic as
well as macroscopic studies is the concept of the renormalization group (RG) as first
developed by Wilson [3]. Here, we discuss its application to the problem of diffusion advection processes in turbulent environments which, basically a phenomenon
of classical nature, still represents an unsolved problem with analytic results on
the field being notoriously difficult to obtain [4–8]. Nevertheless, as shown in the
framework of the Kolmogorov's theory, in turbulent flows scaling invariance emerges
naturally [9]. It is therefore at hand to use the well established tools of RG theory to
obtain results of analytical nature for turbulent flows. Moreover, scaling properties
manifest itself more clearly at higher values of Reynolds number Re [2]. The limit
of Re $\to \infty$ is thus of great importance. Unlike in numerical calculations where
Re $\to \infty$ is difficult to reach, the RG approach allows analytic calculations which in
the end allow to study turbulent phenomena which are independent of microscopic
details of the material as well as of macroscopic flow conditions [10]. RG approach
allows easy incorporation of different scenarios including the advection of various
admixtures [2].

   In this respect, we show how fundamental the RG approach is in connection to
an asymmetric scenario with explicitly broken spatial parity as observed usually in
nature [11–13]. The present analysis generalizes the RG approach from the passive
advection problem, see Refs. [14, 15] for example, to the important scenario of
active admixture where the advected agent influences the underlying turbulent flow.
To this end, we study the general A model of active vector admixture. To avoid
confusion, we stress that all of the results in Refs. [14, 15] are restricted to the
passive scenario and that the name of the present model is completely unrelated to
the classification of Halperin and Hohenberg [16]. The generalization to the active
advection is performed by inclusion of a Lorentz-like force term as it appears in
magnetohydrodynamics (MHD) [17, 18] which corresponds to the choice of $A = 1$
in the general A model defined by Ref. [15] for example. Nevertheless, parameter A
will remain a continuous parameter in complete analogy to passive models discussed
in Refs. [11–13, 15, 19] with detailed definitions given in Sect. 2. The introduction
of a continuous parameter A describes the advection processes of a wide variety of
vector admixture models in a unified manner. In explicit, A is only required to be real
(further restrictions may follow only from the perturbative approach itself, for details
see Ref. [14]) and thus it may take values of $A \in 1, 0, -1$ which correspond to the

usual MHD model, the $A = 0$ model and to the so called linearized Navier-Stokes equations model [15, 19], respectively.

The RG approach to the fully developed turbulence represents a well established and well developed approach which has been used to successfully analyze a wide variety of systems without admixtures being advected [20–26] as well as for advection diffusion processes of several admixtures including passive scalar admixture [13, 27–32], magnetic admixtures [33–36] and also vector admixtures [11, 12, 15, 37–39]. The one-loop techniques for calculation of the turbulent Prandtl number within the A model used here are similar to those of Refs. [33, 34]. In this paper, we incorporate spatial parity breaking into the general A model of vector advection. As shown in Refs. [33, 34] for a special case of $A = 1$, violation of spatial parity leads to a wide variety of new physical phenomena. Most important in this respect is the appearance of the turbulent dynamo via a well known mechanism of spontaneous symmetry breaking, which is commonly used in quantum field theory (QFT). However, authors of Ref. [34] only consider an MHD model which represents a special scenario of $A = 1$ of the general A model considered here. Nevertheless, as argued later in the paper, the same effect also arises for other values of the parameter A with only $A = 0$ being a trivial exception and with $A = -1$ that is a case of linearized Navier-Stokes equation. Taking together, although the general A model has attracted a lot of attention recently, see Refs. [11, 12, 14, 15] for more details, only the case of turbulent environments with passive admixtures has been analyzed so far and we therefore consider here the general helical case. The field theoretic formalism employed in the present paper relies on the mechanism of spontaneous symmetry breaking as first described in Ref. [33] for non-helical environments. Its extension to the helical environment was then performed in Ref. [34].

The paper is structured as follows. In Sect. 2 the A model for the case of active admixture is discussed. The emphasis is laid on the meaning of the parameter A for the structure of interactions. In Sect. 6 we show that all previously defined models posses an additional instability in helical environments. We discuss its physical consequences and show that the appearance of a macroscopic field $\mathbf{B}$ represents a possible mechanism of stabilization of the system at large scales. Inclusion of the symmetry breaking field $\mathbf{B}$ leads then to modifications in the original A model of active admixture advection which are discussed in Sects. 7 and 8. The obtained results are then briefly discussed in Sect. 9.

2   Model A of Active Vector Advection with Spatial Parity Violation

To describe a general active vector admixture $\mathbf{b}$ advected in the turbulent environment of the velocity field $\mathbf{v}$ the framework of the general A model is used. The cross-interaction between $\mathbf{b}$ and $\mathbf{v}$ is defined via a Lorentz-like force term. Thus, the corresponding diffusion-advection equations take the form

$$\partial_t \mathbf{b} = \nu_0 u_0 \triangle \mathbf{b} - (\mathbf{v} \cdot \nabla)\mathbf{b} + A(\mathbf{b} \cdot \nabla)\mathbf{v} - \partial P, \tag{1}$$

$$\partial_t \mathbf{v} = \nu_0 \triangle \mathbf{v} - (\mathbf{v} \cdot \nabla)\mathbf{v} + (\mathbf{b} \cdot \nabla)\mathbf{b} - \partial Q + \mathbf{f}^{\mathbf{v}}, \tag{2}$$

with $\partial_t \equiv \partial/\partial_t$, $\partial_i \equiv \partial/\partial_{x_i}$, $\triangle \equiv \partial^2$, ν_0 denoting the bare viscosity coefficient, u_0 is the bare reciprocal Prandtl number, $P \equiv P(x)$ and $Q \equiv Q(x)$ are the pressure fields and x is a shorthand notation for time and spatial variables, i.e., $x \equiv (t, \mathbf{x})$. We have explicitly considered stochastic forcing term $\mathbf{f}^v$ which is discussed below in the detail. Now, let us stress that A is a real parameter with $A = 1, 0, -1$ representing notably important values. For $A = 1$ the kinematic MHD model is recovered, the $A = 0$ corresponds to the active advection of a vector field and $A = -1$ leads towards the model of the linearized Navier-Stokes equations [15]. The system of Eqs. (1) and (2) is accompanied in this case also by the condition of incompressibility of $\mathbf{v}$, i.e., $\nabla \cdot \mathbf{v} = 0$. Magnetic field obeys standard Maxwell equations and is thus transversal, i.e., $\nabla \cdot \mathbf{b} = 0$.

For further application of QFT formalism we modeled the real flow using the stochastic forcing terms with $\mathbf{f}^v$ being the random force per unit mass. The addition of the Lorentz-like term does not require additional modifications of the model regarding the stochasticity of the velocity field and, as discussed for example in Ref. [34], Eq. (4) does apply for the present model. However, according to Ref. [34], stochastic forcing from Eq. (1) can now be completely omitted. The transverse random force per unit mass $\mathbf{f}^v = \mathbf{f}^v(x)$ is thus taken to model the injection of kinetic energy into the turbulent system at large scales. Therefore, its form must be in accord with the real infrared (IR) energy pumping. We assume it in a power-like form as usual for fully developed turbulence within the RG approach (for more details see Ref. [2]) via

$$D_{ij}^v(x; 0) \equiv \langle f_i^v(x) f_j^v(0) \rangle = \delta(t) \int \frac{d^d \mathbf{k}}{(2\pi)^d} D_0 k^{4-d-2\varepsilon} R_{ij}(\mathbf{k}) e^{i\mathbf{k}\cdot\mathbf{x}}. \tag{3}$$

with d being the spatial dimension, $\mathbf{k}$ the wave number with $k = |k|$, ε the formally small parameter of the present RG calculation, $D_0 \equiv g_0 \nu_0^3 > 0$, and g_0 the coupling constant related to the characteristic ultraviolet (UV) momentum scale Λ by the relation $g_0 \simeq \Lambda^{2\varepsilon}$. The term $R_{ij}(k)$ appearing in Eq. (3) encodes the spatial parity violation of the underlying turbulent environment and is given as

$$R_{ij}(\mathbf{k}) = \delta_{ij} - k_i k_j / k^2 + i\rho \, \varepsilon_{ijl} k_l / k. \tag{4}$$

Here, ε_{ijl} is the Levi-Civita tensor of rank 3, and the real valued helicity parameter ρ satisfies $|\rho| \leq 1$ due to the requirement of positive definiteness of the correlation function. Obviously, $\rho = 0$ corresponds to the fully symmetric (non-helical) case, whereas $\rho = 1$ means that spatial parity is fully broken.

3   Field Theoretic Formulation of the A Model of Active Admixture Without Symmetry Breaking

To transform the set of the differential equations defining the problem, we apply the Martin-Siggia-Rose theorem. Consequently, the system of stochastic differential Eqs. (1) and (2) is equivalent to a field theoretic model of the double set of fields $\Phi = \{\mathbf{v}, \mathbf{b}, \mathbf{v}', \mathbf{b}'\}$ where primed fields are the auxiliary response fields [2, 40]. Such model is defined via the corresponding Dominicis-Janssen action functional $S(\Phi) = S_{free}(\Phi) + S_{vbb'}(\Phi) + S_{vvv'}(\Phi) + S_{bbv'}(\Phi)$, where:

$$S_{free}(\Phi) = -\int dx\, v_i'(\partial_t - \nu_0\Delta)v_i - \int dx\, b_i'(\partial_t - \nu_0 u_0\Delta)b_i$$
$$+ \iint dx\,dy\, \frac{1}{2}v_i'(x)D_{ij}^v(x;y)v_j'(y). \tag{5}$$

where the required summations over dummy indices $i, j \in 1, 2, 3$ are implicitly assumed. The auxiliary fields and their original counterparts $\mathbf{v}, \mathbf{b}$ share the same tensor properties and are transverse. The interaction part of the action, namely S_{int}, is a sum of $S_{vvv'}$, $S_{vbb'}$ and a new term $S_{bbv'}$ which is responsible for all effects connected to the active advection. The corresponding interaction terms reads as

$$S_{vbb}' = -\int dx\, b_i' v_j\, \partial_j\, b_i + A\int dx\, b_i' b_j\, \partial_j\, v_i, \tag{6}$$

$$S_{vvv}' = -\int dx\, v_i' v_j\, \partial_j\, v_i, \tag{7}$$

$$S_{bbv}' = +\int dx\, v_i' b_j\, \partial_j\, b_i. \tag{8}$$

We note that since a quadratic term composed solely of field $\mathbf{b}'$ is missing, the active model will not include any $\langle b_i b_j \rangle_0$ type of a propagator. Thus, in the frequency-momentum representation the following set of bare propagators is obtained:

$$\langle v_i' v_j \rangle_0 = \langle v_i v_j' \rangle_0^* = \frac{P_{ij}(\mathbf{k})}{i\omega + \nu_0 k^2}, \tag{9}$$

$$\langle v_i v_j \rangle_0 = \frac{g_0 \nu_0^3 k^{4-d-2\varepsilon} R_{ij}(\mathbf{k})}{|-i\omega + \nu_0 k^2|^2}. \tag{10}$$

The propagators are represented as usual by the dashed and full lines, where the dashed lines involve the velocity type of fields and full lines represent the vector admixture type fields. The auxiliary fields are denoted using a slash in the corresponding propagators. The general A model also contains three different triple interaction vertices. In the momentum-frequency representation they correspond to $V_{ijl} = i(k_j\delta_{il} - Ak_l\delta_{ij})$ and $W_{ijl} = i(k_l\delta_{ij} + k_j\delta_{il})$. In both cases, momentum $\mathbf{k}$ is flowing into the vertices via the corresponding prime field, i.e., in the former case via field

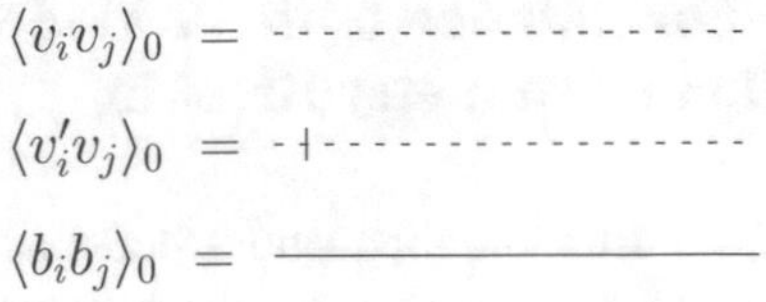

$$\langle v_i v_j \rangle_0 = \text{------------------}$$

$$\langle v_i' v_j \rangle_0 = \text{-|-----------------}$$

$$\langle b_i b_j \rangle_0 = \text{------------------}$$

Fig. 1 The A model of active vector admixture contains only three different propagators, namely $\langle v_i v_j \rangle_0$, $\langle v_i' v_j \rangle_0$ and $\langle b_i' b_j \rangle_0$

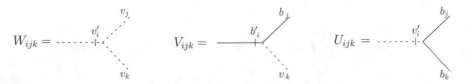

Fig. 2 Three interaction vertices W_{ijk}, V_{ijk} and U_{ijk} do appear in the A model of active vector admixture regardless of the presence of the symmetry breaking mechanism introduced in Sect. 7. The velocity vertex W_{ijk} as well as the vertex, U_{ijk} which originates in the Lorentz force term, are both symmetric and read as $W_{ijk} = U_{ijk} = i \left(\delta_{ij} k_k + \delta_{ik} k_j \right)$. The vertex $V_{ijk} = i \left(\delta_{ij} k_k - A \delta_{ik} k_j \right)$ depends on the parameter A

$\mathbf{b}'$, whereas in the latter via $\mathbf{v}'$. A vertex encoding all active admixture effects, namely $U_{ijk}(k) = i \left(\delta_{ij} k_k + \delta_{ik} k_j \right)$ is present as well. We stress out that signs have been chosen in order to comply with definitions commonly used in the field of field theoretic RG approach to fully developed turbulence. Consequently, calculating symmetry factors corresponding to given Feynman diagram, we have to consider multiplicative factor of $+1/2$ for each W_{ijk} vertex and a factor of $-1/2$ for U_{ijk} vertex which follow from Eqs. (7) and (6) respectively. With such definitions, the expansions of $e^{S_{int}}$ in the perturbative series will produce symmetry factors of either $+1$ or -1 in one loop order. Taking together, the theory contains three different interaction vertices and three different propagators (Figs. 1 and 2).

4   Analysis of Canonical Dimensions

In order to correctly carry out the RG calculation it is required to first perform the analysis of canonical dimensions which allows to determine all relevant UV divergences which in turn identifies all objects (graphs) containing the so called superficial UV divergences. Only such divergences are the relevant ones left for the subsequent RG analysis, for details, see Ref. [2]. The present model contains two scales [2]. Consequently, an arbitrary quantity Q possesses canonical dimension of $d_Q = d_Q^k + 2d_Q^\omega$. From now we will temporarily set $\rho = 0$ but we put forward that in the helical scenario $d_\rho^k = d_\rho^\omega = 0$ with the other canonical dimensions remaining unaltered. The analysis starts by assigning canonical dimensions to the basic variables

Table 1 Canonical dimensions of quantities appearing in the A model of active vector admixture with or without symmetry breaking as discussed in Sect. 7. The quantity $\mathbf{B}$ is present only in the model with incorporated spontaneously broken symmetry

Q	x	t	k	ω	$\mathbf{v}, \mathbf{b}, \mathbf{B}$	$\mathbf{v}', \mathbf{b}'$	m, Λ, μ	ν_0, ν	g_0	g
d_Q^k	-1	0	1	0	-1	$d+1$	1	-2	2ε	0
d_Q^ω	0	-1	0	1	1	-1	0	1	0	0
d_Q	-1	-2	1	2	1	$d-1$	1	0	2ε	0

x, t, k and ω according to the widely used convention [2]. Applying the condition of vanishing canonical dimension of the action functional S we obtain the resulting canonical dimensions in the present model as shown in the Table 1. Remarkably, all coupling constants posses zero canonical dimensions at $\varepsilon = 0$ which means that the investigated A model of active vector admixture is logarithmic for $\varepsilon = 0$. In other words, all UV divergences of the present model are of the form of poles in ε [41] when the minimal subtraction (MS) scheme is employed. Thus, together with the results of Table 1 we define the following renormalized coupling constants:

$$g_0 = g\mu^{2\varepsilon}Z_g. \tag{11}$$

To establish the multiplicative renormalizability of the models we need to show that the action functional by definition given in Eq. (11) has all the necessary structures. We precede that inclusion of the Lorentz-like force term will have profound consequences and the results differ for passive and active scenario. Therefore, the analysis has to be split into the two cases as discussed in the next two sections.

5 Renormalization Group Approach to the A Model of Active Advection

RG approach to the A model of active admixture requires us to investigate the canonical dimension of an arbitrary one-irreducible (1-IR) diagram of the present theory denoted here as $\Gamma_n(k_1, \ldots k_n)$, where k_i with $i \in 1, \ldots n$ are n momenta flowing into the diagram via the external legs. Considering the Table 1 and no further constraints, we arrive at the following canonical dimension δ of an arbitrary diagram with N_Φ being the number of external fields corresponding to the given type of $\Phi \in \{v, v', b, b'\}$:

$$\delta = d + 2 - (N_v + N_b) - (d-1)(N_{v'} + N_{b'}). \tag{12}$$

Nevertheless, the actual degree of superficial UV divergence is lower since all fields present in our theory are transversal. Consequently, every external leg lowers the degree of superficial UV divergence δ by $N_{v'} + N_{b'}$. The actual degree of divergence denoted as δ' reads therefore:

$$\delta' = d + 2 - (N_v + N_b) - d\,(N_{v'} + N_{b'}) \geq 0. \tag{13}$$

Since later in this work, helical effects are considered we require the physical dimension of our theory to be $d = 3$ and avoid the case of $d = 2$ where additional divergences as analyzed in Ref. [33] do appear. We may now easily exploit the actual degree of the superficial UV divergence for diagrams with different number and type of external legs. Doing so systematically, we arrive at the results presented in Table 2.

As a result of the analysis, we notice that all 1-IR diagrams containing four and more external legs of any type are superficially UV convergent. Moreover, 1-IR diagrams with three external legs are superficially UV divergent only when there is no more than one external leg of $\mathbf{v}'$ or $\mathbf{b}'$ nature. For $d = 3$ the same holds also for diagrams with two external legs. The possible diagrams of the tadpole and bubble type are more numerous, however these are actually irrelevant in the theory of turbulence from trivial tensorial reasons discussed, for example, in Ref. [2]. At $d = 3$ we therefore end up with superficial UV divergences contained only in the diagrams $\langle v_i' v_j \rangle_{1-IR}$, $\langle b_i' b_j \rangle_{1-IR}$ and $\langle v_i' b_j b_k \rangle_{1-IR}$ are superficially divergent in $d = 3$. In the dimension of $d = 2$ additional divergences in $\langle v_i' v_j' \rangle_{1-IR}$ and $\langle b_i' b_j' \rangle_{1-IR}$ diagrams do appear as already noticed in Ref. [33]. However, we especially concentrate at the dimension $d = 3$ where helical effects may be incorporated. We also note that diagrams of the types $\langle v_i' v_j \rangle_{1-IR}$ and $\langle b_i' b_j \rangle_{1-IR}$ contain linear divergences in the general case. However, setting $\rho = 0$, as done in this section, causes both types of diagram to contain only logarithmic divergences due to the tensorial reasons. This in turn means that the model is logarithmically divergent which allows subsequent multiplicative renormalization by utilizing the ε expansion.

However, such a benign scenario changes dramatically when helicity is incorporated into the model. Subsequently, diagrams $\langle v_i' v_j \rangle_{1-IR}$, $\langle b_i' b_j \rangle_{1-IR}$ and $\langle v_i' b_j b_k \rangle_{1-IR}$ contain also linear divergences and self-consistent RG analysis is not possible anymore as no suitable counterterms are available in the action functional S which is a scenario discussed in the subsequent sections (Fig. 3).

Table 2 Canonical dimensions of several 1-IR reducible graphs of the A model of active vector advection. The presented results hold for both, the A model with as well as without the mechanism of symmetry breaking. Note that bubble diagrams with $N = 0$, tadpole diagrams with $N = 1$ and also diagrams with $N_{v'} + N_{b'} = 0$ all vanish due to the tensorial reasons [2]

$N_{v'} + N_{b'}$	1	1	1	2	2	2	3	3
$N_v + N_b$	1	2	3	0	1	2	0	1
N	2	3	4	2	3	4	3	4
δ	2	1	0	$3 - d$	$3 - d$	$2 - d$	$5 - 2d$	$4 - 2d$
δ'	1	0	-1	$2 - d$	$1 - d$	$-d$	$2 - 2d$	$1 - 2d$

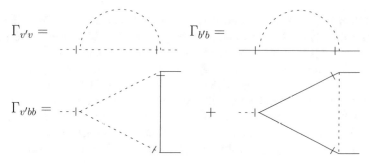

Fig. 3 In the A model without the mechanism of symmetry breaking, there are four different 1-IR diagrams in one-loop order which contain superficial UV divergences as outlined in Sect. 5

6 Helicity Induced Instability

As already discussed in Refs. [14, 34], passing from the non-helical to the general parity broken A model is not straightforward as new type of divergences arises in the theory and the above conclusions are valid only in the non-helical case. Such linear divergences posses the form of $\mathbf{b}' \cdot (\nabla \times \mathbf{b})$ (also referred to as the curl term in the present paper) and appear thus in the 1-irreducible Green function $\langle b_i' b_j \rangle_{1-ir}$. As very well known established, the presence of the curl term for the $A = 1$ case leads to an exponential increase of magnetic fluctuations at large scales with subsequent instabilities emerging in the model [42, 43]. In the steady state, these instabilities are generally attributed to the generation of large-scale magnetic field via the mechanism known as the turbulent dynamo, for details see Refs. [44, 45]. Successful incorporation of such a mechanism into the field theoretical description was already performed by authors of Ref. [34] who introduced a mechanism of spontaneous symmetry breaking into the model. Technically, the original field $\mathbf{b}$, which describes magnetic fluctuations around zero mean magnetic field $\langle \mathbf{b} \rangle \equiv \mathbf{B} = 0$ was substituted by $\mathbf{b} + \mathbf{B}$ with $\mathbf{B}$ allowed to be non-zero. The field $\mathbf{B}$ then provided all necessary means for the elimination of the curl term in the model. However, in Ref. [34] the authors considered only the MHD model which corresponds to the present general A model only when $A = 1$. Nevertheless, the Lorentz term present in the general A model of active vector admixture gives rise to a new interaction term $\mathbf{b}'(\mathbf{b} \cdot \nabla)\mathbf{b}$ in the action functional. Its presence is then of crucial importance for the application of RG methods to systematically treat the curl divergences by employing the mechanism of spontaneous symmetry breaking as described above.

Let us now briefly discuss the extension of the theory valid for $A = 1$ as proposed in Ref. [34] to the arbitrary values of A as considered in the present paper. As a starting point one considers the Lorentz-like term $b_i \partial_i b_j$ in the Navier-Stokes equation (2). The magnitude $|\mathbf{B}|$ of the spontaneously generated field (with an arbitrary direction) follows then from the condition of cancellation of the curl term. Thus, the field **B** depends explicitly on the parameter A in the general case. Moreover, some values of the parameter A may violate the condition of $|\mathbf{B}|$ being positive which may restrict the interval of admissible values of A. All of these facts are easily demonstrated by considering the special case of $A = -1$. Here, no curl terms emerge due to the appearance of symmetrical vertex in the action functional via a mechanism that resembles the vanishing of curl terms in the response function $\langle v'v \rangle$ (for details, see Ref. [34]). Taking together, calculation of the field **B** together with the possible restrictions on A are clearly feasible and are performed in the following section.

7 Field theoretic Formulation of the A Model with Symmetry Breaking

As shown in the previous section, Eqs. (5)–(7) lead to an inherently unstable model due to the $\langle v'v \rangle_{1-IR}$ and $\langle b'b \rangle_{1-IR}$ graphs. However, this is only because we have neglected physical processes that stabilize the advection-diffusion system. To correct this, we follow the procedure presented in Ref. [34] for the special case of $A = 1$ and extend it to arbitrary values of A. To do so, we shift the zero expectation value of the admixture field **b**. In other words, **b** fluctuates around a spontaneously generated non-zero mean value $\mathbf{B} \equiv \mathbf{b}$ with magnitude being dependent on the actual underlying mechanism of symmetry breaking which itself is depended on the parameter A. Moreover, the response field **b′** which corresponds to the admixture type of the field is assumed to have zero mean value. Technically, to introduce spontaneous symmetry breaking into the A model of active vector admixture, we replace the admixture field **b** in Eqs. (5)–(7) by $\mathbf{B} + \mathbf{b}$ and leave all other fields unaltered. Performing such substitution, we notice that the free part of the original action functional S remains the same, while the interacting part contains contributions $S_{vbb'}$ and $S_{bbv'}$ which now give rise to two additional terms which are however quadratic in the fields **v**, **v′**, **b**, **b′**. Therefore, new terms actually belong to the free part of the new action functional $\tilde{S}$ which describes the symmetry broken case. In explicit, the free part of the symmetry broken action functional reads:

$$
\begin{aligned}
S_{\text{free}}(\Phi) = \iint dx\,dy\ & \frac{1}{2} v_i'(x) D_{ij}^v(x; y) v_j'(y) \\
& - \int dx\ v_i'(\partial_t - \nu_0 \Delta) v_i - \int dx\ b_i'(\partial_t - \nu_0 u_0 \Delta) b_i \\
& + A \int dx\ b_i' B_j\, \partial_j\, v_i + \int dx\ v_i' B_j\, \partial_j\, b_i.
\end{aligned}
\tag{14}
$$

The last two terms represent the additional quadratic terms. Since the interaction part of the symmetry broken case posses the same form as S_{int} of the original action functional no additional interaction vertices appear in the symmetry broken model.

However, the free part of the action functional is now much richer than previously and gives rise to the following set of non-zero propagators:

$$\langle v_i v_j \rangle_0 = \frac{\beta(\mathbf{k})\beta^*(\mathbf{k})}{\xi(\mathbf{k})\xi^*(\mathbf{k})} D_v(\mathbf{k}) R_{ij}(\mathbf{k}), \tag{15}$$

$$\langle v_i v_j' \rangle_0 = \langle v_i' v_j \rangle_0^* = \frac{\beta^*(\mathbf{k})}{\xi^*(\mathbf{k})} P_{ij}(\mathbf{k}), \tag{16}$$

$$\langle b_i b_j \rangle_0 = A^2 \frac{(\mathbf{B} \cdot \mathbf{k})^2}{\xi(\mathbf{k})\xi^*(\mathbf{k})} D_v(\mathbf{k}) R_{ij}(\mathbf{k}), \tag{17}$$

$$\langle b_i b_j' \rangle_0 = \langle b_i' b_j \rangle_0^* = \frac{\alpha^*(\mathbf{k})}{\xi^*(\mathbf{k})} P_{ij}(\mathbf{k}), \tag{18}$$

$$\langle b_i v_j \rangle_0 = \langle v_i b_j \rangle_0^* = iA \frac{\beta(\mathbf{k})(\mathbf{B} \cdot \mathbf{k})}{\xi(\mathbf{k})\xi^*(\mathbf{k})} D_v(\mathbf{k}) R_{ij}(\mathbf{k}), \tag{19}$$

$$\langle v_i b_j' \rangle_0 = \langle b_i' v_j \rangle_0^* = i \frac{(\mathbf{B} \cdot \mathbf{k})}{\xi^*(\mathbf{k})} P_{ij}(\mathbf{k}), \tag{20}$$

$$\langle b_i v_j' \rangle_0 = \langle v_i' b_j \rangle_0^* = iA \frac{(\mathbf{B} \cdot \mathbf{k})}{\xi^*(\mathbf{k})} P_{ij}(\mathbf{k}), \tag{21}$$

where momenta are always flowing from the first to the second field and the parameters α, β, ξ, D_v are given as follows

$$\alpha(\mathbf{k}) = i\omega_k + \nu k^2, \qquad\qquad \beta(\mathbf{k}) = i\omega_K + u\nu k^2, \tag{22}$$

$$\xi(\mathbf{k}) = A(\mathbf{B} \cdot \mathbf{k})^2 + \alpha(\mathbf{k})\beta(\mathbf{k}), \qquad D_v(\mathbf{k}) = g_0 \nu_0 k^{4-d-2\varepsilon}. \tag{23}$$

Graphical depiction of propagators is shown in Fig. 4.

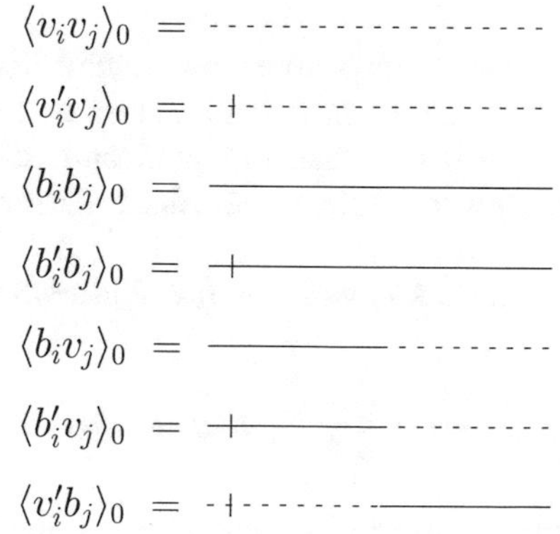

$$
\begin{aligned}
\langle v_i v_j \rangle_0 &= \text{-----------------}\\
\langle v_i' v_j \rangle_0 &= \text{-|----------------}\\
\langle b_i b_j \rangle_0 &= \text{———————}\\
\langle b_i' b_j \rangle_0 &= \text{+———————}\\
\langle b_i v_j \rangle_0 &= \text{————— - - - - - -}\\
\langle b_i' v_j \rangle_0 &= \text{+———— - - - - - -}\\
\langle v_i' b_j \rangle_0 &= \text{-|- - - - - - ———}
\end{aligned}
$$

Fig. 4 The A model of active vector admixture with incorporation of the mechanism of spontaneous symmetry breaking induced by the macroscopic field **B**. It contains seven different propagators as shown in the picture above. The momentum in the propagators on the picture is considered to flow from left to the right

Setting the field **B** to zero, one immediately recovers the free propagators of the original model. We notice that the A model of active admixture with incorporated symmetry breaking posses three propagators $\langle v_i'v_j \rangle_0$, $\langle v_i v_j \rangle_0$, $\langle b_i'b_j \rangle_0$ of the original theory but they are modified to include the symmetry broken field **B** in their corresponding definitions. Moreover, four principally new propagators $\langle b_i b_j \rangle_0$, $\langle v_i b_j \rangle_0$, $\langle v_i b_j' \rangle_0$ and $\langle v_i'b_j \rangle_0$ do emerge. All propagators of the spontaneously broken model depend explicitly on the parameter A. The interaction vertices in the symmetry broken theory remain completely unaltered.

8 Renormalization Group Analysis of the A Model with Symmetry Breaking for Helical Environments

Although the A model with incorporated symmetry breaking has different Feynman rules and its corresponding action functional is significantly different, the analysis of the canonical dimensions is the same as previously due to **B** having the same canonical dimension as the field **b**. Consequently, as before only $\langle v_i'v_j \rangle_{1-IR}$, $\langle b_b'b_j \rangle_{1-IR}$ and $\langle v_i'b_jb_k \rangle_{1-IR}$ diagrams contain superficial UV divergences. The corresponding renormalized action functional reads therefore

$$
\begin{aligned}
S^R = &- \int dx \; v_i'(\partial_t - Z_1 \nu \triangle)v_i - \int dx \; b_i'(\partial_t - Z_2 \nu u \triangle)b_i \\
&+ \iint dx \, dy \frac{1}{2} v_i'(x) D_{ij}^v(x;y) v_j'(y) - \int dx \; b_i' v_j \, \partial_j \, b_i + A \int dx \; b_i' b_j \, \partial_j \, v_i \\
&+ A \int dx \, b_i' B_j \, \partial_j \, v_i + \int dx \; v_i' B_j \, \partial_j \, b_i + \int dx \, v_i' Z_3 \, b_j \, \partial_j \, b_i - \int dx \, v_i' v_j \, \partial_j \, v_i,
\end{aligned}
$$

$$(24)$$

where Z_1, Z_2 and Z_3 are the corresponding renormalization constants which are discussed later in the detail. Let us now rather stress that the renormalization of the symmetry broken model is nevertheless significantly altered as the instability problem discussed in Sect. 6 does not appear here due to the presence of the stabilizing background field **B**.

The renormalization constants Z_i with $i \in 1, 2, 3$ have the following form in the MS scheme:

$$
Z_i = 1 + \frac{g}{\varepsilon} z_i^{(1)} + higher \; order,
\tag{25}
$$

where the inscription *higher order* means terms beyond the one loop order and the subscript of coefficients $z_i^{(1)}$ with $i \in 1, 2, 3$ denotes the corresponding renormalization constant while the superscript denotes the loop order. In Eq. (25) only the renormalized variables are inserted which leads to a divergence free 1-irreducible Green's functions $\langle v_i'v_j \rangle_{1-ir}$ and $\langle b_i'b_j \rangle_{1-ir}$ which are associated to the corresponding

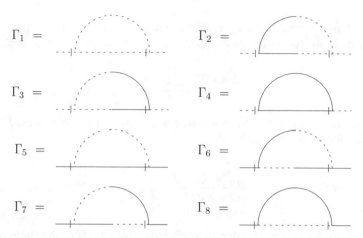

Fig. 5 Due to the presence of spontaneous symmetry breaking, the A model of active vector admixture contains additional diagrams for $\langle v'_i v_j \rangle$ and $\langle b'_i b_j \rangle$

self-energy operators $\Sigma_{v'v}$ and $\Sigma_{b'b}$. These in turn are given as the sum of corresponding one loop Feynman diagrams. Writing down the corresponding perturbative series one obtains the following expressions:

$$\Sigma_{v'v} = \sum_{i \in 1,2,3,4} S_i \Gamma_i, \quad \Sigma_{b'b} = \sum_{i \in 5,6,7,8} S_i \Gamma_i, \tag{26}$$

where the Feynman diagrams Γ_i are shown in Fig. 5 and the corresponding symmetry factors S_i are obtained by considering the definitions of vertices and a proper perturbative expansions. One obtains $S_1 = S_3 = S_5 = S_6 = 1$ and $S_2 = S_4 = S_7 = S_8 = -1$. The graphs under the discussion contain terms which are linear as well as quadratic in **p** and a complete RG analysis would require thus to determine both types of divergences. However, in the present paper we are only interested in the value of the spontaneously generated background field **B** which as shown in Ref. [34] is related to the above discussed linear divergences in ρ. These also turn to be the most dangerous divergences in the present model as discussed later.

Omitting unnecessary technical details, the isolation of linear divergences in ρ is quite straightforward to perform in both $\langle v_i v_j \rangle_{1-IR}$ and $\langle b_i b_j \rangle_{1-IR}$ graphs. In the case of $\langle v_i v_j \rangle_{1-IR}$ graphs, the tensorial structure of the problem ensures that no such divergences exist at all as already noticed, for example, in Ref. [14] for a corresponding passive advection limit of the present model. In explicit, one obtains:

$$\Gamma_1^{\rho,+} = \Gamma_2^{\rho,+} = \Gamma_3^{\rho,+} = \Gamma_4^{\rho,+} = 0, \tag{27}$$

where the superscript ρ denotes the corresponding parts linear in ρ and the superscript $+$ was added to remind that $A > 0$ case is considered. However, the previous result is also valid for the case of $A < 0$. As previously, we employ an analogous notation for the terms linear in ρ for $\langle b_i b_j \rangle_{1-IR}$ graphs and limit ourselves to the case $A > 0$. With such notation, the results after a straightforward ω integration obtain the following form:

$$\Sigma^\rho_{b'b,+} = \sum_{i \in 5,6,7,8} S_i \, \Gamma^\rho_{i,+} \tag{28}$$

where $i \in \{5, 6, 7, 8\}$, S_i are the aforementioned symmetry factors and the $\Gamma^\rho_{i,+}$ terms posses the following integral form:

$$\Gamma^\rho_{i,+} = i\rho \, p_\gamma \epsilon_{\gamma\alpha\delta} \frac{g\mu^{2\varepsilon}\nu A(1+A)}{4(1+u)^2(2\pi)^d} \int\limits_m^\Lambda d^d k \, F_{i,+} |k|^{4-d-2\varepsilon} k_\beta k_\delta, \tag{29}$$

where α is the corresponding coordinate placed on the field $\mathbf{b}'$ of the outer incoming leg while β corresponds to the coordinate of the field $\mathbf{b}$ of the outer outgoing leg of the $\langle b_i b_j \rangle_{1-IR}$ graph and we stress that summations over dummy indices are implicitly understood. The functions $F_i^{\rho,+}$ for $i \in 5, 6, 7, 8$ are functions of $\tilde{\mathbf{B}}$ and the analytic structure of $\Sigma^{\rho,+}_{bb'}$ allows then a cancellation of the otherwise unrenormalizable linear divergences in ρ. In explicit, we obtain:

$$\Gamma^\rho_{5,+} = -\frac{A|\beta|^2\alpha}{|\xi|^2\xi} D_v k_q p_j H_{js}(1+A), \quad \Gamma^\rho_{6,+} = -\frac{A^2\beta^*(Bk)^2}{|\xi|^2\xi} D_v k_q p_j H_{js}(1+A),$$

$$\Gamma^\rho_{7,+} = -\frac{A^2\beta(Bk)^2}{|\xi|^2\xi} D_v p_j H_{sj}(1+A), \quad \Gamma^\rho_{8,+} = \frac{A^2\beta(Bk)^2}{|\xi|^2\xi} D_v p_j H_{sj}(1+A).$$

Summing all this contributions together, we finally get:

$$\Sigma^\rho_{b'b,+} = -\frac{A\beta^*}{|\xi|^2} D_v k_q p_j H_{js}(1+A) + 2A^2 \frac{(Bk)^2}{|\xi|^2\xi} 2u\nu k^2 D_v k_q p_j H_{js}(1+A).$$

Performing the following substitution

$$\mathbf{B}' = \sqrt{\frac{|A|}{u\nu^2}} \mathbf{B} \tag{30}$$

allows us to further simplify the result to

$$\Sigma^\rho_{b'b,+} = i\rho \frac{A(1+A)g\mu^{2\varepsilon}\nu}{2(1+u)} \int \frac{d^d k}{(2\pi)^d} \frac{-k^8 + k^4(B'k)^2}{k^5(k^4 + (B'k)^2)^2} k^{4-d-2\varepsilon} k_q p_j k_l \varepsilon_{jsl}.$$

The momentum integral is then easily expressed as

$$\int d^d k \frac{-k^8 + k^4(Bk)^2}{k^5(k^4 + (Bk)^2)^2} k^{4-d-2\varepsilon} k_q k_l \equiv F_1 \delta_{ql} + F_2 \frac{B_q B_l}{B^2} \equiv J_{ql}.$$

The magnitude of $|\mathbf{B}|'$ follows now from the requirement of the UV divergence cancellation contained in

$$F_1 \sim J_{qq} - B'_q J_{ql} B'_l / B'^2.$$

Using d-dimensional spherical coordinates we obtain for $\varepsilon = 0$ (standard approach to RG calculations):

$$F_1 \sim \int_m^\Lambda dk \int_0^1 dt \, t^{-1/2} (1-t)^{d/2-1/2} \frac{-k^2 + B'^2 t}{(k^2 + B'^2 t)^2} k^2$$

$$= \int_0^1 dt \, t^{-1/2} (1-t)^{d/2-1/2} (-\Lambda + 2B' \sqrt{t}),$$

from which finally the magnitude of the symmetry breaking field is obtained as

$$B' = \frac{\Lambda}{\sqrt{\pi}} \frac{\Gamma(d/2 + 3/2)}{\Gamma(d/2 + 1)}. \tag{31}$$

The remaining divergences in the present model are linear dependent with time which is in contrast to the present one which grows exponentially and is thus physically the dominant one. For the physical dimension $d = 3$ we obtain

$$B' = \frac{8\Lambda}{3\pi}. \tag{32}$$

9 Conclusion

In this work, we have discussed the general A model of active admixture. As already shown in Ref. [14], active nature of the admixture is required to consistently renormalize the theory in the presence of helical divergences which are linear in ρ. However, considering just a simple generalization of the general A model to incorporate active nature of advection is insufficient and one must explicitly consider spontaneous generated symmetry breaking. Subsequently, we discussed the aspects of UV renormalization and show their non-trivial dependence on the magnitude of the field $\mathbf{B}$ which is generated by the mechanism of spontaneous breaking. Mechanism of spontaneous breaking is thus shown to allow complete renormalization of the general A model which is the object of an ongoing work.

Acknowledgements The work was supported by VEGA Grant No. 1/0345/17 of the Ministry of Education, Science, Research and Sport of the Slovak Republic.

References

1. M.E. Peskin, D.V. Schroeder, *An Introduction to Quantum Field Theory, The Advanced Book Program, Perseus Books* (Reading, Massachusetts, 1995)
2. A.N. Vasil'ev, *Quantum-Field Renormalization Group in the Theory of Critical Phenomena and Stochastic Dynamics* (Chapman & Hall/CRC, Boca Raton, 2004)
3. K.G. Wilson, Renormalization group and critical phenomena. I. Renormalization group and the Kadanoff scaling picture. Phys. Rev. B, **4**(9), 3174–3183 (1971); K.G. Wilson, Renormalization group and critical phenomena. II. Phase-space cell analysis of critical behavior. Phys. Rev. B **4**(9), 3184–3205 (1971)
4. A. Yoshizawa, S.-I. Itoh, K. Itoh, *Plasma and Fluid Turbulence: Theory and Modelling* (IoP, Bristol, 2003)
5. D. Biskamp, *Magnetohydrodynamic Turbulence* (Cambridge University Press, Cambridge, 2003)
6. A.S. Monin, A.M. Yaglom, *Statistical Fluid Mechanics*, vol. 2 (MIT Press, Cambridge, MA, 1975)
7. W.D. McComb, *The Physics of Fluid Turbulence* (Clarendon, Oxford, 1990)
8. B.I. Shraiman, E.D. Siggia, Rev. Nat. **405**, 639–646 (2000)
9. U. Frisch, *Turbulence: The Legacy of A. N. Kolmogorov* (Cambridge University Press, Cambridge, 1996)
10. L.T. Adzhemyan, N.V. Antonov, A.N. Vasil'ev, *The Field Theoretic Renormalization Group in Fully Developed Turbulence* (Gordon & Breach, London, 1999)
11. N.V. Antonov, N.M. Gulitskiy, Phys. Rev. E **91**, 013002 (2015)
12. N.V. Antonov, N.M. Gulitskiy, Phys. Rev. E **92**, 043018 (2015)
13. L.T. Adzhemyan, J. Honkonen, T.L. Kim, L. Sladkoff, Phys. Rev. E **71**, 056311 (2005)
14. M. Hnatič, P. Zalom, Phys. Rev. E **89**, 043023 (2016)
15. H. Arponen, Phys. Rev. E **79**, 056303 (2009)
16. P.C. Hohenberg, B.I. Halperin, Rev. Mod. Phys. **49**, 435 (1977)
17. S.N. Shore, *Astrophysical Hydrodynamics: An Introduction* (Wiley-VCH Verlag GmbH & KGaA, Weinheim, 2007)
18. Y. Zhou, W.H. Matthaeus, P. Dmitruk, Rev. Mod. Phys. **76**, 1015 (2004)
19. L.T. Adzhemyan, N.V. Antonov, A. Mazzino, P. Muratore-Ginanneschi, A.V. Runov, Europhys. Lett. **55**, 801 (2001)
20. L.T. Adzhemyan, A.N. Vasilev, Y.M. Pismak, Theor. Math. Phys. **57**, 1131 (1983)
21. L.T. Adzhemyan, N.V. Antonov, M.V. Kompaniets, A.N. Vasil'ev, Int. J. Mod. Phys. B **17**, 2137 (2003)
22. L.T. Adzhemyan, J. Honkonen, M.V. Kompaniets, A.N. Vasil'ev, Phys. Rev. E **68**, 055302(R) (2003)
23. L.T. Adzhemyan, A.N. Vasil'ev, M. Gnatich, Theor. Math. Phys. **74**, 180–191 (1988)
24. L.T. Adzhemyan, N.V. Antonov, A.N. Vasil'ev, Usp. Fiz. Nauk **166**, 1257 (1996) [Phys. Usp. **39**, 1193 (1996)]
25. L.T. Adzhemyan, J. Honkonen, M.V. Kompaniets, A.N. Vasilev, Phys. Rev. E 71, 036305 (2005)
26. L.T. Adzhemyan, J. Honkonen, T.L. Kim, M.V. Kompaniets, L. Sladkoff, A.N. Vasil'ev, J. Phys. A: Math. Gen. **39** 7789 (2006)
27. L.T. Adzhemyan, A.N. Vasil'ev, M. Gnatich, Theor. Math. Phys. **58**, (1983)
28. L.T. Adzhemyan, N.V. Antonov A.N. Vasilev, Phys. Rev. E **58**, 1823 (1998)
29. L.T. Adzhemyan, N.V. Antonov, V.A. Barinov, Y.S. Kabrits, A.N. Vasilev, Phys. Rev. E **63**, 025303(R) (2001); L.T. Adzhemyan, N.V. Antonov, V.A. Barinov, Y.S. Kabrits, A.N. Vasilev, Phys. Rev. E **64**, 019901 (2001)
30. C. Pagani, Phys. Rev. E **92**, 033016 (2015)
31. S.V. Novikov, Theor. Math. Phys. **136**, 936 (2003)
32. N.V. Antonov, M.M. Kostenko, Phys. Rev. E **90**, 063016 (2014)

33. L.T. Adzhemyan, A.N. Vasil'ev, M. Gnatich, Theor. Math. Phys. **64**, 777 (1985)
34. L.T. Adzhemyan, A.N. Vasilev, M. Hnatich, Theor. Math. Phys. **72**, 940 (1987)
35. N.V. Antonov, M.M. Kostenko, Phys. Rev. E **92**, 053013 (2015)
36. E. Jurčišinová, M. Jurčišin, R. Remecký, P. Zalom, Phys. Rev. E **87**, 043010 (2013)
37. L.T. Adzhemyan, N.V. Antonov, P.B. Goldin, M.V. Kompaniets, J. Phys. A: Math. Theor. **46**, 135002 (2013); N.V. Antonov, N.M. Gulitskiy, Theor. Math. Phys. **176**, 851 (2013)
38. H. Arponen, Phys. Rev. E **81**, 036325 (2010)
39. S.V. Novikov, J. Phys. A: Math. Gen. **39**, 8133 (2006)
40. P.C. Martin, E.D. Siggia, H.A. Rose, Phys. Rev. A **8**, 423 (1973); C. De Dominicis, J. Phys. (Paris), Colloq **37**, C1–247 (1976); H.K. Janssen, Z. Phys. B **23**, 377 (1976); R. Bausch, H.K. Janssen, H. Wagner, *ibid.* **24**
41. J.C. Collins, *Renormalization: An Introduction to Renormalization, the Renormalization Group and the Operator-Product Expansion* (Cambridge University Press, Cambridge, 1986)
42. S.S. Moiseev, R.Z. Sagdeev, A.V. Tur, G.A. Khomenko, V.V. Yanovskii, Z. Eksp, Teor. Fiz. **85**, 1979 (1983)
43. A. Pouquet, J.D. Fournier, P.L. Sulem, J. Phys. Lett. (Paris) **39L**, 199 (1978)
44. S.I. Vainshtein, Y.B. Zel'dovich, A.A. Ruzmaikin, *The Turbulent Dynamo in Astrophysics [in Russian]* (Nauka, Moscow, 1980); S.I. Vainshtein, *Magnetic Fields in Space [in Russian]* (Nauka, Moscow, 1983)
45. H.K. Moffat, *Magnetic Field Generation in Electrically Conducting Fluids* (Cambridge University Press, Cambridge, 1978); (Wiley & KGaA, Weinheim, 2007); (Wiley & KGaA, Weinheim, 2007). Fluid Mech. **13**, 82–85 (1962); (Wiley & KGaA, Weinheim, 2007)

Modeling Turbulence via Numerical Functional Integration

Ilja Honkonen and Juha Honkonen

Abstract We investigate the possibility of modeling turbulence via numerical functional integration. Our approach is based on the functional integral widely used in the theory of critical phenomena and stochastic dynamics. A detailed argument for this choice is given. By transforming the incompressible stochastic Navier-Stokes equation into a functional integral we are able to calculate equal-time spatial correlation of system variables using standard methods of multidimensional integration. In contrast to direct numerical simulation, our method allows for simple parallelization of the problem as the value of the integral at any point is independent of other points. Thus the entire problem does not have to fit into available memory of any one computer but can be distributed even onto several supercomputers and the cloud. We present the mathematical background of our method and its numerical implementation. The free and open source implementation is composed of a fast serial program for evaluating the integral over a given volume and a Python wrapper that divides the problem into subvolumes and distributes the work among available processes. We use Monte-Carlo integrators of the GNU Scientific Library for integrating subvolumes. We show first results obtained with our method and discuss its pros, cons and future developments.

Keywords Stochastic Navier-Stokes equation · Numerical functional integration

I. Honkonen (✉)
Finnish Meteorological Institute, PO Box 503, 00101 Helsinki, Finland
e-mail: ilja.honkonen@fmi.fi

J. Honkonen
Department of Military Technology, National Defence University,
PO Box 7, 00861 Helsinki, Finland
e-mail: juha.honkonen@mil.fi

© Springer Nature Switzerland AG 2019
C. H. Skiadas and I. Lubashevsky (eds.), *11th Chaotic Modeling
and Simulation International Conference*, Springer Proceedings
in Complexity, https://doi.org/10.1007/978-3-030-15297-0_11

1 Introduction

Understanding turbulence is likely relevant for phenomena of any scale, from particle collisions in an accelerator [1] and human blood circulation [2] to atmospheric and oceanic circulation, solar wind [3] and even galaxy clusters [4]. Most commonly used methods for studying turbulence involve solving the Navier-Stokes (NS) equation for the velocity field **v** of incompressible fluid

$$\partial_t \mathbf{v} + \mathbf{v} \cdot \nabla \mathbf{v} = \nu \nabla^2 \mathbf{v} - \nabla p + \mathbf{f}, \qquad \nabla \cdot \mathbf{v} = 0, \tag{1}$$

where $\mathbf{v}(t, \mathbf{x})$ is the divergenceless velocity field, ν the kinematic viscosity, p the pressure and $\mathbf{f}$ the external force, which may be fixed or random depending on the setup of the problem.

In large-eddy simulation (LES) the velocity field is divided into a sum of large-scale modes $\bar{u}(t, x)$ and small-scale modes $v'(t, x)$

$$v(t, x) = \bar{u}(t, x) + v'(t, x)$$

and the latter $v'(t, x)$ are filtered out directly in the NS equation. Filtering out of small-scale modes gives rise to equations for the large-scale modes $\bar{u}(t, x)$ containing correlations of the former. These correlations cannot be calculated exactly and the main modeling problem is how to take into account these correlations. It is worth mentioning that in the functional-integral approach filtering out the small-scale modes amounts to integration out of the corresponding variables – an approach that is easy to formalize and rather popular in field theories of critical phenomena.

Implicit LES methods do not include a term for viscosity in NS equation, but numerical errors due to finite accuracy of floating point numbers act implicitly as an artificial viscosity.

Direct numerical simulation (DNS) methods use the full forced NS equation which allows, in principle, to fully describe turbulent flow. Solution of NS equations with different initial conditions provides statistical data, from which conclusions about the unknown probability distribution of the turbulent velocity field are inferred through calculation of correlation functions of the velocity field. To arrive at true steady state and accumulate enough data for reliable determination of correlation functions requires large amounts of numerical data. In practice DNS methods are computationally very expensive if one is to describe both large and small spatial scales and their interaction, and require a powerful supercomputer. The problem is exacerbated by the fact that, for example, doubling the Reynolds number increases the amount of memory required by at least an order of magnitude [5], making it currently impossible to model many systems with realistic Reynolds numbers.

We present a different approach for modeling turbulence based on the functional-integral representation of the generating function of correlation functions of turbulent velocity field [6]. In principle, the generating function contains all statistical information of the stochastic problem which is thus available in an analytical form at the outset, contrary to simulations. By transforming the generating function of correlation functions of Burgers' equation into a functional integral, we calculate equal-time spatial correlation functions of system variables using standard methods of multidimensional integration. In contrast to direct numerical simulation, our method allows for simple parallelization of the problem as the value of the integral within any region can be calculated separately from others. Thus the calculations required for obtaining one correlation data set can be distributed to several supercomputers and/or the cloud simultaneously.

2 Functional Integral for the Generating Function

We start from the functional integral used in the perturbative analysis of critical phenomena and stochastic transport equations (see, e.g., [7–9]). This representation is based on the solution of the generic Langevin equation with additive noise f:

$$\frac{\partial \varphi}{\partial t} = V(\varphi) + f := -K\varphi + U(\varphi) + f, \tag{2}$$

with the (white) noise statistics

$$\langle f(t, \mathbf{x}) f(t', \mathbf{x}') \rangle = \delta(t - t') D(\mathbf{x} - \mathbf{x}'), \quad \langle f \rangle = 0. \tag{3}$$

Standard procedures [7, 9] give rise to functional representation of the generating function of correlation functions of the solution of the Langevin equation

$$G(J) = \langle \exp\{\varphi[f]J\} \rangle = \int \mathcal{D}f \int \mathcal{D}\varphi \int \mathcal{D}\tilde{\varphi}$$

$$\exp\left\{-\theta(0)U' - \frac{1}{2}fD^{-1}f + \tilde{\varphi}\left[-\partial_t \varphi - K\varphi + U(\varphi) + f\right] + \varphi J\right\}, \tag{4}$$

where $\theta(0)$ is the (unspecified by this procedure) value of the temporal step function at the origin arising from the diagonal value of the Green function of the free Langevin equation and U' is the functional derivative of the nonlinear term of the right side of (2). In representation (4) a shorthand notation is used in which integrals over space an time as well as sums over indices are implied. It should be emphasized that the perturbation expansion of correlation functions brought about by generating function (4) is independent of the choice of $\theta(0)$. However, when the functional integral is calculated by other means the effect of the ambiguous value of $\theta(0)$ is an open problem.

To test the proposed approach, here we study the stochastic problem generated by Burgers equation for the (one-dimensional) velocity field u:

$$\frac{\partial u}{\partial t} = -u\frac{\partial u}{\partial x} - \nu\nabla^2 u + f, \tag{5}$$

where the pressure gradient is included in the random force f. The nonlinear part of the right side of the stochastic differential equation (5) is a total derivative (cf. notation (2))

$$U(u) = -u\frac{\partial u}{\partial x} = -\frac{1}{2}\frac{\partial}{\partial x}u^2.$$

Therefore, the ambiguous term in the integrand of (4) in detailed representation is of the form

$$\theta(0)U' = \theta(0)\int dt \int dx \frac{\partial u(t,x)}{\partial x}$$

and vanishes as the integral of a derivative over the whole space (or with periodic boundary conditions). Therefore, we see that the generic ambiguity of the functional representation (4) is actually absent in the stochastic Burgers problem! This turns out to be the case in our ultimate goal – the stochastic NS problem of incompressible fluid – as well, since the nonlinear term in (1) is also a total derivative:

$$v_j\partial_i v_j = \partial_i\left(\frac{1}{2}v^2\right).$$

The goal is calculation of velocity correlations. Therefore, we integrate out excessive fields $\tilde{\varphi}$ and f in (4) to obtain the generating function determined by the action functional $S[\varphi]$ in the form

$$G(J) = \int \mathcal{D}\varphi \exp\{S[\varphi] + \varphi J\}$$

$$= \int \mathcal{D}\varphi \exp\left\{-\frac{1}{2}[-\partial_t\varphi - K\varphi + U(\varphi)]D^{-1}[-\partial_t\varphi - K\varphi + U(\varphi)] + \varphi J\right\}. \tag{6}$$

The ambiguous term has been omitted here in view of the properties of hydrodynamic equations discussed above.

It is customary to analyze the statistical properties of the turbulent system in terms of single-time correlation functions. It should be noted that the generating function (6) yields temporal and spatial statistical description of the system and the factor $\exp\{S[\varphi]\}$ in the integrand has the meaning of the probability density function of fluctuations of the field φ. For the purposes of calculation of the single-time statistical properties (6) is excessive and is actually an implicit representation of the probability density function for the single-time fluctuations. Although expression (6) allows to write a (functional) differential equation for the generating function of the single-time correlation functions [9], an explicit closed functional-integral representation for it is not known, hence we use (6) as the quantity containing the full statistical description of the random system.

Integral (6) is then calculated on a lattice with periodic boundary conditions with finite time steps for evolution. The functional-integral representation is devised to describe steady state, thus we impose periodic BC with respect to time as well in the discrete integral. To test the numerical implementation we have used the simplest case of uncorrelated in space random force, i.e. $D(x - x') = D\delta(x - x')$. On one-dimensional lattice with spacing a and evolution time step τ we arrive at the multidimensional integral determined by the action

$$S[v] = -\frac{a}{2D\tau} \sum_{l=1}^{L} \sum_{m=1}^{M} \left(v(t_{l+1}, x_m) - v(t_l, x_m) \right.$$
$$+ \left\{ v(t_l, x_m) \left[\frac{v(t_l, x_{m+1}) - v(t_l, x_{m-1})}{2a} \right] \right.$$
$$\left. \left. - \nu \left[\frac{v(t_l, x_{m+1}) - 2v(t_l, x_m) + v(t_l, x_{m-1})}{a^2} \right] \right\} \tau \right)^2. \quad (7)$$

The periodic conditions imposed on the velocity field in time and space have not been written explicitly in (7) for economy of notation. A change of variables to dimensionless velocity $v \to v\sqrt{D\tau/a}$ demonstrates that the coupling constant of the model is $\sqrt{D\tau a}/\nu$.

3 Numerical Implementation

We use the HDIntegrator program [10] for evaluating the functional integrals in parallel on the Finnish Meteorological Institute's Cray XC 40 supercomputer. The integration volume is subdivided into smaller and smaller subvolumes until one or more of user-defined criteria for convergence is reached. Convergence of the solution for each subregion is checked by evaluating the integral twice, where the second evaluation uses some factor of more samples decided by the user. Currently the solution is defined as converged when one more of the following criteria are satisfied: (1) the absolute relative difference between results within a subvolume is smaller than some factor, or (2) the absolute difference between results is smaller than some value, or (3) the maximum absolute value of results is less than some value. Listing 1 shows an example invocation and output of hdintegrator for calculating (half of) the volume of a unit sphere in 4 dimensions with a 3-dimensional integrand.

Listing 1 Example invocation and output of the parallel Python wrapper for calculating (half of) the volume of a unit sphere in 4 dimensions using a 3-dimensional integrand over the interval $[-1, 1]$

```
$ mpiexec −n 5 ./hdintegrator.py \
>          —integrand integrands/N−sphere \
>          —dimensions 3 \
>          —min−extent −1 \
>          —max−extent 1
2.465467073965016 0.002690919078211793
```

We implement the integrand for functional integrals using the Monte Carlo integration algorithms of the GNU Scientific Library [11]. HDIntegrator communicates with the integrand via standard input and output in ASCII format. Every line of input to the integrand consists of the number of samples and the extent of integration volume in every dimension. Every line of output from the integrand consists of the result, absolute error and a suggestion in which dimension to split the subvolume in case convergence is not achieved. Listing 2 shows an example invocation of the integrand for evaluating an integral and is essentially how the integrand is executed also by the parallel Python wrapper.

Listing 2 Example invocation and output of a integrand for calculating (half of) the volume of a unit sphere in 3 dimensions using a 2-dimensional integrand with 10^7 samples over the interval $[-1, 1]$

```
$ echo 1e7 −1 1 −1 1 | ./integrands/N−sphere
2.093776991699273e+00 4.355051371078182e−04 1
```

4 Single-Time Velocity Correlation of Burgers Equation

The actual multi-dimensional integral to represent the functional integral for the two-point correlation function is of the form

$$\langle v(t_i, x_k)v(t_i, x_n) \rangle = C \int \prod_{l=1}^{L} \prod_{m=1}^{M} dv(t_l, x_m)\, v(t_i, x_k)v(t_i, x_n)$$

$$\times \exp\left[-\frac{1}{2D\tau} \sum_{l=1}^{L} \sum_{m=1}^{M} \left(v(t_{l+1}, x_m) - v(t_l, x_m) \right. \right.$$

$$\left. + \left\{ v(t_l, x_m) \left[\frac{v(t_l, x_{m+1}) - v(t_l, x_{m-1})}{2a} \right] \right. \right.$$

$$\left. \left. - \nu \left[\frac{v(t_l, x_{m+1}) - 2v(t_l, x_m) + v(t_l, x_{m-1})}{a^2} \right] \right\} \tau \right)^2 \right], \quad (8)$$

where C is the normalization constant. In (8) the lattice constant is $a = x_n - x_{n-1} = 1$, the time step $\tau = t_l - t_{l-1} = 1$, the (kinematic) viscosity is $\nu = 1$ and the variance of the random force is $D = 1$. For economy of notation we have not written explicitly in (8) the periodic conditions imposed on the velocity field in time and space.

We calculate the single-time velocity correlation of Burgers equation on a lattice of 10 spatial and 2 temporal points for which $L = 2, M = 10$ in (8) and transform the integration range from $\pm\infty$ to ± 1. We examine the convergence of the result by calculating the integral using different convergence criteria when the number of samples within each subvolume is doubled from 10^6: (1) the absolute relative difference between results is at most between 2.5 and 8.5 % (1.025 and 1.085). (2) the absolute difference between results is at most between 0.5 and 2. (3) the maximum absolute value of either result is at most between 0.5 and 2.

Figure 1 shows the normalized single-time velocity correlation of Burgers equation in a 10×2 space and time lattice as a function of correlation distance in number of spatial points. For each correlation distance, two results are shown for different convergence criteria used. The results with strictest convergence criteria are shown in cyan. Each value is normalized by calculating the integral with the velocity factors in (8) in front of the exponential set to 1.

Figure 2 shows a histogram of integration subvolume centres as a function of distance from origin for an integral of correlation distance of 5. The distances from origin are normalized by $1/\sqrt{D}$ where D is the number of dimensions. Integration proceeds to smaller and smaller subvolumes until convergence is reached and one can see that the largest number of subvolumes, and hence the worst convergence, is concentrated around a shell at a distance of approximately 0.55 from origin.

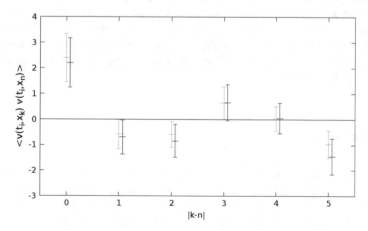

Fig. 1 Normalized single-time velocity correlation of Burgers equation as a function of correlation distance. Results with strictest convergence criteria are shown in cyan

Fig. 2 Histogram of number of centres of integration subvolumes as a function of normalized distance from origin. Most subvolumes, and hence worst convergence, is concentrated on a spherical shell at a distance of approximately 0.55 from origin.

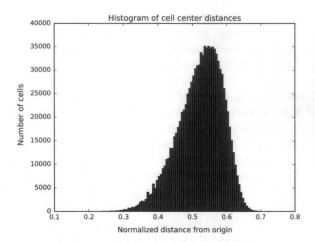

5 Discussion

The smallest computational resources required for obtaining one point in Fig. 1 used approximately 500 core hours of computational time and required a total of less than 10 GB of memory while the largest resources for one result required on the order of 10^4 core hours and 100 GB of memory. The required memory can most likely be decreased significantly by further optimizing its use in the parallel Python wrapper as well as switching from a Python implementation to e.g. C++. Substantial gains in required computational time will probably require a different integrand from (8). In this regard potential optimizations include switching from cartesian to spherical coordinates in order to better concentrate resolution where it is needed (cf. Fig. 2) and/or directly calculating the Fourier spectrum of the single-time velocity correlation. Utilizing GPUs and/or cloud computing is also an option worth exploring in the future.

6 Conclusion

We present a method for modelling turbulence via numerical functional integration. As a first step, we study single-time velocity correlation of Burgers' equation by transforming it into a functional integral that we solve with a parallel Python wrapper over the Monte Carlo integrators of GNU Scientific Library. We have used the functional integral widely used in the theoretical analysis of dynamic critical phenomena and stochastic dynamics with the subsequent transformation to discrete variables with respect to space and time.

In contrast to direct numerical simulation, our method allows for simple parallelization of the problem as even a single integral can be evaluated independently over different subvolumes. Here we evaluated each point in Fig. 1 separately but in parallel using 140–280 Intel Haswell cores and at most our calculation used 1540 cores simultaneously of the Cray XC40 supercomputer installed at the Finnish Meteorological Institute.

The results show that, as a first step, our method is promising but there still remains substantial work in developing an optimized approach for integration most suitable to this particular problem.

Funding The work of IH was funded by the Emil Aaltonen foundation.

References

1. J.P. Blaizot, E. Iancu, Y. Mehtar-Tani, Medium-induced QCD cascade: democratic branching and wave turbulence. Phys. Rev. Lett. **111**, 052001 (2013)
2. H.N. Sabbah, P.D. Stein, Turbulent blood flow in humans: its primary role in the production of ejection murmurs. Circ. Res. **38**(6), 513–525 (1976)
3. M.L. Goldstein, R.T. Wicks, S. Perri, F. Sahraoui, Kinetic scale turbulence and dissipation in the solar wind: key observational results and future outlook. Phil. Trans. R. Soc. A **373**, 20140147 (2015)
4. I. Zhuravleva, E. Churazov, A.A. Schekochihin, S.W. Allen, P. Arévalo, A.C. Fabian, W.R. Forman, J.S. Sanders, A. Simionescu, R. Sunyaev, A. Vikhlinin, N. Wernerl, Turbulent heating in galaxy clusters brightest in X-rays. Nature **515** (2014)
5. W.D. McComb, *The Physics of Fluid Turbulence* (Clarendon, Oxford, 1990)
6. I. Honkonen, J. Honkonen, Modelling turbulence via numerical functional integration using Burgers' equation (2018), arXiv:1803.07560
7. J. Zinn-Justin, *Quantum Field Theory and Critical Phenomena*, 4th edn. (Oxford University Press, Oxford, 2002)
8. L.T. Adzhemyan, N.V. Antonov, A.N. Vasil'ev, *The Field Theoretic Renormalization Group in Fully Developed Turbulence* (Gordon & Breach, London, 1999)
9. A.N. Vasil'ev, *The Field Theoretic Renormalization Group in Critical Behavior Theory and Stochastic Dynamics* (Chapman Hall/CRC, Boca Raton, 2004)
10. I. Honkonen, High-dimensional integrator. J. Open Sour. Softw. **2**, 437 (2017). https://doi.org/10.21105/joss.00437
11. M. Galassi, J. Theiler et al., *GSL—GNU Scientific Library, Version 2.4* (2017), https://www.gnu.org/software/gsl/

Nonlinear Noise Reduction on *TESS* Simulated Data

N. Jevtić, S. Shaffer and P. Stine

Abstract The high quality *Kepler Space Telescope* light curves have allowed us to deepen our understanding of nonlinear time series analysis and develop novel applications. Since simulated data represent the epitome of signal pre-processing, two simulated *TESS* stars were analyzed, one with a noisy light curve and one with a periodic light curve, in order to explore one of the tenets of nonlinear time series analysis: That pre-processed data do not lend themselves to these tools.

Keywords Nonlinear noise reduction · Kepler Space Telescope · *TESS* simulated light curves · Variable stars

1 Introduction

1.1 Nonlinear Time Series Analysis

Time-delay phase space reconstruction is used to analyze stellar light curves. A one-dimensional string of numbers yields a multi-dimensional representation. This embedding is a surrogate for the phase-space representation that would be obtained if we knew the equations governing the behavior at the source. The data itself, without any a priori assumptions, yields the time delay and the dimension of the phase space. The time delay is obtained using the information theory counterpart of the autocorrelation function, average mutual information (AMI). The optimal time delay is chosen at the first minimum of AMI [2]. The smallest accommodating dimension of the phase space, which is related to the number of degrees of freedom of the process, is obtained by the method of false nearest neighbors (FNN) [8]. For a detailed introduction to the methodology please see Kantz and Schreiber [7] and Schrebier [10]. For the use of nonlinear time series analysis on stellar data relevant here please see Jevtić et al. [5, 6].

N. Jevtić (✉) · S. Shaffer · P. Stine
Bloomsburg University of Pennsylvania, Bloomsburg, PA 17815, USA
e-mail: njevtic@bloomu.edu

© Springer Nature Switzerland AG 2019
C. H. Skiadas and I. Lubashevsky (eds.), *11th Chaotic Modeling and Simulation International Conference*, Springer Proceedings in Complexity, https://doi.org/10.1007/978-3-030-15297-0_12

123

1.2   Data Requirements

For the methodology to be successful in identifying chaotic systems the following are required:

- The observable should couple all the active degrees of freedom (energy, power, brightness, luminosity)
- Continuous data
- Uniform sampling
- Long data sets
- Fine digitization
- Access to a large dynamic range
- As little additive noise as possible
- No filtering or averaging.

In this paper we explore the validity of the last restriction by examining the pre-processed simulated astrophysical limited length over-sampled noisy light curves [4].

2   Kepler Versus TESS as Data Sources

Both of the two NASA planet-hunter space telescopes, the venerable *Kepler* [1] and the newly launched Transiting Exoplanet Survey Satellite (*TESS*) [9] detect exoplanets using the transit method by small drops in the brightness of a star as the planet gets between Earth and the star. Both produce continuous, uniformly sampled light curves. The two telescopes are, however, of very different design and operation. *Kepler* is in a heliocentric orbit while *TESS* utilizes a 2:1 lunar resonant orbit. *Kepler* has a mirror objective while *TESS* has an ingenious set of four lenses. During it's nominal mission, *Kepler* (K1) observed a region in Cygnus-Libra of only 0.28% of the sky to a depth of 3000 ly. *TESS* will conduct an all-sky transiting exoplanet survey of both the southern and northern hemispheres of an area about 400 greater. It will observe brighter stars to a depth of 300 ly. *TESS* targets will be observed for shorter times, many for only 27 days as opposed to the *Kepler* multi-year light curves. *Kepler* has 4 arcsec pixels; *TESS's* pixels are much larger covering 21 arcsec. This results in greater "crowding." Moreover, the depth of *TESS* pixels is six times greater. Therefore, where a cosmic ray in *Kepler* could impact only a single pixel, in *TESS* it will produce a trail which impacts many pixels. The photometric precision for a 10th magnitude star is estimated to be about 200 ppm based on 1 h of data collection, considerably higher than for *Kepler*. Thus, the analysis of *TESS* light curves poses different challenges.

3 The Simulated Data

The simulated data are common domain and available at https://archive.stsci.edu/tess/ete-6.html. The data were processed through the *TESS* Science Processing Operations Center pipeline [4].

The data were simulated to allow the stellar community to prepare their own analysis software and follow-up observation strategies for the *TESS* Mission in advance of receiving data now scheduled for the end of 2018 or the beginning of 2019.

Simulated 2-min cadence data consist of 20,610 data points (for many applications, this is oversampled).

Stellar variability is modeled after the Q10 *Kepler* data observed for the corresponding KIC ID.

In addition to stellar variability, signals have been injected for transiting planets and eclipsing binaries.

In principle, shorter light curves are noisier. Since *TESS* will observe a significant fraction of stars for only 27 days or two orbits, noise will be increased. The simulation length was slightly longer than two orbits which will be referred to as orbits 1 and 2.

Other issues include the fact that the much larger 21 arcsec pixels will result in more crowding. Since *TESS* pixels are double the *Kepler* width and six times thicker, cosmic ray effects will be more significant.

4 The Processing Command-Line Sequence in TISEAN

The analyses reported here were conducted using the TISEAN (Time Series Analysis) package by Hegger, Kantz and Schreiber at the Max-Planck-Institut fur Physics komplexer Systeme [3].

The analyses consist of the following command-line sequence and strategic decisions at each step:

mutual—used to find the first minimum of AMI for the initial optimal delay
false_nearest—used to find the phase-space dimension
ghkss—used for noise reduction
delay—used to produce a reconstructed phase-space portrait.

We analyze simulated light curves for two stars: one with a noise-like and the other with a periodic light curve.

5 Example I—Nonlinear Analysis of Simulated Data for a Noise-like Light Curve

The simulated data for *TIC* 89039049 (*TESS* input catalog number) were chosen as an example of a noise-like light curve. From its Data Validation Report Summary this star has a single planet candidate with a period of 10.707 days. The star has a radius 10% larger than the Sun and a surface temperature of 6060 K (Fig. 1).

5.1 Choice of Delay and Embedding Dimension

The first minimum of AMI (Average Mutual Information), Fig. 2, is at a delay of 150. The false nearest neighbors (FNN) curve (Fig. 3) drops slowly, even at a dimensions of 10 not getting down to zero.

Fig. 1 Orbit 1 *TIC* 89039049 light curve

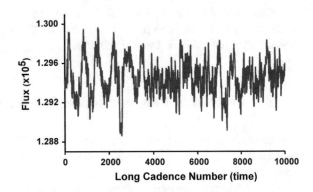

Fig. 2 Average mutual information, with a global minimum at a delay of 150

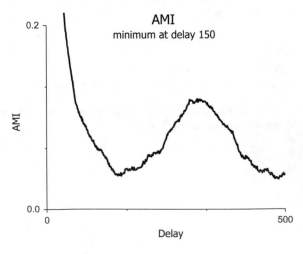

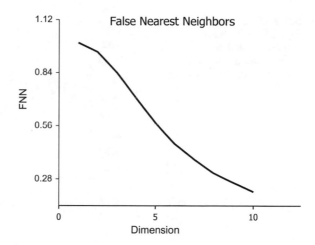

Fig. 3 The significant False Nearest Neighbor fraction reflecting the noise-like light curve

In view of the false nearest neighbor result, the significant false nearest neighbor fraction even at a large hyper-dimension of 10, reflecting the noise content of the curve, noise was reduced in phase-space using the orthogonal nonlinear projective method from a hyper-dimension of 5 to a phase-space dimension of two. Though the global minimum is at a delay of 150, the standard delay of one resulted in good noise reduction. The results are presented for the 6th iteration for a delay of 1.

5.2 Power Spectra and Light Curve

The power spectra for one *TIC* 89039049 orbit without and with noise reduction are shown in Fig. 4.

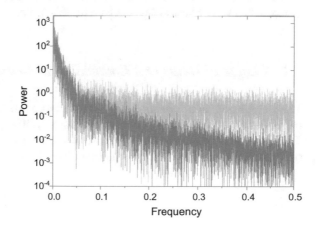

Fig. 4 *TIC* 89039049 Orbit 1 power spectra of prior to and after noise reduction for a delay of 1 of a factor ~100

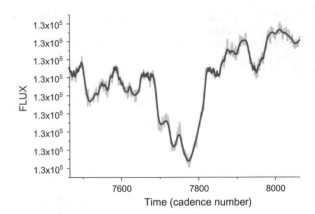

Fig. 5 Orbit 1 section of light curve with transit without and with noise reduction for a delay of 1 from 5D to 2D (6th iteration)

The spectrum prior to noise reduction has a significant white-noise tail. With the noise reduction, at the highest frequencies, noise has been reduced by a factor of about 100.

A section of the *TESS* 89039049 light curve with a planetary transit in orbit 1, prior to and after nonlinear projective noise reduction is shown in Fig. 5. A comparison of the nonlinear projective noise-reduced light curves for the same region for orbits 1 and 2 is shown in Fig. 6 (top). These results are to be compared with simulated data results (bottom) as published by NASA that show the phased orbit 1 and 2 transits and average for both.

6 Example II—Nonlinear Analysis of Simulated Data for a Periodic Light Curve

TIC 18114057 was chosen as an example of a regular, periodic light curve as seen in Fig. 7. No star or period data is given for this system.

6.1 Choice of Delay and Embedding Dimension

The average mutual information (Fig. 8) forecasts a power spectrum that reflects the periodic nature of the light curve and forecasts splitting of lines into components. The false nearest neighbor results imply that a phase-space reconstruction dimension of 5 is sufficient to accommodate the portrait (Fig. 9).

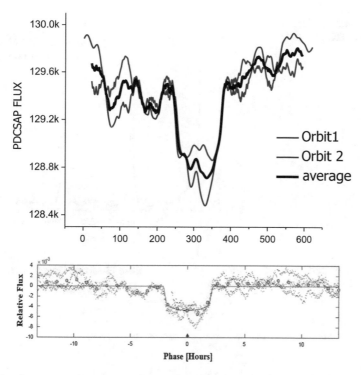

Fig. 6 Transit reconstruction after nonlinear noise reduction (top) compared with simulated data results (bottom) for *TIC* 89039049 as published by NASA in the respective validation report. The phased orbit 1 and 2 transits and average are shown for both

Fig. 7 *TIC* 18114057 Orbit 1 light curve after noise reduction

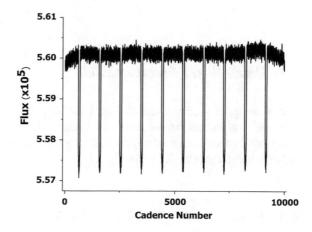

Fig. 8. Average mutual
information TIC 18114057

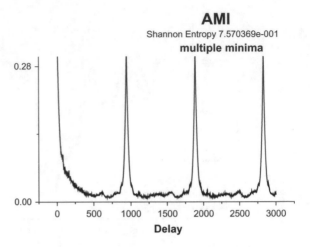

Fig. 9 False nearest
neighbors TIC 18114057

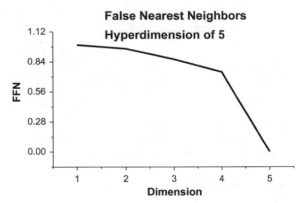

6.2 Phase Space Reconstruction

The phase space portrait shown in Fig. 10 in 3D is for a delay of 81. This is the delay
for which the portrait is most unfolded. It results in the most efficient noise reduction
due to the best definition of surfaces that the trajectories lie on. The residuals for the
noise-reduced data are shown in Fig. 11.

6.3 Power Spectra for TIC 18114057

The comparison of power spectra of *TIC* 18114057 prior to (top) and after nonlinear
noise reduction (bottom) is given in Fig. 12 (left). On the right is a blow-up of the
lower-frequency region. We access twice the frequency range in the noise-reduced
power spectrum.

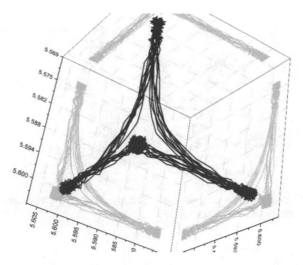

Fig. 10 Phase-space portrait for a delay of 81 with noise reduction from a hyper dimension of 6 to a dimension of 2

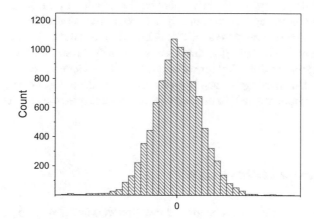

Fig. 11 Residuals for noise reduced data

7 Discussion: Choice of Parameters for Nonlinear Analysis of Tess Simulated Data

To test whether nonlinear time series analysis can yield meaning-full results for simulated *TESS* short, oversampled, noisy light curves, we investigated two simulated star systems, one whose light curve is noisy and looks random (*TIC* 89039049) and one with a very periodic light curve (*TIC* 18114057). After nonlinear projective noise reduction in phase space, we reconstructed their phase-space portraits, obtained their power spectra, and identified the transit of a planet in the noisy light curve.

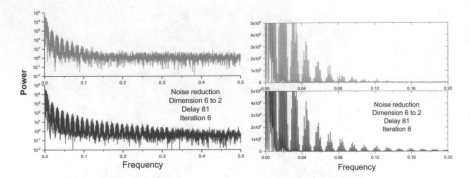

Fig. 12 (left) Comparison of power spectra prior to (top) and after nonlinear noise reduction (bottom). (right) Enlarged lower-frequency region

7.1   Time Delay

For theoretical systems, or slightly noisy but long enough time series, τ, the time delay for the reconstruction does not matter. Whatever reasonable delay we chose, the reconstruction retains it's shape. This delay is found at the first minimum of AMI. The noise reduction is most efficient if we can identify the surface the trajectory lies on. Thus a search for the delay for which the trajectories spread out (unfold) most to yield the most supporting surface is justified. This value is most often on the order of the 1st minimum of AMI. In the two cases above this gave the most efficient noise reduction.

7.2   Phase-Space Dimension

The theoretically sufficient embedding dimension **m** is not always optimal. For local projective noise reduction, the redundancy of an embedding with small τ and large embedding dimension **m** most often allows for better noise reduction. For our examples this is not the case. Even for *TESS* 18114057, for which the FFN goes to 0 at a dimension of 5, a higher phase-space dimension of 6 yielded better results.

8   Conclusion: Even for Pre-processed Data Nonlinear Noise Reduction Works!

Despite the fact that the data were pre-processed, noise was successfully reduced in both the examples presented allowing us to conclude that pre-processed data do lend themselves to nonlinear time series analysis! However, for pre-processed data and if the time series are short and noisy, most of the rules for the optimal time delay and embedding dimension have to be used with care, only as guidelines. Our results for TIC 89039049 and TIC 18114057, with two totally different light curves, show this. For the noisy TIC 89039049 light curve, with just the four commands, we were able to reproduce the planet transit with veracity. For the TIC 18114057 periodic light curve we were able to gain access to twice the frequency range in it's power spectrum. One possibility is that the restriction dates from the early work focused on detecting chaotic systems, whereas we are now using the toolkit for the more general and ubiquitous category of nonlinear systems.

Acknowledgements We wish to thank: The entire *Kepler* team for the development and operations of this outstanding mission and all involved in KASC for the handling of the data. Their efforts have allowed us to conduct the theoretical work on nonlinear time series methodology presented here. Funding for the *Kepler Mission* is provided by NASA's Science Mission Directorate.
TESS Science Processing Operations Center (SPOC) Pipeline at NASA Ames Research Center.
The Max Planck Institute for Complex Systems for the TISEAN software: R. Hegger, H. Kantz, and T. Schreiber, *Practical implementation of nonlinear time series methods: The TISEAN package*, CHAOS, 9, 413 (1999) and the book it became.

References

1. Borucki et al., IAUS **249**, 17 (2008)
2. A.M. Fraser, H.L. Swinney, Phys. Rev. A **33**, 1134 (1986)
3. R. Hegger, H. Kantz, T. Schreiber, Practical implementation of nonlinear time series methods: the TISEAN package. Chaos **9**, 413 (1999)
4. J.M. Jenkins et al., Proc. SPIE **9913**, 99133E (2016)
5. N. Jevtić et al., ApJ **635**, 527 (2005)
6. N. Jevtić et al., AN **333**(10), 983 (2012)
7. H. Kantz, T. Schreiber, *Nonlinear Time Series Analysis*, 2nd edn. (Cambridge University Press, Cambridge, 2004)
8. M.B. Kennel et al., Phys. Rev. A **45**, 3403 (1992)
9. Rickert et al., JATIS **1**, 1 (2015)
10. T. Schrebier, Phys. Rep. **308**(1), 1 (1999)

A Phase-Space Approach to Non-stationary Nonlinear Systems

Vladimir L. Kalashnikov and Sergey L. Cherkas

Abstract A phase-space formulation of non-stationary nonlinear dynamics including both Hamiltonian (e.g., quantum-cosmological) and dissipative (e.g., dissipative laser) systems reveals an unexpected affinity between seemly different branches of physics such as nonlinear dynamics far from equilibrium, statistical mechanics, thermodynamics, and quantum physics. One of the key insights is a clear distinction between the "vacuum" and "squeezed" states of a non-stationary system. For a dissipative system, the "squeezed state" (or the coherent "condensate") mimics vacuum one and can be very attractable in praxis, in particular, for energy harvesting at the ultrashort time scales in a laser or "material laser" physics including quantum computing. The promising advantage of the phase-space formulation of the dissipative soliton dynamics is the possibility of direct calculation of statistical (including quantum) properties of coherent, partially-coherent, and non-coherent dissipative structure without numerically consuming statistic harvesting.

Keywords Weil-Wigner-Moyal representation of stochastic dynamics · Coherent and turbulent dissipative structures · The vacuum states of nonlinear systems

1 Introduction

The study of the dynamics of self-organized dissipative systems could bridge the alas different shores of our knowledge, and it has to be based on an understanding of a multiscale nature of underlying phenomena. Here, we shall try to demonstrate as the most general and, nevertheless, outwardly disjoined concepts can contribute productively to the study of nonlinear dynamics of nonequilibrium nonlinear sys-

V. L. Kalashnikov (✉)
Institute of Photonics, Vienna University of Technology, Vienna, Austria
e-mail: vladimir.kalashnikov@tuwien.ac.at

S. L. Cherkas
Institute for Nuclear Problems, Belarus State University, Minsk, Belarus
e-mail: cherkas@inp.bsu.by

© Springer Nature Switzerland AG 2019
C. H. Skiadas and I. Lubashevsky (eds.), *11th Chaotic Modeling and Simulation International Conference*, Springer Proceedings in Complexity, https://doi.org/10.1007/978-3-030-15297-0_13

tems. The keystone here is a phase-space formulation of a problem which reveals the intrinsic affinity between both classical and quantum Hamiltonian as well as non-Hamiltonian systems. Such affinity promises a breakthrough in the study and practical mastering of scalable coherent structures in the midst of noisy dissipative environment. The application area ranges from neurophysiology to quantum computing and high-energy laser physics.

Here, we intend to illustrate the Weyl-Wigner-Moyal approach to the construction of the phase-space representation of seemly dissimilar systems ranging from quantum cosmology to ultrafast laser physics. The statistical mechanics and the theory of turbulence phenomena are the bearings in this enterprise.

There is a deep and physically relevant analogy between the evolutional laws for a mixed state of a quantum system (whether "closed" or "open") and the statistical mechanics. The Hamiltonian formulation of classical mechanics reveals this elegant and genuine kinship.

Let us remind the von Neumann law for the density matrix $\rho \equiv \sum_i P_i |\psi_i\rangle \langle \psi_i|$ evolution [1]:

$$\frac{\partial \rho(t)}{\partial t} = \frac{i}{\hbar} [\rho(t), H(t)], \tag{1.1}$$

where $H(t)$ is the time-dependent Hamiltonian of a system, including, in the general case, the "environment" ("basin") and the interactional parts ([∗, ∗] denotes a commutator). This equation is a direct analog to the famous Liouville equation for the evolution of a phase-space distribution function ρ in the statistical mechanics:

$$\frac{\partial \rho}{\partial t} = -\{\rho(t), H(t)\}, \tag{1.2}$$

({∗, ∗} denotes the Poisson bracket) and to the law of evolution of a dynamical variable $A(t)$ within the frameworks of Hamiltonian formulation of classical mechanics:

$$\frac{dA(t)}{dt} = \{H(t), A(t)\}. \tag{1.3}$$

However, the conceptual difference is that the phase space in the quantum mechanics is the operator space, and these operators can be noncommitting in the general case. A study of this space is a mathematically challenging issue, and, we face the interpretation challenges additionally. The instance of such problem, which is relevant to our work, is the practically useful definition of *vacuum state of a time-dependent quantum system* (e.g., the time-dependent quantum oscillator) and its distinguish from a so-called "squeezed state." The classical definition implying the vacuum state $|0\rangle$ as a "zero space" of annihilation operator $\hat{a}|0\rangle = 0$ is not practically useful in many cases. The important insight of the Hamiltonian minimization $\langle 0|\hat{H}|0\rangle$ is closely related to the situations, which will be considered below. At last, the asymptotical "uncertainty

minimization" criterium $\left\langle \left| \hat{p} - \langle \hat{p} \rangle \right|^2 \left| \hat{x} - \langle \hat{x} \rangle \right|^2 \right\rangle = \frac{1}{4} + \sigma^2 = \frac{1}{4}\left(1 + \langle 0|\hat{x}\hat{p} + \hat{p}\hat{x}|0\rangle^2\right)$
[2] is relevant to important both quantum and classical problems.

2 A Time-Dependent (Driven) Quantum Oscillator

An issue of time-dependent (driven) oscillator arises naturally in some fields of the theoretical physics. In particular, it has an application in cosmology and astrophysics, where the scalar, fermion, gravitational, and other quantum fields evolve in an expanding Universe. Nevertheless, the definition of the ground (vacuum) state remains to be obscure. It would be desirable to define vacuum state without appealing to the adiabatic series or analytical solution that can be impossible in praxis. This issue is addressed in the suggested method, which allows finding the true vacuum state numerically if such a state exists.

Let us remind the problem in more detail. The Hamiltonian of a time-dependent oscillator has the following form:

$$H = \frac{1}{2}\left(\dot{x}^2 + \omega^2(t)x^2\right). \tag{2.1}$$

The standard commutation relations for the momentum and coordinate operators are:

$$\left[\hat{p}(t), \hat{x}(t)\right] = \left[\hat{\dot{x}}(t), \hat{x}(t)\right] = -i. \tag{2.2}$$

The mean value of the kinetic and potential energies difference is expressed as

$$\langle 0|\frac{1}{2}\hat{p}^2(t) - \frac{1}{2}\omega(t)\hat{x}^2(t)|0\rangle = \dot{\sigma}(t). \tag{2.3}$$

Here

$$\sigma = \frac{1}{2}\langle 0|\hat{x}(t)\hat{p}(t) + \hat{p}(t)\hat{x}(t)|0\rangle \tag{2.4}$$

has a sense of the additional uncertainty arising in the Heisenberg uncertainty relation:

$$\left\langle \left| \hat{p} - \langle \hat{p} \rangle \right|^2 \left| \hat{x} - \langle \hat{x} \rangle \right|^2 \right\rangle > \frac{1}{4} + \sigma^2 \tag{2.5}$$

and ⟨ l, l ⟩ are arbitrary states. For a family of the squeezed states, including a true vacuum, the inequality (2.5) becomes equality.

The straightforward computation shows that σ satisfies the nonlinear equation

$$\left(4\sigma\omega^2 + \ddot{\sigma}\right)\left(4\sigma\omega^3 + \ddot{\sigma}\omega - 2\dot{\sigma}\dot{\omega}\right) - \omega\dot{\omega}^2\left(4\sigma^2 + 1\right) = 0, \tag{2.6}$$

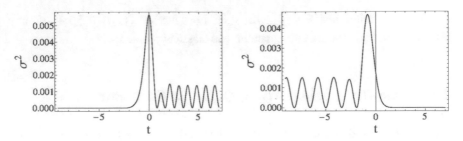

Fig. 1 The examples of the σ^2-function behavior for the in- and the out-vacuum states

for the states belonging to a family of the squeezed vacuum states including the true vacuum ones. Thus, one has the nonlinear Eq. (2.6) for choosing the true vacuum state from a family of the squeezed states. The nonlinearity in (2.6) arises from (2.2). We suggest that a true vacuum state corresponds to the monotonic time-dependence of $\sigma(t)$.

Since the criterium of a monotonic behavior of the $\sigma(t)$-function within a time interval $\{t_1, t_2\}$ is chosen, one may use the minimization of the functional

$$Z(r, \delta) = \int_{t_1}^{t_2} \left(\frac{\mathrm{d}}{\mathrm{d}t} \sigma(t, r, \delta) \right)^2 \mathrm{d}t, \qquad (2.7)$$

where r, δ are the parametrization parameters for the whole family of the squeezed states. In a non-steady case, the vacuum state is a conditional notion for the in-vacuum $t_1 \rightarrow -\infty$ and the out-vacuum $t_2 \rightarrow \infty$ states. As may see, the nonlinear equation appears even in a linear quantum problem for determining a true vacuum state of the time-dependent oscillator.

The examples of the $\sigma(t)$—behavior for the in- and the out-vacuum states are shown in Fig. 1.

Other insight bridging the quantum and classical systems could regard to a decreasing of the dispersion of the dynamical variables mean values. An example is the cosmological mini-superspace model. The Hamiltonian in this model is simultaneously a constraint condition $H = 0$ which should be satisfied alongside with the equations of motion.

Let us consider the toy model with a massless scalar field ϕ and the "by hand" introduced decrease of the cosmological constant V_0 [3]. The Hamiltonian of the model has the form:

$$H = -\frac{p_a^2}{2a} + \frac{p_\phi^2}{2a^3} + V_0 \frac{a^3}{1 + \beta a^3}, \qquad (2.8)$$

where p_a and p_ϕ are the momentums associated with the Universe scale-factor a and the scalar field ϕ, respective, and β is some constant. This Hamiltonian assumes a modification of the gravity theory with a cosmological constant in a sense that this "constant" $V_0 \frac{a^3}{1+\beta a^3}$ is non-zero at the small-scale factors and decreases as $\propto a^{-3}$ at the large scale-factors (i.e., it is a model of the terminating inflation).

The corresponding equations of motion are:

$$\ddot{\alpha} + \frac{3}{2}(\dot{\alpha}^2 + \dot{\phi}^2) - \frac{3V_0}{(1 + \beta e^{3\alpha})^2} = 0, \quad \ddot{\phi} + 3\dot{\alpha}\dot{\phi} = 0, \tag{2.9}$$

where $\alpha \equiv \ln a$.

For quantization, one should consider the Hamiltonian constraint as a condition for a state vector $|\Psi\rangle$: $\hat{H}|\Psi\rangle = 0$. As a result, we come to the Wheeler-DeWitt equation [3, 4]:

$$\left(\frac{1}{2a^2} \frac{\partial}{\partial a} a \frac{\partial}{\partial a} - \frac{1}{2a^3} \frac{\partial^2}{\partial \phi^2} + V_0 \frac{a^3}{1 + \beta a^3} \right) \Psi(a, \phi) = 0. \tag{2.10}$$

The paradox is that there is no explicit time-variable in this equation, which manifests the so-called "problem of time" in the quantum cosmology [5]. Formally, the Hamiltonian is the field equation constraint in the general theory of relativity. That means that the total energy of the gravitational field and the matter vanishes. Thus, all states form the Hamiltonian "null-space" after canonical quantization [that results in the Wheeler-DeWitt equation (2.10)]. That is all quantum states are "vacuum states" (the Hamiltonian "annihilates" them). But the Hamiltonian provides a time-evolution. Thus, there is no time-evolution in the quantum cosmology.

However, this is rather a pseudo-problem, since the time-evolution remains in the equations of motion (2.9) so that one could only write "hats" over $\hat{\alpha}$ and $\hat{\phi}$ to consider them as the quasi-Heisenberg operators and Eq. (2.9) as the operator equations [6]. The commutation rules for these operators follow from the Dirac brackets for a constraint system. They can be evaluated explicitly at the initial moment of time then the system allows evolving in accordance with the equations of motion. The Hilbert space for the quasi-Heisenberg operators is built on the basis of an asymptotical solution of the Wheeler-DeWitt equation (2.10).

The results of the calculation are shown in Fig. 2. One can see that the Universe becomes "classical" after the inflation end. It means that the sufficiently quick decrease of the cosmological constant causes suppressing the dispersion of the scale factor logarithm.

3 The Relation with the Solitonic and Statistical Physics

Here we invent the connection with the solitonic physics [7] based on the idea that the classical states are the result of the quantum evolution of a nonlinear system evolving to the state with the small dispersions of the mean values of observables. Thus, the

Fig. 2 The mean value of
the logarithm of the scale
factor $\langle \alpha \rangle$ and its dispersion
$\sigma(\alpha)$ for the model (2.8)
with the cosmological
constant $V_0 = 1$, $\beta = 0$
(dashed curves), and with the
decreasing cosmological
constant $V_0 = 1$, $\beta = 10^{-8}$
(solid curves)

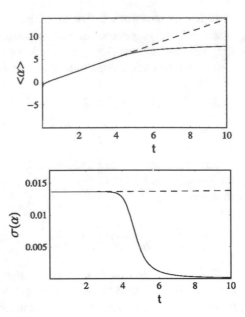

nonlinear equations arise in a quantum linear physics when one tries constructing a vacuum state. On the other hand, one may see that the quantum systems tend to classics ones in some cases. Thereby, the solitonic physics can be incorporated into the field of quantum physics including both linear and nonlinear phenomenon.

More specifically, a soliton can be interpreted as a coherent structure formed in the self-interacting bosonic system, i.e., as a classical analog of the Bose-Einstein condensate [8–10]. Such a coherent condensate is defined by the two-point correlation function in the momentum $p-$ space: $\left\langle A_p(t) A_{p'}^*(t) \right\rangle = n_p \delta(p - p')$, where $A_p(t) = \frac{1}{\sqrt{2\pi}} \int \psi(t, x) e^{-ipx} dx$, $\psi(t, x)$ is a field amplitude, and n_p is a "particle number" distribution characterizing the soliton "shape." The "condensation" means a flow of energy to zero wavenumbers $p \to 0$ that is the increase of long-range correlations and the suppression of fluctuations in direct analogy with minimization of dispersion of a quantum system transiting to a classical state (Fig. 2) [11]. Simultaneously, that results in the minimization of the Hamiltonian $H(\psi)$ defined as

$$H(\psi) = \int \left(\left| \frac{\partial \psi}{\partial x} \right|^2 \mp |\psi|^4 \right) dx \tag{3.1}$$

for the well-known (1+1)-dimensional cubic nonlinear Schrödinger equation which describes an evolution of slowly varying wave in a nonlinear medium [9, 12]:

$$i\frac{\partial \psi}{\partial t} = \frac{\delta H}{\delta \psi^*} \text{ or } i\frac{\partial \psi}{\partial t} + \frac{\partial^2 \psi}{\partial x^2} \pm |\psi|^2 \psi = 0. \tag{3.2}$$

Fig. 3 The Langmuir
dispersion relation $\lambda \propto p^2$
(black curve) and the
Rayleigh-Jeans equilibrium
distribution for a DS or the
turbulence (red curve). The
condensation in the vicinity
of $p = 0$ is illustrated by
shading

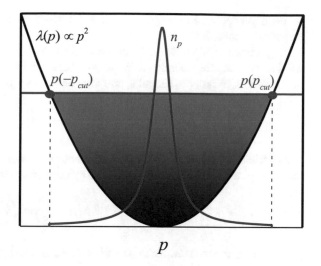

Such a coherent condensate (i.e., a soliton) minimizing the Hamiltonian and exist-
ing as a steady "ground" state (i.e., $\psi(t, x) = \phi(x)\mathrm{e}^{-i\lambda t}$) allows treating as an analog
of the vacuum state of the nonlinear system far from the thermodynamic equilibrium.

A soliton (i.e., a coherent "condensate") has the minimal entropy so that the rest
of entropy concentrates in the small-scale fluctuations with large $|\psi_x|^2$ outside the
condensate [8, 13]. As a result, the condensate evolves toward the Rayleigh-Jeans
equilibrium distribution [11]:

$$n_p \propto \frac{1}{p^2 - \mu}\Theta\left(p_{\text{cut}}^2 - p^2\right) \tag{3.3}$$

which obeys two correlation scales: a long-range one defined by a negative "chemical
potential" μ, and a short-range one defined by a momentum cut-off at p_{cut} which
is caused by the nonlinear and dissipative effects (Θ is the Heaviside function, see
Fig. 3).

Thus, a bridge to the statistical mechanics is in the offing, and such invention
is relevant to the description not only coherent solitons but also to the study of the
dissipative solitons (DS) and the turbulent phenomena [11, 14].

4 Phase-Space Representation of Nonlinear Dynamics

As was demonstrated above, the phase-space (Hamiltonian) description of nonlinear
dynamical systems in both quantum and classical mechanics provides with a guide-
line in the solution and interpretation of the complex problems that entwines the

seemingly disjointed concepts ranging from quantum cosmology to solitonics and statistical mechanics.

Regarding the quantum mechanics operating in a linear operator Hilbert space, we need associating the operator $\hat{A}$ in the x-representation with an appropriate function in the Weyl's, Wigner's, Moyal's, and Groenewold's style [15]:

$$\tilde{A}(x, p) = \int e^{-ipq/\hbar} \left\langle x + \frac{q}{2} \middle| \hat{A} \middle| x - \frac{q}{2} \right\rangle dq, \tag{4.1}$$

and to relate the quantum density with the so-called Wigner function $W(x, p)$ which has a direct association with the probability density operator $\hat{\rho} = |\psi\rangle\langle\psi|$ [16]:

$$W(x, p) \propto \int e^{-ipq/\hbar} \psi\left(x + \frac{q}{2}\right) \psi^*\left(x - \frac{q}{2}\right) dq \tag{4.2}$$

that provides the measurable expectation value of a $\hat{A}$-operator:

$$\langle A \rangle = \iint W(x, p) \tilde{A}(x, p) dx dp. \tag{4.3}$$

Finally, we have to associate the noncommutativity of operators with some ordering rule, e.g., in the Weyl's style:

$$\hat{p}^2\hat{x} \rightarrow \frac{1}{3}\left(\tilde{p}^2\tilde{x} + \tilde{p}\tilde{x}\tilde{p} + \tilde{x}\tilde{p}^2\right). \tag{4.4}$$

Returning to nonlinear optics, the nonlinear Schrödinger Eq. (3.2) with a potential $U(x)$ allows the phase-space representation through the Wigner transformation

$$W(x, p) \propto \int e^{-ipq} \psi\left(x + \frac{q}{2}\right) \psi^*\left(x - \frac{q}{2}\right) dq \tag{4.5}$$

resulting in [17]:

$$\frac{\partial W(x, p)}{\partial t} + \frac{\beta}{2} p \frac{\partial W(x, p)}{\partial x}$$
$$+ \sum_{s=0}^{\infty} \frac{(-1)^s}{(2s+1)! 2^{2s}} \frac{\partial^{2s+1} U(x)}{\partial x^{2s+1}} \frac{\partial^{2s+1} W(x, p)}{\partial p^{2s+1}} = 0, \tag{4.6}$$

where β is a group-delay parameter ("kinetic energy" term), and the self-interaction potential is $U(x) = \sqrt{\frac{1}{2\pi}} \int W(x, p) dp$.

Two problems are that the resulting Eq. (4.6) contains the infinite expansion term, and it is the integrodifferential equation in (2+1)-dimensions. Nevertheless, our calculations demonstrated that the geometrical optics approximation $\Delta x \Delta p \gg 1$ is

Fig. 4 The Wigner function on (x,p)-plane for the dimensionless dispersion $\beta = 1$ after the 6 dimensionless nonlinear propagation lengths t

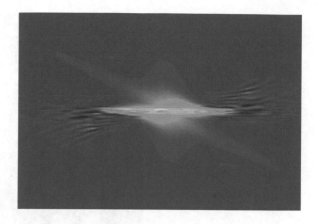

well-working even for the "true vacuum" (not only "squeezed one," see below) states and can be modeled by the Vlasov's equation [$s = 0$ in Eq. (4.6)] [18, 19]:

$$\frac{\partial W(x, p)}{\partial t} \pm p\frac{\partial W(x, p)}{\partial x} + \frac{\partial U(x)}{\partial x}\frac{\partial W(x, p)}{\partial p} = 0, \qquad (4.7)$$

re-interpreting the Wigner function as a probability distribution function:

$$W(x, p) \propto \int e^{-ipq}\left\langle \psi\left(x + \frac{q}{2}\right)\psi^*\left(x - \frac{q}{2}\right)\right\rangle \mathrm{d}q \qquad (4.8)$$

where $\langle \bullet \rangle$ denotes a statistical average. Equation (4.7) describes the quasi-particles statistics in the effective self-consistent potential, and we may interpret a soliton as a self-organized ensemble of interacting quasi-particles ("internal modes") and use the methods of statistical mechanics.

Figures 4, 5, 6 and 7 demonstrate the evolution of a Wigner function in the so-called anomalous dispersion regime $\beta > 0$, where the classical soliton exists [20]:

$$i\frac{\partial \psi}{\partial t} + \frac{\beta}{2}\frac{\partial^2 \psi}{\partial x^2} + |\psi|^2\psi = 0. \qquad (4.9)$$

When the nonlinearity prevails over the dispersion, the initial pulse inevitable collapses (Fig. 4). But a compensation of pulse squeezing due to nonlinearity by dispersion results in the soliton formation, which is stable, perfectly localized and coherent structure (Fig. 5). When the dispersion prevails over nonlinearity, the pulse spreads in the time domain but with the conservation of its spectral width. It is an example of squeezing described by the so-called "chirp" parameter θ, that is a slope of the Wigner function in our case (Figs. 6 and 7).

In the normal dispersion regime ($\beta < 0$), the tendency to collapse is arrested, so that the energy concentration at zero wave-number (carrier frequency) results in a squeezing state with a huge "chirp" (Fig. 8).

Fig. 5 The Wigner function
for the dimensionless
dispersion $\beta = 2$

Fig. 6 The Wigner function
for the dimensionless
dispersion $\beta = 3$

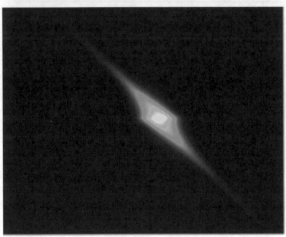

However, this state is not steady. It tends to disappear in the "fluctuation sea."
One may propose a way out of this problem: let's make our system dissipative that
could provide an inverse energy cascade outward the zero wave-number but without
the coherency loss. The example of such open system is a laser with the linear and
nonlinear gain, loss (μ, κ), and the spectral dissipation (α).

$$i\frac{\partial \psi}{\partial t} + \frac{\beta}{2}\frac{\partial^2 \psi}{\partial x^2} + |\psi|^2\psi = i\left(\mu\psi + \alpha\frac{\partial^2 \psi}{\partial x^2} + \kappa|\psi|^2\psi\right). \tag{4.10}$$

The combination of these factors provides a right energy redistribution $E = 2\left(\mu|\psi|^2 + \kappa|\psi|^4 - \alpha\left|\frac{\partial \psi}{\partial x}\right|\right) + \alpha\frac{\partial^2 |\psi|^2}{\partial x^2}$ (Fig. 9) that stabilizes the DS coherent structure.

Fig. 7 The Wigner function
for the dimensionless
dispersion $\beta = 4$

Fig. 8 The Wigner function
for the dimensionless
dispersion $\beta = -4$

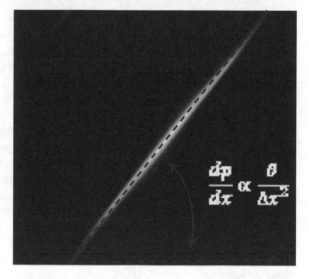

Here, we deal with a DS with a nontrivial internal structure providing a huge chirp without spectral squeezing that allows a coherent energy harvesting. This huge chirp validates the lowest-order term approximation in the Weyl-Wigner Eq. (4.7) and we reveal with surprise that a dissipative soliton is a self-organized "ensemble" of self-interacting quasi-particles, somewhat like an elementary "community," and the methods of statistical mechanics could allow the description of such metaphorical "community" without a direct statistic gathering of the "individual fates."

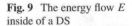

Fig. 9 The energy flow E
inside of a DS

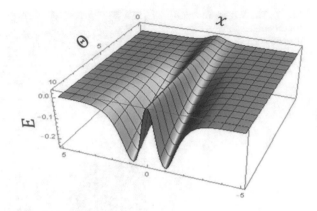

5 Conclusions

Phase-space formulation of non-stationary nonlinear systems reveals an affinity
between seemly different branches of physics such as dynamics of nonlinear systems
far from equilibrium, statistical mechanics, thermodynamics, and quantum physics.
One of the key insights is a clear distinction between the "vacuum" and "squeezed"
states of a system. A soliton can be treated as a "vacuum state" of a closed nonlinear
system, and such low-entropy state minimizes a Hamiltonian so that the second law
of thermodynamics needs an entropy concentration in small-scaled (down to quantum level) fluctuations. The "squeezed states" (or coherent "condensates") mimic
vacuum ones and can be very attractable in praxis, in particular, for energy harvesting at ultrashort time scales. However, such states are not steady-state in a closed
system. The stabilization of such coherent structure is possible in an open, i.e., dissipative system. That means a DS formation. The phase-space analysis demonstrates
a close analogy between DS and turbulence phenomena in plasma and condensed
media that allows formulating the statistical mechanics and quantum approaches to
the extremely broad diapason of nonlinear phenomena. In particular, the promising
advantage of the phase-space formulation of the DS dynamics is the possibility of
direct calculation of statistical (including quantum) properties coherent, partially-
coherent, and non-coherent dissipative structure without numerically consuming
statistic harvesting.

References

1. H.-P. Breuer, F. Petruccione, *The Theory of Open Quantum Systems* (Oxford University Press,
   New York, 2002)
2. S.V. Anischenko, S.L. Cherkas, V.L. Kalashnikov, Functional minimization method addressed
   to the vacuum finding for an arbitrary driven quantum oscillator. Nonlinear Phenom. Complex
   Syst. **12**(1), 16–26 (2009). (arXiv:0806.1593 [quant-ph])

3. S.L. Cherkas, V.L. Kalashnikov, Quantum evolution of the universe in the constrained quasi-Heisenberg picture: from quanta to classics? J. Gravit. Cosmol. **12**(2–3), 126–129 (2006). (arXiv:gr-qc/0512107)
4. B.S. DeWitt, Dynamical theory in curved spaces. I. A review of the classical and quantum action principles. Rev. Mod. Phys. **29**, 377–397 (1957)
5. M. Bojowald, *The Universe: A View from Classical and Quantum Gravity* (Wiley-VCH, Weinheim, 2013)
6. S.L. Cherkas, V.L. Kalashnikov, Quantum mechanics allows setting initial conditions at a cosmological singularity: Gowdy model example. Theor. Phys. **2**(3), 124–135 (2017). (arXiv: 1504.06188 [gr-qc])
7. V.L. Kalashnikov, Optics and Chaos: Chaotic, Rogue and Noisy Optical Dissipative Solitons, in *Handbook of Applications of Chaos Theory*, ed. by ChH Skiadas, Ch. Skiadas (Chapman and Hall, Boca Raton, 2016), pp. 587–626
8. R. Jordan, B. Turkington, C.L. Zirbel, A mean-field statistical theory for the nonlinear Schrödinger equation. Physica D **137**(3–4), 353–378 (2000)
9. S. Dyachenko, A.C. Newell, A. Pushkarev, V.E. Zakharov, Optical turbulence: weak turbulence, condensates and collapsing filaments in the nonlinear Schrödinger equation. Physica D **57**(1–2), 96–160 (1992)
10. Y. Lai, H.A. Haus, Quantum theory of solitons in optical fibers. II. Exact solution. Phys. Rev. A **40**(2), 854–866 (1989)
11. V.L. Kalashnikov, E. Sorokin, Self-Organization, Coherence and Turbulence in Laser Optics, in *Complexity in Biological and Physical Systems*, ed. by R. Lopez-Ruiz (IntechOpen, London, 2018), pp. 97–112
12. D.J. Benney, A.C. Newell, The propagation of nonlinear wave envelopes. J. Math. Phys. **4**, 133–139 (1967)
13. C. Connaughton, Ch. Josserand, A. Picozzi, Y. Pomeau, S. Rica, Condensation of classical nonlinear waves. Phys. Rev. Lett. **95, 26**, 263901 (2005)
14. V.L. Kalashnikov, E. Sorokin, Turbulence of Optical Dissipative Solitons, in *Chaotic Modeling and Simulation* (2018), April Issue, pp. 125–137
15. G.S. Agarwal, E. Wolf, Calculus for functions of noncommuting operators and general phase-space methods in quantum mechanics. I. Mapping theorems and ordering of functions of noncommuting operators. Phys. Rev. D **2**(10), 2161–2186 (1970)
16. W.B. Case, Wigner functions and Weyl transforms for pedestrians. Am. J. Phys. **76**(10), 937–946 (2008)
17. D. Dragoman, Wigner distribution function in nonlinear optics. Appl. Opt. **35**, 4142–4146 (1996)
18. B. Hall, M. Lisak, D. Anderson, R. Fedele, V.E. Semenov, Phys. Rev. E **65**, 035602(R) (2002)
19. J. Garnier, M. Lisak, A. Picozzi, Toward a wave turbulence formulation of statistical nonlinear optics. J. Opt. Soc. Am. B **29**, 2229–2242 (2012)
20. G.P. Agrawal, *Nonlinear Fiber Optics* (Elsevier, Amsterdam, 2013)

Spatial Extent of an Attractor

Avadis Hacınlıyan and Engin Kandıran

Abstract Lyapunov exponents characterize the rate of approach or recession of nearby trajectories in a dynamical system defined by differential equations or maps. They are usually taken as indicators of chaotic behavior. The density of orbits in the state space or equivalently, the Poincare map is usually taken as another such indicator. Although these indicators usually give correct results, there are instances in which they can lead to confusing or misleading information. For instance, a system of three linear differential equations can have three positive eigenvalues λ_i leading to a solution $\exp \lambda_i t$. The Wolf-Benettin algorithm [4] would report three positive Lyapunov exponents, in spite of the fact that the system is not chaotic. Another example is the Khomeriki model [1] or even the usual Bloch equations that would report a spectrum of all negative Lyapunov exponents but produce completely full state space plots, if the AC field is sufficiently strong. We will consider the class of systems proposed by Sprott [3] consisting of three-dimensional ODE's with at most two quadratic non-linearities as examples. Many of them obey two scenarios one of which is Lorenz model like behavior where an unstable linearized fixed point is surrounded by two stable fixed points so that the unstable fixed-point leads to a throw and catch behavior. The other is Rössler-like behavior whereas the system moves away from a weakly unstable linearized fixed point, nonlinear terms return it to equilibrium with a spiral out catch in mechanism. Since the presence of an attractor may involve structural stability, these two mechanisms are expected to produce different spatial extents for the attractor. Although Lyapunov exponents indicate time dependent behavior, spatial extent would complement this as a spatial measure of localization, thus comple-

A. Hacınlıyan
Department of Physics and Department of Information Systems and Technologies,
Yeditepe University, Ataşehir, İstanbul, Turkey
e-mail: avadis@yeditepe.edu.tr

E. Kandıran (✉)
Department of Information Systems and Technologies,
Yeditepe University, Ataşehir, İstanbul, Turkey
e-mail: engin.kandiran@yeditepe.edu.tr

E. Kandıran
The Institute for Graduate Studies in Sciences and Engineering,
Yeditepe University, İstanbul, Turkey

© Springer Nature Switzerland AG 2019
C. H. Skiadas and I. Lubashevsky (eds.), *11th Chaotic Modeling and Simulation International Conference*, Springer Proceedings in Complexity, https://doi.org/10.1007/978-3-030-15297-0_14

menting the Lyapunov exponents that characterize horizon of predictability. Direct numerical simulation and where feasible, the normal form approach will be used to investigate selected examples of the three degree of freedom systems.

Keywords Lyapunov exponents · Sprott systems · Poincare map · Normal form · Simulation · Chaotic simulation

1 Introduction

In order to determine if a system is chaotic, the usual first step is calculating the Liapunov spectrum. Since the calculation of largest Liapunov exponent is usually easier than the calculation of whole spectrum, many studies only take the largest exponent into consideration. Numerical algorithms can be applied to an arbitrary dynamical system, but they are limited by some problems:

- Computational stability
- Convergence and truncation errors.

In a dynamical system, a zero Liapunov exponent may serve as an indicator for conserved quantities, because a conserved quantity could be used to lower the degree of freedom by one. The normal form expansion and a solution of the normal form equations is available in the vicinity of equilibrium points. Time and higher powers of residual terms in radial variable, if available, can be eliminated from the solution to give an approximately conserved quantity.

It is an attractive idea to replace a system by a locally equivalent, simpler system and a polynomial transformation. It is hoped that such a decomposition would approximate chaos by stretching and folding the solution of the simpler system with the polynomial transformation. The method of normal forms achieves this by a systematic procedure. If there are no resonances, the simpler system is linear, if there are resonances, the nonlinear terms remain, but its truncations can often be integrated or used to estimate extent of the attractor.

Usually, the three dimensional systems leading to an attractor (Sprott) have the $(+,0,-)$ or $(a \pm ib, c)$ with either a or c positive Lyapunov spectrum. However, there are at least two different kinds of such systems, the Lorentz system that uses the throw and catch mechanism (an unstable central fixed point surrounded by two stable fixed points) and the Rössler system that uses the spiral out, then fall in mechanism. Both will have comparable Lyapunov spectra but lead to different dynamics.

In this study, we use the normal form approach to find simpler forms of the three Sprott systems given in [4] and look for extent of the attractors and their conserved quantities. In Sect. 2, formal description of normal form analysis is given. In the Sect. 3, three Sportt systems are defined and analyzed using normal form analysis. Finally, in Sect. 4, a final discussion is given.

2 Normal Form of a Dynamical System

Normal form expansions give the local properties of chaotic behavior correctly, but have convergence problem. Aim of the normal form is to reduce the system to a simpler system, whose structure is determined by resonance dictated by the eigenvalues of the linearized system near an equilibrium point. It may sometimes be desirable to change the resonant structure of the base system by applying nonpolynomial transformations, or using nonlinear transformations that introduce a manifold over which the system survives in long time and scales in a way analogous to the center manifold.

Consider an ODE:

$$\dot{x} = Ax + F_2(x) + F_3(x) + \cdots + F_N(x) + \cdots \tag{1}$$

where $x \in \Re^n$ and A is a linear operator in Jordan Canonical Form, $F_i, i = 1, \ldots, N$. are homogeneous polynomials of degree N (Note: initial values $x(t = 0)$ are given). Let $\lambda = (\lambda_1, \ldots, \lambda_N)$ the eigenvalues of A.

Suppose system (1) is transformed by the near identity transformation:

$$x_i = y_i + h_{i2}(y_k) + \cdots + h_{iM}(y_k) + \cdots \tag{2}$$

where $y_k \in \Re^n, h_{iM}(y_k) : \Re^n \to \Re^n$ are also homogeneous polynomials of degree M. These polynomials can are determined order by order so that F_i are eliminated. The condition for elimination of ith degree term is:

$$Dh_i(y)Ay - Ah_i(y) = F_i(y) \tag{3}$$

These equations are linear in the over complete space spanned by the eigenvectors of the linear part $\times$ monic monomials of appropriate degree. They can be solved if the eigenvalues λ_j of the linearized part do not satisfy the resonance condition:

$$\lambda_j = \sum_k m_k \lambda_k \tag{4}$$

where m_k are positive integers or zero and $\sum_k m_k$ sum is called as **order of resonance**. This polynomial normal form expansion fails to converge if the system does not admit an additional symmetry however nicely illustrates the stretch and fold approach.

We will form A into a special form that will ease our calculation.

$$\lambda_1 = \gamma \quad \text{and} \quad \lambda_{2,3} = \alpha \pm i\beta \tag{5}$$

Let corresponding eigenvectors be v_1, v_2, v_3. Let T be a transformation matrix [2]:

$$T = \begin{bmatrix} \vdots & \vdots & \vdots \\ v_1 & v_2 & v_3 \\ \vdots & \vdots & \vdots \end{bmatrix} \tag{6}$$

which is invertable. Then the transformation matrix is used to transform the linear part of the system into diagonal or Jordan canonical form:

$$T^{-1}x = \tilde{x} \tag{7}$$

3 Analysis of Sprott Systems by Normal Form Approach

In this study, we have studied the nth order normal form of the following systems:

- Sprott D system
- Sprott E system
- Sprott F system.

These systems are examples of possible situations where the size of attractor and lyapunov exponents can be estimated using normal forms and where this is not possible.

3.1 Sprott D System

The Sprott D system is defined as:

$$\dot{x}_1 = -x_2 \tag{8a}$$
$$\dot{x}_2 = x_1 + x_3 \tag{8b}$$
$$\dot{x}_3 = x_1 x_3 + 3x_2^2 \tag{8c}$$

The system has only one fixed point which is origin. Since the origin is the fixed point we can apply the normal form procedure directly where A_0.

$$A_0 = \begin{bmatrix} 0 & -1 & 0 \\ 1 & 0 & 1 \\ 0 & 0 & 0 \end{bmatrix} \tag{9}$$

Fig. 1 Phase portrait in 3-D

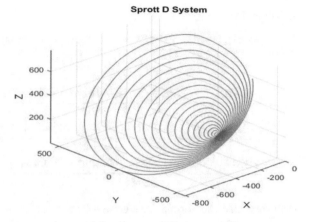

Fig. 2 Phase portrait in $x_1 - x_2$ plane

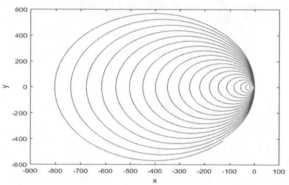

The system shows chaotic behavior with following Liapunov exponents $(0.1027, 0, -1.3198)$. Phase portrait of the system for initial values $(x_0, y_0, z_0) = (-0.2, 0, 1)$ is given in Figs. 1 and 2.

Eigenvalues of A_0 are $\lambda_{2,3} = \pm i$ and $\lambda_1 = 0$. From the eigenvalues, we can understand that the sytem is resonant in every order. The corresponding eigenvectors are:

$$\begin{bmatrix} -1 \\ 0 \\ 1 \end{bmatrix} \quad \text{and} \quad \begin{bmatrix} \pm i \\ 1 \\ 0 \end{bmatrix} \tag{10}$$

The transformation matrix T is:

$$T = \begin{bmatrix} -1 & i & -i \\ 0 & 1 & 0 \\ 1 & 0 & 1 \end{bmatrix} \tag{11}$$

After applying the transformation to our system we have calculated the normal form of the system. The normal form expansion through the sixth order in cylindrical coordinate is:

$$\dot{\rho} = \frac{\rho u (456753\rho^4 + 557072\rho^2 u^2 - 27456\rho^2 + 389988u^4 - 57024u^2 + 13824)}{27648}$$

(12a)

$$\dot{\theta} = \frac{(-14261\rho^4 - 30698\rho^2 z^2 - 2016\rho^2 - 50598u^4 + 9504u^2 + 6912)}{6912}$$

(12b)

$$\dot{u} = \frac{(-1214543\rho^6 - 2973264\rho^4 u^2 + 119808\rho^4 + 2660040\rho^2 u^4 - 15552\rho^2 u^2}{82944}$$

$$\frac{+ 82944\rho^2 - 393840u^6 + 207360u^4 - 82944u^2)}{82944}$$

(12c)

Unfortunately, the normal form equations are not trivial to draw a conclusion about the extent of the attractor. One would need an equilibrium like value for ρ or u. Although the normal form equations can be set to 0, there is no nontrivial equilibrium value, $\dot{\rho}$ and $\frac{d\rho}{d\theta}$ are not only functions of θ. Although numerical simulation gives a bounded region for the attractor, its extent is not independent of the initial conditions.

3.2   Sprott E System

The Sprott E System is given as:

$$\dot{x}_1 = x_2 x_3 \tag{13a}$$

$$\dot{x}_2 = -x_1^2 - x_2 \tag{13b}$$

$$\dot{x}_3 = 1 - 4x_1 \tag{13c}$$

The system has only one equilibrium point which is at $(\frac{1}{4}, \frac{1}{16}, 0)$. When this point is moved to the origin, system takes the following form:

$$\dot{\bar{x}}_1 = \frac{\bar{x}_3}{16} + \bar{x}_2 \bar{x}_3 \tag{14a}$$

$$\dot{\bar{x}}_2 = \frac{\bar{x}_1}{2} - \bar{x}_2 + \bar{x}_1^2 \tag{14b}$$

$$\dot{\bar{x}}_3 = -4\bar{x}_1 \tag{14c}$$

The normal form of the Sprott E System has the following form in cylindrical coordinates

$$\dot{\rho} = \frac{1}{3200}\rho^3\left(23 + \frac{4273991}{468000}\rho^2 - \frac{5732625917801}{7008768000000}\rho^4\right) \tag{15a}$$

$$\dot{\theta} = \frac{1}{2} + \frac{1}{9600}\rho^2\left(113 - \frac{14470361}{1872000}\rho^2 - \frac{30250848101177}{2336256000000}\rho^4\right) \tag{15b}$$

$$\dot{w} = -w - \frac{1}{800}\rho^2 w\left(23 + \frac{5584999}{468000}\rho^2 - \frac{2063305495801}{7008768000000}\rho^4\right) \tag{15c}$$

This system undergoes subcritical Hopf bifurcation. Radius of the chaotic attractor can be estimated from $\dot{\rho} = 0$. This yields $\rho = 3.6445$. Using averaging over the attractor it can be estimated that

$$< \lambda\rho - f(\rho) > \approx 0 \tag{16}$$

where

$$f(\rho) = \frac{d\rho}{d\theta} = (\frac{23}{1600}\rho^3 - \frac{8041177}{1497600000}\rho^5 + \frac{6890526323021}{11214028800000000}\rho^7) + O(\rho^9) \tag{17}$$

This gives an estimate for $\lambda = 0.0512$.

Its order of magnitude agrees with the numerically calculated largest Liapunov exponent 0.078 with error 0.000085 although this averages the theta and not the time derivative. At that point we should be averaging over the attractor (Figs. 3 and 4).

3.3 Sprott F System

Governing equations of Sprott F system are:

$$\dot{x}_1 = x_2 + x_3 \tag{18a}$$

$$\dot{x}_2 = -x_1 + ax_2 \tag{18b}$$

$$\dot{x}_3 = x_1^2 - x_3 \tag{18c}$$

where a is bifurcation parameter. For $a = 0.5$, Sprott shows that system has no chaotic behavior. For $a = 0.5$ eigenvalues of the linearized system are $(\pm i, -1)$.

Fig. 3 Phase space of
Sprott E

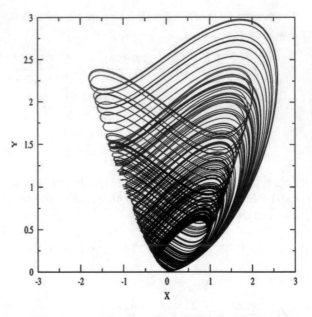

Fig. 4 The time
development original system
versus the normal form

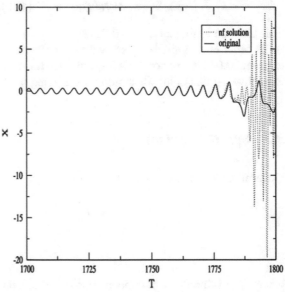

The normal form expansion through seventh order:

$$\dot{\rho} = -\frac{1850797}{88128000}\rho^7 - \frac{251}{900}\rho^5 - \frac{1}{20}\rho^3 \tag{19a}$$

$$\dot{\theta} = \frac{3499}{1088000}\rho^4 + \frac{1}{60}\rho^2 + \frac{12692663}{440640000}\rho^6 + 1 \tag{19b}$$

$$\dot{w} = \frac{44446901}{550800000}\rho^6 w + \frac{497}{4500}\rho^4 + \frac{1}{5}\rho^2 w - w \tag{19c}$$

According to these equations ρ and w have a stable equilibrium point at the origin and system goes super-critical Hopf bifurcation. The Liapunov exponents of the system are $(0, 0, -1.234)$. By the Descartes sign rule there is no positive root for ρ. Solution of $\dot{\rho}$ gives the trivial solution $\rho = 0$ and the four complex roots with positive real parts:

$$\rho_{2,3} = 0.459095557571 \pm 0.52646646306i$$
$$\rho_{4,5} = -0.459095557571 \pm 0.52646646306i$$

Thus the normal form does not estimate the *extent* of the attractor.

4 Conclusion

Although the linearized form of a dynamical system near its equilibrium point can give a partial understanding of the Liapunov spectrum and the extent of the attractor, more powerful techniques than simple scenarios based on the linearized system are needed. Developing such techniques is a difficult and system specific task. Benchmark systems with three variables and quadratic nonlinearities proposed by Sprott have been studied by the NF technique. Algebraic techniques also require an estimate of the attractor's extent. Such a criterion would represent a novel criterion to chaotic behavior, since it would represent a balancing action between the folding and stretching that produces the attractor.

Algebraic techniques also may furnish an estimate of the attractor's extent. Such a criterion would represent a novel criterion to chaotic behavior, since it would represent a balancing action between the folding and stretching that produces the attractor. We have illustrated a case where it works, a case where it fails and a case that requires more study. Furthermore, non uniqueness of the NF transformation, and clarification of the resonant system in the problemmatic case where the linearized system has the eigenvalue spectrum (0;i) leading to a more complicated NF merit further study.

References

1. R. Khomeriki, Route to and from the NMR chaos in diamagnets. Euro. Phys. J. B Condens. Matter Complex Syst. **10**(1), 99–103 (1999)
2. N. Perdahci, A. Hacinliyan, Normal forms and nonlocal chaotic behavior in sprott systems. Int. J. Eng. Sci. **41**, 1085–1108 (2003)
3. J.C. Sprott, Some simple chaotic flows. Phys. Rev. E **50**(2), R647–R650 (1994)
4. Alan Wolf, Jack B. Swift, Harry L. Swinney, John A. Vastano, Determining lyapunov exponents from a time series. Physica D Nonlinear Phenom. **16**(3), 285–317 (1985)

The FitzHugh-Nagumo Model and Spatiotemporal Fractal Sets Based on Time-Dependent Chaos Functions

Shunji Kawamoto

Abstract It is presented firstly that a one-dimensional (1-D) time-dependent logistic map for population growth is derived from the chaos solution consisting of a time-dependent chaos function, and the logistic map has the dynamics of coherence and incoherence in time, which are the so-called chimera states discussed in the field of complex systems, by introducing the bifurcation diagram and a time-dependent system parameter for the 1-D map as one of non-equilibrium open systems. Secondly, the 2-D time-dependent solvable chaos map corresponding to the FitzHugh-Nagumo model for neural phenomena is obtained on the basis of time-dependent chaos functions, and the 2-D map is shown to have chimera states in time under the assumption of a time-dependent system parameter, and to find spatiotemporal fractal sets defined by initial values as the dynamic stability region for neural cells.

Keywords Chaos function · Time-dependent logistic map · Time-dependent chaos function · Time-dependent system parameter · Non-equilibrium open system · Chimera states · FitzHugh-Nagumo model · 2-D solvable chaos map · Spatiotemporal fractal set · Dynamic stability region

1 Introduction

As is known, there have been many advances in the field of nonlinear dynamics, such as soliton, chaos and fractals, which appear in physical, biological, chemical, mechanical and social sciences. After the theory for shallow water waves, soliton has been a self-reinforcing solitary wave that maintains the shape during the travel with constant speed, and arises as the solution to weekly nonlinear dispersive partial differential equations for physical system [1]. At the same time, it has been shown that the one-dimensional (1-D) nonlinear difference equations possess a rich spectrum of dynamical behavior as chaos in many respects, and a family of shapes and irregular

S. Kawamoto (✉)
Osaka Prefecture University, Sakai, Osaka, Japan
e-mail: kawamoto@eis.osakafu-u.ac.jp

© Springer Nature Switzerland AG 2019
C. H. Skiadas and I. Lubashevsky (eds.), *11th Chaotic Modeling and Simulation International Conference*, Springer Proceedings in Complexity, https://doi.org/10.1007/978-3-030-15297-0_15

159

patterns called fractals have been proposed for the geometric representation [2, 3]. Recently, the chaos theory has been widely extended to biology, medicine, optics, pattern recognition and human sciences [4]. Here, it is important to emphasize that as the nonlinearity makes the nonlinear differential equations difficult to solve analytically, the personal computer with mathematical software has played an important role to calculate nonlinear equations and to draw the figures.

On the other hand, the coexistence of coherence and incoherence in non-locally coupled phase oscillators has been presented and called chimera states which display remarkable spatiotemporal patterns [5, 6]. After that, the chimera states are described as chaotic transients forming a self-organized pattern in a population of non-locally coupled oscillators and coupled chaotic systems, such as the logistic map, the Rössler system, the Lorenz system and the FitzHugh-Nagumo oscillators for complex spatiotemporal patterns [7–9]. Moreover, it has been studied numerically that chimera states can be stable even without taking the continuous limit [10]. In addition, chimera states have been considered to happen in modular neural networks [11], and it is investigated if the states can be synchronized across different interacting networks [12].

In the meantime, 2-D and 3-D chaos maps have been proposed for the analysis of population growth, mechanical vibration, electrical oscillation, atmospheric convection and chemical reaction [13]. Moreover, the time-dependent chaos functions have been introduced and discussed briefly for the application to engineering with the nonlinear time series expansion [14]. Then, a 2-D solvable chaos map corresponding to the FitzHugh-Nagumo (FHN) model [15, 16] is obtained, and the discrete properties have been considered numerically [17].

This paper is organized as follows: Firstly, a 1-D time-dependent logistic map for population growth is derived in Sect. 2 from the chaos solution consisting of a time-dependent chaos function, and the logistic map is shown to have the dynamics of coherence and incoherence in time, which are the so-called chimera states discussed in the field of complex systems, by introducing the bifurcation diagram and a time-dependent system parameter for the 1-D map, as one of non-equilibrium open systems. In Sect. 3, the forced Van der Pol oscillator [18] is transformed into a 2-D map which corresponds to the FHN model for neural phenomena, on the basis of time-dependent chaos functions. Finally, Sect. 4 presents the numerical calculation of the 2-D map with the bifurcation diagram and a time-dependent system parameter for the chimera states in time, and the spatiotemporal fractal sets defined by initial values are obtained as the dynamic stability region for neural cells. The last section is devoted to conclusions.

2   A Time-Dependent Logistic Map

Firstly, we introduce a logistic function;

Fig. 1 The bifurcation diagram of the logistic map (2) with (3)

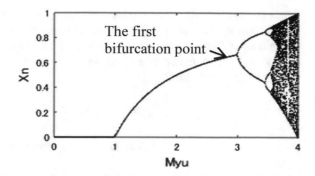

The first bifurcation point

$$P(t) = a/\left(b + e^{-ct}\right) \qquad (1)$$

with the time $t > 0$ and real constants $\{a \neq 0, b > 0, c \neq 0\}$ for population growth [19]. By differentiating (1), and by applying the difference method and variable transformations, we find the well-discussed logistic map;

$$x_{n+1} = \mu x_n(1 - x_n), \qquad (2)$$

$$0 \leq \mu \equiv 1 + c(\Delta t) \leq 4, \qquad (3)$$

which is a 1-D chaotic map, and is known to have chaotic dynamics, where the system parameter μ given by (3) consists of the real constant c of (1) and the time step Δt of the difference method. The map (2) with (3) has been known to exhibit stable and unstable fixed points by bifurcation processes [20]. Then, the processes have been formulated for a large class of recursion relations exhibiting infinite bifurcation [21], and the bifurcation diagram has been obtained experimentally for a nonlinear electric circuit [22]. The diagram for (2) with (3) is shown in Fig. 1, and we find that the first bifurcation point arises at $\mu = 3.0$, and the chaotic dynamics end at $\mu = 4.0$, as discussed in [20]. For Fig. 1, we carry out 200 iterations of the map (2) with (3), and drop the first 150 iterations to illustrate the remaining 50 subsequent values of x_n. Here, it is important to note that if the parameter μ increases in time, then the solution x_n would transit from coherent state into incoherent state in time.

On the other hand, it has been shown recently that from the time-dependent chaos function given by

$$x_n(t) = \sin^2(2^n t), \; t \neq \pm m\pi/2^l, \qquad (4)$$

the time-dependent chaos map is derived as

$$x_{n+1}(t) = 4x_n(t)(1 - x_n(t)), \; t \neq \pm m\pi/2^l, \qquad (5)$$

where $\{l, m\}$ are finite positive integers [13, 14], that is, the map (2) with (3) has the chaos solution (4) at $\mu = 4.0$. We assume here that the system parameter μ of (2) depends on a change of the environment for population growth in time, such as the climate and the economy of society, and then we find the time-dependent logistic map from (2) to (5) and Fig. 1;

$$x_{n+1}(t) = \mu(t)x_n(t)(1 - x_n(t)), \tag{6}$$

$$1 \leq \mu(t) \leq 4, \ t \neq \pm m\pi/2^l, \tag{7}$$

where the μ is assumed to be an increasing function of time t, and the map (6) has the chaos solution (4) at $\mu(t) = 4.0$.

The map (6) and the condition (7) are rewritten for the numerical calculation as

$$x_{n+1}(t_{i+1}) = \mu(t_i)x_n(t_i)(1 - x_n(t_i)), \tag{8}$$

$$1 \leq \mu(t_i) \leq 4, \tag{9}$$

where we can choose the discrete time t_i satisfying (7) [17]. Under the function $\mu = \mu(t_i)$, we introduce the following three cases of $\mu(t_i)$ with the time interval $t_{i=0} \leq t_i \leq t_{i=120}$ as shown in Fig. 2a; Case 1: $1 \leq \mu(t_{i=0-120}) \leq 4.0$ (blue line), Case 2: $1 \leq \mu(t_{i=0-120}) \leq 3.6$ (red line) and Case 3: $1 \leq \mu(t_{i=0-120}) \leq 2.8$ (black line). Then, we have the solution $x_n(t_i)$ obtained by iterating the map (8) with $\mu(t_i)$ in Fig. 2b as the orbit and as the time-dependent sequence of points in Fig. 2c, which are the chimera states or the chaotic transients considered in [5–12]. Then, it is found that Cases 1 and 2 (blue and red) show the coexistence of coherent state and incoherent state in time, and Case 3 (black line) gives a coherent state, because the first bifurcation point in Fig. 1 does not arise for $\mu(t_{i=0-120}) < 3.0$. The MATLAB program for Fig. 2 is shown in Appendix.

Moreover, the system parameter $\mu(t_{i=0-240})$ is assumed in Fig. 3a as an increasing and decreasing function of $t_{i=0-240}$ for a long time interval $t_{i=0} \leq t_i \leq t_{i=240}$ given by the following three cases; Case 1: $1 \leq \mu(t_{i=0-240}) \leq 4.0$ (blue line), Case 2: $1 \leq \mu(t_{i=0-240}) \leq 3.6$ (red line) and Case 3: $1 \leq \mu(t_{i=0-240}) \leq 2.8$ (black line). Then, we obtain the solution $x_n(t_{i=0-240})$ in Fig. 3b, and as the sequence of points in Fig. 3c, respectively. It is found that Cases 1 and 2 (blue and red) evolve coherently, incoherently and coherently in time, which are chaotic during $\mu(t_{i=0-240}) \geq 3.0$, and Case 3 (black) presents a coherent state for $\mu(t_{i=0-240}) < 3.0$. Here, it is interesting to note that the map (8) with (9) has the following two fixed points;

$$x^* = 0, \ (\mu(t_i) - 1)/\mu(t_i), \tag{10}$$

where the system parameter is a function of time, and therefore the fixed point is movable in time.

Fig. 2 The time-dependent logistic map (8) for the interval $t_{i=0} \leq t_i \leq t_{i=120}$; **a** three cases (blue, red, black) of $\mu(t_{i=0-120})$, **b** the solution $x_n(t_{i=0-120})$, and **c** the sequence of points $\{x_n(t_{i=0-120})\}$

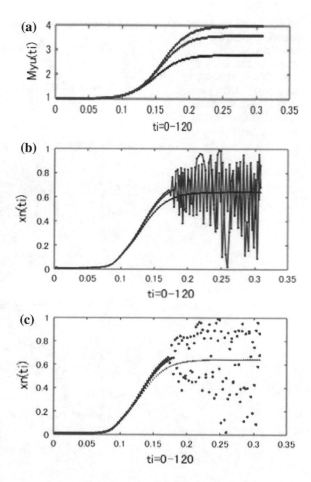

3 The FitzHugh-Nagumo Model and 2-D Time-Dependent Chaos Map

The forced Van der Pol oscillator is given by

$$\ddot{x} - \varepsilon(1 - x^2)\dot{x} + x = E_0 \sin(\omega t), \tag{11}$$

which represents a model for a simple vacuum tube oscillator circuit with a nonlinear damping term, and describes the heart beat as a relaxation oscillation, where ε is the damping coefficient, and E_0 and ω are the strength and the frequency of the periodic external forcing, respectively [18]. We introduce the Liénard transformation

$$y \equiv \varepsilon(x - \frac{1}{3}x^3) - \dot{x}, \tag{12}$$

Fig. 3 The time-dependent logistic map (8) for a long time interval $t_{i=0} \leq t_i \leq t_{i=240}$; **a** three cases (blue, red, black) of $\mu(t_{i=0-240})$, **b** the solution $x_n(t_{i=0-240})$, and **c** the sequence of points $\{x_n(t_{i=0-240})\}$

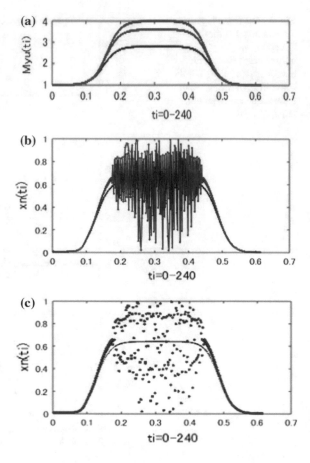

and find the 2-D model from (11) and (12) as

$$\dot{x} = \varepsilon\left(x - \frac{1}{3}x^3\right) - y, \tag{13}$$

$$\dot{y} = x - E_0\sin(\omega t), \tag{14}$$

which is known to have chaotic behaviors in the equivalent circuit with sinusoidal forcing [23]. As is well-discussed, the FitzHugh-Nagumo (FHN) model [15, 16] is a 2-D simplification of the Hodgkin-Huxley model [24] of spike generation in squid giant axons, and by adding terms $\{a, by(t), z(t)\}$ with constants $\{a, b\}$ and a stimulus external current $z(t)$ to the model (13) and (14) with $E_0 = 0$, we have the FHN model;

$$\dot{x} = \varepsilon\left(x - \frac{1}{3}x^3\right) - y + z, \tag{15}$$

$$\dot{y} = x + (a - by), \tag{16}$$

which has the coefficient ε of the nonlinear damping term, and it is found that setting $\varepsilon = 0$ in (11) and (15) gives a linear differential equation, respectively. Here, by a transformation $\varepsilon y \rightarrow y$ in the Liénard transformation $y \equiv x - x^3/3 - \dot{x}/\varepsilon$ used in [15], we find the same coefficient ε of nonlinear term in the 2-D model (13) and (14), and can regard the ε as the system parameter of the FHN model (15) and (16).

Recently, the following 2-D time-dependent chaos system corresponding to the FHN model (15) and (16) has been proposed as

$$x_n(t_i) = a_1 \sin^2\left(2^n t_i\right) + b_1(t_i), \tag{17}$$

$$y_n(t_i) = a_2 \cos\left(2^n t_i\right), \tag{18}$$

$$\frac{1}{a_1}(x_n(t_i) - b_1(t_i)) + \frac{1}{a_2^2} y_n^2(t_i) = 1, \tag{19}$$

$$x_{n+1}(t_{i+1}) = \frac{\varepsilon}{a_2^2}(x_n(t_i) - b_1(t_i))y_n^2(t_i) + b_1(t_i), \tag{20}$$

$$y_{n+1}(t_{i+1}) = -\left(\frac{2a_2}{a_1}\right)x_n(t_i) + a_2 + \left(\frac{2a_2}{a_1}\right)b_1(t_i), \tag{21}$$

where (17) and (18) are the time-dependent chaos solutions consisting of chaos functions with nonzero coefficients $\{a_1, a_2\}$ and the external force function $b_1(t_i)$ of discrete time t_i. The condition (19) derived from the solutions (17) and (18) gives a discrete quadratic curve, and plays a key role for the FHN model (20) and (21) on the $x_n(t_i) - y_n(t_i)$ plane. Therefore, the 2-D map (20) and (21) at $\varepsilon = 4$ has the chaos solutions (17) and (18), and the orbit of neural cells has been discussed numerically by iterating the chaos solutions (17) and (18) based on chaos functions with initial values [17].

The bifurcation diagram is illustrated in Fig. 4 with $a_1 = 1.0$, $a_2 = 1.0$ and the external force term $b_1(t_i) = 0$ in the 2-D map (20) and (21). The first bifurcation point $(x^*, y^*) = (0.15, 0.7)$ arrises at $\varepsilon = 2.2$, and the chaotic dynamics end at $\varepsilon = 4.0$. Here, we carry out 200 iterations of the 2-D map, and drop the first 150 iterations to show the remaining 50 subsequent values of $\{x_n, y_n\}$. The orbit and the sequence of points (collision points) at $\varepsilon = 0.5, 1.5, 2.5, 3.5$ for the 2-D map (20) and (21) are shown in Fig. 5a1, a2, a3, a4, and in Fig. 5b1, b2, b3, b4, respectively. At $\varepsilon = 0.5$ and $\varepsilon = 1.5$, the orbit has a coherent state with the interval $0 \leq \varepsilon \leq 1.0$ and $1.0 \leq \varepsilon \leq 2.2$ as shown in Fig. 4, and the orbit converges to the stable fixed point from the initial point in time as illustrated in Fig. 5a1, b1, a2, b2. However, at $\varepsilon = 2.5$, the orbit shows an incoherent state in Fig. 5a3 with the interval $2.2 \leq \varepsilon \leq 4.0$ of Fig. 4. Then, it is found that the sequence of points shows a discrete limit cycle in Fig. 5b3, and the orbit converges to the limit cycle from the initial point, which is discussed for the Van der Pol oscillator [25] and the FHN model [26]. Moreover, at $\varepsilon = 3.5$, which is in the chaotic region of Fig. 4, the orbit is incoherent as shown in Fig. 5a4, and the sequence of points is illustrated in Fig. 5b4. For $\varepsilon = 4.0$, it has been presented that the orbit and the sequence of points give chaos solutions (17) and (18),

Fig. 4 The bifurcation diagram for the 2-D map (20) and (21) with $a_1 = 1.0$, $a_2 = 1.0$ and $b_1(t_i) = 0$

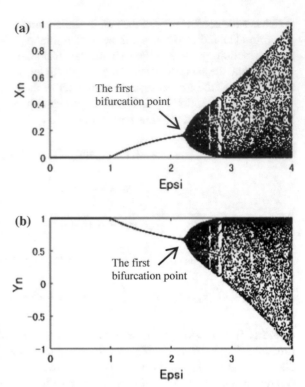

and the quadratic curve (19) on the $x_n - y_n$ plane, by the numerical calculation of (17) and (18) without the accumulation of round-off error caused by the numerical iteration as discussed in [17].

Here, it is interesting to note that the 2-D map (20) and (21) has the following three fixed points;

$$(x^*, y^*) = (0, 1), \left(\frac{1}{2} \left(1 \pm \frac{1}{\sqrt{\varepsilon}} \right), \mp \frac{1}{\sqrt{\varepsilon}} \right), \tag{22}$$

where ε is the system parameter. Therefore, it is found that if ε is a function of time, then the fixed points are movable in time.

4 Spatiotemporal Fractal Sets

As is discussed in Sect. 3, we assume that the ε in the 2-D map (20) and (21) depends on a change of the environment, such as the temperature and the external force of neural phenomena in time, that is, the ε is a function of time $\varepsilon(t_i)$. Then, we have the following map from the 2-D map (20) and (21) as

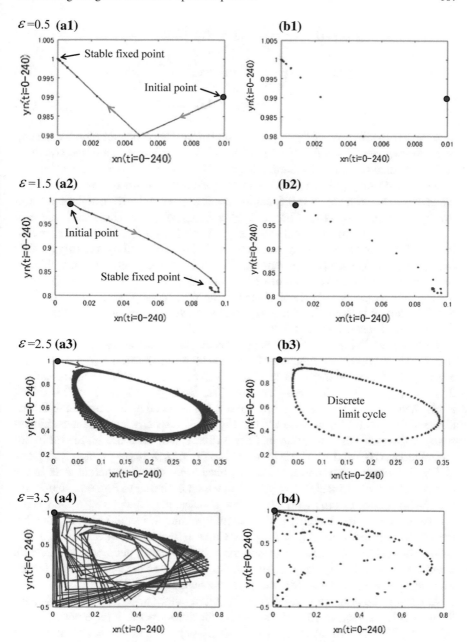

Fig. 5 The orbit: **a1**, **a2**, **a3** and **a4**, and the sequence of points (collision points): **b1**, **b2**, **b3** and **b4** for the 2-D map (20) and (21) with the initial point $(x_0, y_0) = (0.01, 0.99)$ at $\varepsilon = 0.5, 1.5, 2.5, 3.5$, respectively

$$x_{n+1}(t_{i+1}) = \frac{\varepsilon(t_i)}{a_2^2}(x_n(t_i) - b_1(t_i))y_n^2(t_i) + b_1(t_i), \tag{23}$$

$$y_{n+1}(t_{i+1}) = -\left(\frac{2a_2}{a_1}\right)x_n(t_i) + a_2 + \left(\frac{2a_2}{a_1}\right)b_1(t_i), \tag{24}$$

$$1 \le \varepsilon(t_i) \le 4.0, \tag{25}$$

here the 2-D chaotic map (23) and (24) is found to have the chaos solutions consisting of time-dependent chaos functions (17) and (18) at $\varepsilon(t_i) = 4.0$, and the condition (25) is obtained from the bifurcation diagram Fig. 4 as the interval of $\varepsilon(t_i)$.

As the 2-D map (23) and (24) with (25) includes the time-dependent system parameter $\varepsilon(t_i)$ with the first bifurcation point at $\varepsilon = 2.2$, we introduce the following three cases of $\varepsilon(t_i)$; Case 1: $1 \le \varepsilon(t_{i=0-120}) \le 4.0$ (blue line), Case 2: $1 \le \varepsilon(t_{i=0-120}) \le 3.6$ (red line) and Case 3: $1 \le \varepsilon(t_{i=0-120}) \le 2.0$ (black line) in Fig. 6a. Then, we have the solutions $x_n(t_i)$ and $y_n(t_i)$ obtained by iterating the 2-D map (23) and (24) with (25) as shown in Fig. 6b, c, respectively, and find chimera states in time with the exception of Case 3 (black line). In addition, the orbit of collision points $(x_n(t_i), y_n(t_i))$ are presented on the $x_n(t_{i=0-120}) - y_n(t_{i=0-120})$ plane of Fig. 6d, where it is found that Case 3 (black line) gives a coherent state in the region $0 \le x_n < 0.15$ and $0.7 < y_n \le 1.0$.

Furthermore, in Fig. 7a, the system parameter $\varepsilon(t_i)$ of the 2-D map (23) and (24) with (25) is assumed as a function of t_i with a long time interval $t_{i=0} \le t_i \le t_{i=240}$ given by the following three cases; Case 1: $1 \le \varepsilon(t_{i=0-240}) \le 4.0$ (blue line), Case 2: $1 \le \varepsilon(t_{i=0-240}) \le 3.6$ (red line) and Case 3: $1 \le \varepsilon(t_{i=0-240}) \le 2.0$ (black line). Then, we find the solutions $x_n(t_i)$ and $y_n(t_i)$ obtained by iterating the 2-D map (23) and (24) with (25) as illustrated in Fig. 7b, c respectively, and have chimera states in time with the exception of Case 3 (black line). In addition, the orbits for the solutions $x_n(t_i)$ and $y_n(t_i)$ are presented on the $x_n(t_{i=0-240}) - y_n(t_{i=0-240})$ plane of Fig. 7d, where it is found that Case 3 (black line) gives a coherent state in the region $0 \le x_n < 0.15$ and $0.7 < y_n \le 1.0$, which is similar to Fig. 6d. That is, for the system parameter $\varepsilon(t_i)$ with a long time $t_{i=0-240}$, the solutions $x_n(t_i)$ and $y_n(t_i)$ evolve coherently, incoherently and coherently in time. It is interesting to note that the $\varepsilon(t_i)$ is symmetric about $t_i = t_{i=120}$ for Cases 1, 2 and 3. However, the solutions $x_n(t_i)$ and $y_n(t_i)$ evolve linearly during the coherent states, but are not symmetric about $t_i = t_{i=120}$ during the incoherent states, as same as the case of $\mu(t_i)$ for the time-dependent logistic map (8) with (9) for population growth.

Moreover, the orbits in Fig. 7d with the initial point $(x_0, y_0) = (0.01, 0.99)$ for the three cases of $\varepsilon(t_{i=0-240})$ are illustrated as the time-dependent sequence of points (blue, red, black) on the $x_n(t_{i=0-240}) - y_n(t_{i=0-240})$ plane in Fig. 8a. Then, for the following five initial points as $(x_0, y_0) = (0.01, 0.99), (0.01, 0.98), (0.01, 0.97), (0.01, 0.96)$ and $(0.01, 0.95)$, we have the sequence of points (blue, red, black) shown in Fig. 8b, respectively. It is found that the boundary curve of blue points is given by the discrete quadratic curve (19) constructed at $\varepsilon(t_i) = 4.0$ of the 2-D map (23) and (24), and the region of blue points is numerically obtained from the curve (19) as

Fig. 6 The time-dependent
2-D map (23) and (24) for the
interval $t_{i=0} \le t_i \le t_{i=120}$;
a three cases of $\varepsilon(t_{i=0-120})$,
b the solution $x_n(t_{i=0-120})$,
c the solution $y_n(t_{i=0-120})$,
and **d** the orbit of points
$(x_n(t_{i=0-120}), y_n(t_{i=0-120}))$

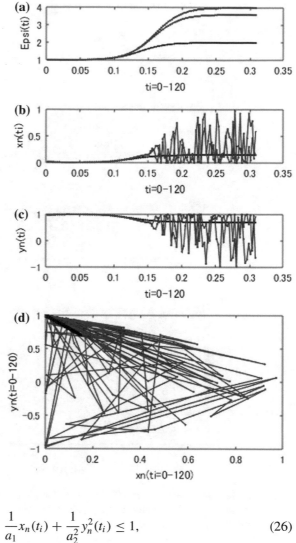

$$\frac{1}{a_1}x_n(t_i) + \frac{1}{a_2^2}y_n^2(t_i) \le 1, \tag{26}$$

which gives a dynamic stability region without the external force $b_1(t_i)$ for the following fractal set M defined by initial values $\{x_0, y_0\}$ [17, 27];

$$M = \{x_0, y_0 \in \mathbf{R} \mid \lim_{n \to \infty} x_n, y_n < \infty\}, \tag{27}$$

where the $\varepsilon(t_i)$ in (23) plays an important role as the system parameter in the 2-D map. It is found that for the region of red points in the Case 2 (red) of Fig. 8a, b, the red quadratic boundary curve is equivalent to the Case 1 (blue) under the condition; $1 \le \varepsilon(t_{i=0-240}) \le 4.0$ or $1/2 \le a_2(t_{i=0-240}) \le 1.0$ in the

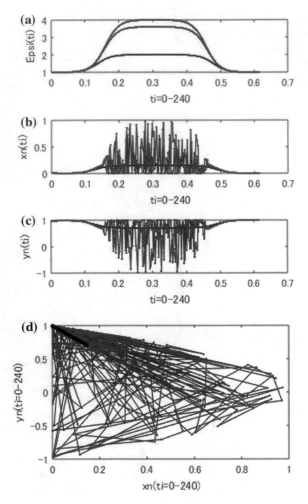

Fig. 7 The time-dependent 2-D map (23) and (24) for a long interval $t_{i=0} \le t_i \le t_{i=240}$; **a** three cases of $\varepsilon(t_{i=0-240})$, **b** the solution $x_n(t_{i=0-240})$, **c** the solution $y_n(t_{i=0-240})$, and **d** the orbit of points $(x_n(t_{i=0-240}), y_n(t_{i=0-240}))$

2-D map (23) and (24), and the red points give a dynamic stability region with the same initial values as the blue points. The two regions include black points corresponding to the Case 3 (black) in Fig. 8. Here, it should be emphasized for the fractal sets that an initial point (x_0, y_0) is firstly given, and we have points $\{(x_1, y_1), \ldots, (x_{240}, y_{240})\}$ by iterating the 2-D map, that is, if we start the iteration with the point (x_1, y_1) as a new initial point $(\tilde{x}_0, \tilde{y}_0)$, then we find a new sequence of points $\{(\tilde{x}_1, \tilde{y}_1) \equiv (x_2, y_2), \ldots, (\tilde{x}_{239}, \tilde{y}_{239}) \equiv (x_{240}, y_{240})\}$, and so on. Therefore, the dynamic stability regions consisting of blue points and red points in Fig. 8 can be regarded as spatiotemporal fractal sets for the 2-D map (23) and (24) with (25).

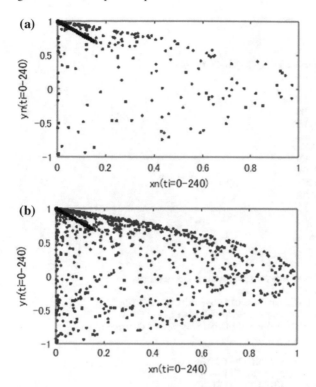

Fig. 8 Spatiotemporal fractal sets for three cases (blue, red, black) with **a** one initial point, and with **b** five initial points

5 Conclusions

In this paper, we have obtained firstly the time-dependent logistic map for population growth by introducing the time-dependent chaos function, and have presented numerically the coexistence of coherence and incoherence in time, under the assumption of a time-dependent system parameter caused by the change of environment as one of non-equilibrium open systems. Secondly, we have considered the 2-D solvable chaos map corresponding to the FHN model, and have discussed the coherent and incoherent dynamics in time, under the assumption of a time-dependent system parameter. Then, we obtain spatiotemporal fractal sets numerically as the coexistence of coherent and incoherent states. The result may describe the essential nonlinear dynamics of the 1-D time-dependent logistic map for population growth and the 2-D map corresponding to the FHN model for neural phenomena.

Appendix

```
% MATLAB program for Fig. 2 by S. Kawamoto
% initial conditions
T=zeros(1, 120);
MYU1=zeros(1, 120);
MYU2=zeros(1, 120);
MYU3=zeros(1, 120);
X1=zeros(120, 120);
X2=zeros(120, 120);
X3=zeros(120, 120);
XX1=zeros(1, 120);
XX2=zeros(1, 120);
XX3=zeros(1, 120);
c=50;
L=1;
PR=1223;
X10=0.01;
X20=0.01;
X30=0.01;
% time-dependent logistic map and system parameter μ (ti)
for I=1:120, T(I)=I.*L.*pi./PR; end
for I=1:120, MYU1(I)=0.999+3.001/(1.0+3000*exp(-c.*T(I)));
          MYU2(I)=0.999+2.601/(1.0+2600*exp(-c.*T(I)));
          MYU3(I)=0.999+1.801/(1.0+1800*exp(-c.*T(I))); end
for I=1:120
    for N=1
        X1(I, N)=MYU1(I).*X10.*(1.0-X10);
        X2(I, N)=MYU2(I).*X20.*(1.0-X20);
        X3(I, N)=MYU3(I).*X30.*(1.0-X30);
    end
    for N=2:I
        X1(I, N)=MYU1(I).*X1(I, N-1).*(1.0-X1(I, N-1));
        X2(I, N)=MYU2(I).*X2(I, N-1).*(1.0-X2(I, N-1));
        X3(I, N)=MYU3(I).*X3(I, N-1).*(1.0-X3(I, N-1));
    end
end
for I=1:120
    XX1(I)=X1(I, I);
    XX2(I)=X2(I, I);
    XX3(I)=X3(I, I);
end
% figures (a) – (b)
figure('Position', [100 100 350 100])
plot(T, MYU3, '-k.', 'MarkerFaceColor', 'k', 'MarkerSize', 5); hold on
plot(T, MYU2, '-r.', 'MarkerFaceColor', 'r', 'MarkerSize', 5); hold on
plot(T, MYU1, '-b.', 'MarkerFaceColor', 'b', 'MarkerSize', 5); hold off
xlabel('ti=0-120'); ylabel('MYU(ti)')
figure('Position', [100 100 350 200])
plot(T, XX3, '-k.', 'MarkerFaceColor', 'k', 'MarkerSize', 5); hold on
plot(T, XX2, '-r.', 'MarkerFaceColor', 'r', 'MarkerSize', 5); hold on
plot(T, XX1, '-b.', 'MarkerFaceColor', 'b', 'MarkerSize', 5); hold off
xlabel('ti=0-120'); ylabel('xn(ti)')
figure('Position', [100 100 350 200])
plot(T, XX3, 'k.', 'MarkerFaceColor', 'k', 'MarkerSize', 4); hold on
plot(T, XX2, 'r.', 'MarkerFaceColor', 'r', 'MarkerSize',7); hold on
plot(T, XX1, 'b.', 'MarkerFaceColor', 'b', 'MarkerSize', 7); hold off
xlabel('ti=0-120'); ylabel('xn(ti)')
```

References

1. A. Scott, *Nonlinear Science* (Routledge, London, 2005)
2. H. Peitgen, H. Jurgens, D. Saupe, *Chaos and Fractals—New Frontiers of Science* (Springer, New York, 1992)
3. F.C. Moon, *Chaotic and Fractal Dynamics* (Wiley, New York, 1992)
4. C. H. Skiadas, C. Skiadas (eds.), in *Handbook of Application of Chaos Theory* (Chapman and Hall/CRC Press, 2016)
5. Y. Kuramoto, D. Battogtokh, Coexistence of coherence and incoherence in nonlocally coupled phase oscillators. Nonlinear Phenom. Complex Syst. **5**, 380–385 (2002)
6. D.M. Abrams, S.H. Strogatz, Chimera states for coupled oscillators. Phys. Rev. Lett. **93**, 174102 (2004)
7. I. Omelchenko, Y. Maistrenko, P. Hövel, E. Schöll, Loss of coherence in dynamical networks: spatial chaos and chimera states. Phys. Rev. Lett. **106**, 234102 (2011)
8. I. Omelchenko, B. Riemenschneider, P. Hövel, Y. Maistrenko, E. Schöll, Transition from spatial coherence to incoherence in coupled chaotic systems. Phys. Rev. E **85**, 026212 (2012)
9. I. Omelchenko, A. Provata, J. Hizanidis, E. Schöll, P. Hövel, Robustness of chimera states for coupled FitzHugh-Nagumo oscillators. Phys. Rev. E **91**, 022917 (2015)
10. Y. Suda, K. Okuda, Persistent chimera states in nonlocally coupled phase oscillators. Phys. Rev. E **92**, 060901(R) (2015)
11. J. Hizanidis, N.E. Kouvaris, G. Zamora-López, A. Díaz-Guilera, C.G. Antonopoulos, Chimera-like states in modular neural networks. Nat. Sci. Rep. **6**, 19845 (2016)
12. R.G. Andrzejak, G. Ruzzene, I. Malvestio, Generalized synchronization between chimera states. Chaos **27**, 053114 (2017)
13. S. Kawamoto, 2-D and 3-D solvable chaos maps. Chaotic Model. Simul. (CMSIM) **1**, 107–118 (2017)
14. S. Kawamoto, Chaotic time series by time-discretization of periodic functions and its application to engineering. Chaotic Model. Simul. (CMSIM) **2**, 193–204 (2017)
15. R. FitzHugh, Impulses and physiological states in theoretical models of nerve membrane. Biophys. J. **1**, 445–466 (1961)
16. J. Nagumo, S. Arimoto, S. Yoshizawa, An active pulse transmission line simulating nerve axon. Proc. IRE. **50**, 2061–2070 (1962)
17. S. Kawamoto, The FitzHugh-Nagumo model and 2-D solvable chaos maps. Chaotic Model. Simul. (CMSIM) **3**, 269–283 (2018)
18. B. Van der Pol, J. Van der Mark, Frequency demultiplication. Nature **120**(3019), 363–364 (1927)
19. P.F. Verhulst, Mathematical researches into the law of population growth increase. Nouveaux Mémoires de l'Académie Royale des Sciences et Belles-Lettres de Bruxelles **18**, 1–42 (1845)
20. R.M. May, Simple mathematical models with very complicated dynamics. Nature **261**, 459–467 (1976)
21. M.J. Feigenbaum, Quantitative universality for a class of nonlinear transformations. J. Stat. Phys. **19**(1), 25–52 (1978)
22. P. Bryant, C. Jaffries, Bifurcations of a forced magnetic oscillator near points of resonance. Phys. Rev. Lett. **53**(3), 250–253 (1984)
23. K. Tomita, Periodically forced nonlinear oscillators, in *Chaos*, ed. by A.V. Holden (Manchester University Press, Manchester, 1986), pp. 213–214
24. A.L. Hodgkin, A.F. Huxley, A quantitative description of membrane current and its application to conduction and excitation in nerve. J. Physiol. **117**, 500–544 (1952)
25. T. Kanamaru, Van der Pol oscillator. Scholarpedia **2**(1), 2202 (2007)
26. E.M. Izhikevich, R. FitzHugh, FitzHugh-Nagumo model. Scholarpedia **1**(9), 1349 (2006)
27. N. H. Tuan Anh, D. V. Liet, S. Kawamoto, Nonlinear dynamics of two-dimensional chaotic maps and fractal sets for snow crystals, in *Handbook of Application of Chaos Theory*, ed. by C. H. Skiadas, C. Skiadas (Chapman and Hall/CRC Press, 2016), pp. 83–91

Stochastic Navier-Stokes Equation for a Compressible Fluid: Two-Loop Approximation

Michal Hnatič, Nikolay M. Gulitskiy, Tomáš Lučivjanský,
Lukáš Mižišin and Viktor Škultéty

Abstract A model of fully developed turbulence of a compressible fluid is briefly reviewed. It is assumed that fluid dynamics is governed by a stochastic version of Navier-Stokes equation. We show how corresponding field theoretic-model can be obtained and further analyzed by means of the perturbative renormalization group. Two fixed points of the RG equations are found. The perturbation theory is constructed within formal expansion scheme in parameter y, which describes scaling behavior of random force fluctuations. Actual calculations for fixed points' coordinates are performed to two-loop order.

Keywords Stochastic Navier-Stokes equation · Anomalous scaling ·
Field-theoretic renormalization group · Compressibility

M. Hnatič · L. Mižišin
Bogoliubov Laboratory of Theoretical Physics, Joint Institute for Nuclear Research, 141980
Dubna, Russian Federation
e-mail: hnatic@saske.sk

L. Mižišin
e-mail: lukas.mizisin@gmail.com

M. Hnatič · L. Mižišin
Institute of Experimental Physics, Slovak Academy of Sciences, Watsonova 47,
040 01 Košice, Slovakia

M. Hnatič · T. Lučivjanský (✉)
Faculty of Sciences, Šafárik University, Moyzesova 16, 040 01 Košice, Slovakia
e-mail: tomas.lucivjansky@upjs.sk

N. M. Gulitskiy
Department of Physics, Saint-Petersburg State University,
7/9 Universitetskaya nab., St. Petersburg 199034, Russian Federation
e-mail: n.gulitskiy@spbu.ru

V. Škultéty
School of Physics and Astronomy, The University of Edinburgh,
Peter Guthrie Tait Road, Edinburgh EH9 3FD, UK
e-mail: viktoroslavs@gmail.com

C. H. Skiadas and I. Lubashevsky (eds.), *11th Chaotic Modeling
and Simulation International Conference*, Springer Proceedings
in Complexity, https://doi.org/10.1007/978-3-030-15297-0_16

1 Introduction

Many natural phenomena are concerned with hydrodynamic flows. Ranging from microscopic up to macroscopic spatial scales fluids can exist in profoundly different states. Especially intrigued behavior is observed in case of turbulent flows. Such flows are ubiquitous in nature and are more common than generally believed [1, 2]. Despite a substantial amount of effort that has been put into investigation of turbulence, the problem itself remains unsolved.

Most of studies are devoted to the case of incompressible fluid. However, particularly in an astrophysical context we have to deal with a compressible fluid rather than incompressible one [3]. In recent years there has also been an increased research activity of compressible turbulence in magnetohydrodynamic context [4–8]. In this work, our aim is to study compressible turbulence [9, 10], partially motivated by previous studies [11–14]. In case of a compressible medium, we are in fact examining system in which sound modes are generated. In fact, any compression leads to acoustic (sound) waves that are transmitted through the medium and serve as the prime source for dissipation. So the problem of the energy spectrum (and dissipation rate) of a compressible fluid is essentially one of stochastic acoustics.

The investigation of such behavior as anomalous scaling requires a lot of thorough analysis to be carried out. The phenomenon manifests itself in a singular power-like behavior of some statistical quantities (correlation functions, structure functions, etc.) in the inertial-convective range in the fully developed turbulence regime [1, 2, 15]. A quantitative parameter that describes intensity of turbulent motion is so-called Reynolds number Re that represents a ratio between inertial and dissipative forces. For high enough values of Re $\gg$ 1 inertial interval is exhibited in which just transfer of kinetic energy from outer L (input) to microscopic l (dissipative) scales take place.

A very useful and computationally effective approach to the problems with many interacting degrees of freedom on different scales is the field-theoretic renormalization group (RG) approach which can be subsequently accompanied by the operator product expansion (OPE); see the monographs [16–20]. One of the greatest challenges is an investigation of the Navier-Stokes equation for a compressible fluid, and, in particular, a passive scalar field advection by this velocity ensemble. The first relevant discussion and analysis of passive advection emerged a few decades ago for the Kraichnan's velocity ensemble [21–24]. Further studies developed its more realistic generalizations [25–32]. The RG+OPE technique was also applied to more complicated models, in particular, to the compressible case [11, 33–45]. Our aim here is to improve existing (one-loop) results on compressible stochastic Navier-Stokes equation and determine relevant physical quantities to two-loop order. Note that in contrast to static phenomena transition from one-loop to two-loop approximation pose in stochastic dynamics much more demanding task.

The paper is a continuation of previous works [12–14] and it is organized as follows. In the introductory Sect. 2 we give a brief overview of the model and we reformulate stochastic equations into field-theoretical language. Section 3 is devoted to the renormalization group analysis. In Sect. 4 we present the fixed points' structure, describe possible scaling regimes and calculate critical dimensions. The concluding Sect. 5 is devoted to a short discussion and future plans.

2 Model

Let us start with a discussion of a model for compressible velocity fluctuations. The dynamics of a compressible fluid is governed by the stochastic Navier-Stokes equation [9] taken in the form

$$\rho \nabla_t v_i = \nu_0 [\delta_{ik} \partial^2 - \partial_i \partial_k] v_k + \mu_0 \partial_i \partial_k v_k - \partial_i p + f_i^v, \tag{1}$$

where the operator ∇_t stands for an expression $\nabla_t = \partial_t + v_k \partial_k$, also known as a Lagrangian (or convective) derivative. Further, $\rho = \rho(t, x)$ is a fluid density field, $v_i = v_i(t, x)$ is the velocity field, $\partial_t = \partial/\partial t$ is a time derivative, $\partial_i = \partial/\partial x_i$ is a i-th component of spatial gradient, $\partial^2 = \partial_i \partial_i$ is the Laplace operator, $p = p(t, x)$ is the pressure field, and f_i^v is the external force, which is specified later. In what follows we employ a condensed notation in which we write $x = (t, x)$, where a spatial vector variable x equals $(x_1, x_2, \ldots, x_d)$ with d being a dimensionality of space. Although it is possible to consider d as additional free parameter [14], in this work spatial dimension d implicitly takes most physically relevant value 3. Two parameters ν_0 and μ_0 in Eq. (1) are two viscosity coefficients [9]. Summations over repeated vector indices (Einstein summation convention) are always implied in this work.

Let us make two important remarks regarding the physical interpretation of Eq. (1). First, this equation should be regarded as an dynamic equation only for a fluctuating part of the total velocity field. In other words, it is assumed that the mean (regular) part of the velocity field has already been subtracted [1, 2]. Second, the random force f_i^v mimics not only an input of energy, but to some extent it is responsible for neglected interactions between fluctuating part of the velocity field and the mean part [16, 19]. In reality, the latter interactions are always present and their mutual interplay generates turbulence [2].

Let us note that stochastic theory of turbulence is similar to a fluctuation theory for critical phenomena [16, 46]. The main difference is lack of Hamilton-like operator for turbulence. Nevertheless, it is still possible to take advantage of well-established theoretical tools borrowed from quantum field theory and employ them on turbulence [16, 18].

To complete the theoretical set-up of the model, Eq. (1) has to be augmented by additional two relations. They are a continuity equation and a certain thermodynamic relation [9]. The former one can be written in the form

$$\partial_t \rho + \partial_i (\rho v_i) = 0 \tag{2}$$

and the latter we choose as follows

$$\delta p = c_0^2 \delta \rho, \tag{3}$$

where δp and $\delta \rho$ describe deviations from the equilibrium values of pressure field and density field, respectively.

Viscous terms in Eq. (1) characterize dissipative processes in the system and in a turbulent state it is expected their relevance at small length scales. Without a continuous input of energy, turbulent processes would eventually die out because of dissipation and the flow would eventually become regular. There are various possibilities for modeling of energy input [19]. For translationally invariant theories it is convenient to specify properties of the random force f_i in time-momentum representation

$$\langle f_i(x) f_j(x') \rangle = \frac{\delta(t - t')}{(2\pi)^d} \int_{k>m} d^d k \, D_{ij}^v(k) e^{ik \cdot (x-x')}, \tag{4}$$

where the delta function in time variable ensures Galilean invariance of the model [19]. The integral in Eq. (4) is infrared (IR) regularized with a parameter $m \sim L_v^{-1}$, where L_v denotes outer scale, i.e., scale of the biggest turbulent eddies. More details can be found in the literature [19, 47]. The kernel function $D_{ij}^v(k)$ is now assumed in the following form

$$D_{ij}^v(k) = g_0 \nu_0^3 k^{4-d-y} \left\{ P_{ij}(k) + \alpha Q_{ij}(k) \right\}, \tag{5}$$

where g_0 is a coupling constant, $k = |k|$ is the wave number, y is a suitable scaling exponent, and α is a free dimensionless parameter. Parameter α basically measures intensity with which energy flows into a system via longitudinal modes.

Further, the projection operators P_{ij} and Q_{ij} in the momentum space read

$$P_{ij}(k) = \delta_{ij} - \frac{k_i k_j}{k^2}, \quad Q_{ij} = \frac{k_i k_j}{k^2}, \tag{6}$$

and correspond to the transversal and longitudinal projector, respectively.

Due to its functional form with respect to momentum dependence, function (5) corresponds to a non-local term in ensuing field theoretic action. However, physical and plausible mathematical considerations [16] justify this choice. One of the reasons is a straightforward modeling of a steady input of energy into the system from outer scales. In what follows we attack the problem with the RG approach. The value of the scaling exponent y in Eq. (5) describes a deviation from a logarithmic behavior (that is obtained for $y = 0$). In the stochastic theory of turbulence the main interest is in the limit behavior $y \to 4$ that yields an ideal pumping from infinite spatial scales [19].

Let us make a brief remark about possible generalization of the model. Although, we present our results with a general spatial dimension d, we have always implicitly in mind its most realistic value $d = 3$. However, it would be possible to generalize the model [14] and consider d as additional small parameter, similar to the well-known φ^4−theory in critical statics [18, 20]. Usually the spatial dimension d plays a passive role and is considered only as an independent parameter. However, Honkonen and Nalimov [48] showed that in the vicinity of space dimension $d = 2$ additional divergences appear in the model of the incompressible Navier-Stokes ensemble and these divergences have to be properly taken into account. Their procedure also naturally leads into improved perturbation expansion [49, 50]. As can be seen from the RG discussion in the next section a similar situation occurs for the model (1) in the vicinity of space dimension $d = 4$. In this case an additional divergence appears in the 1-irreducible Green function $\langle v'v' \rangle_{1\text{-ir}}$. Utilizing this feature one can employ a double expansion scheme, in which the formal expansion parameters are y, and $\varepsilon = 4 - d$, i.e., a deviation from the space dimension $d = 4$ [32, 48].

Our main theoretical tool is the renormalization group theory. Its proper application requires a proof of a renormalizability of the model, i.e., a proof that only a finite number of divergent structures exists in a diagrammatic expansion [17, 18]. As was shown in [51], this requirement can be accomplished by the following procedure: first the stochastic equation (1) is divided by density field ρ, then fluctuations in viscous terms are neglected, and, finally using the expressions (2) and (3) the problem is formulated into a system of two coupled differential equations

$$\nabla_t v_i = \nu_0 [\delta_{ik} \partial^2 - \partial_i \partial_k] v_k + \mu_0 \partial_i \partial_k v_k - \partial_i \phi + f_i, \tag{7}$$

$$\nabla_t \phi = -c_0^2 \partial_i v_i, \tag{8}$$

where a new field $\phi = \phi(x)$ has been introduced for convenience. This new field is related to the density fluctuations via the relation $\phi = c_0^2 \ln(\rho/\overline{\rho})$ [11, 40]. A parameter c_0 denotes the adiabatic speed of sound, $\overline{\rho}$ is the mean value of density field ρ, and $f_i = f_i(x)$ is the external force normalized per unit mass.

According to the general theorem [16, 18], the stochastic problem given by Eqs. (7) and (8), is tantamount to the field theoretic model with a doubled set of fields $\Phi = \{v_i, v_i', \phi, \phi'\}$ and given De Dominicis-Janssen action functional. The latter can be written in a compact form as a sum of two terms

$$\mathcal{S}_{\text{total}}[\Phi] = \mathcal{S}_{\text{vel}}[\Phi] + \mathcal{S}_{\text{den}}[\Phi], \tag{9}$$

where the first term describes a velocity part

$$\mathcal{S}_{\text{vel}}[\Phi] = \frac{v_i' D_{ij}^v v_j'}{2} + v_i' \left[-\nabla_t v_i + \nu_0 (\delta_{ij} \partial^2 - \partial_i \partial_j) v_j + u_0 \nu_0 \partial_i \partial_j v_j - \partial_i \phi \right], \tag{10}$$

and the second term is given by the expression

$$S_{\text{den}}[\Phi] = \phi'[-\nabla_t\phi + v_0\nu_0\partial^2\phi - c_0^2(\partial_i v_i)]. \tag{11}$$

Here, D_{ij}^v is the correlation function (5). Note that we have introduced a new dimensionless parameter $u_0 = \mu_0/\nu_0 > 0$ and a new term $v_0\nu_0\phi'\partial^2\phi$ with another positive dimensionless parameter v_0, which is needed to ensure multiplicative renormalizability [16, 40].

Further, we employ a condensed notation, in which integrals over the spatial variable x and the time variable t, as well as summation over repeated indices, are not explicitly written, for instance

$$\phi'\partial_t\phi = \int dt \int d^dx \, \phi'(t, x)\partial_t\phi(t, x),$$

$$v_i'D_{ik}v'_k = \sum_{ik} \int dt \int d^dx \int d^dx' \, v_i(t, x)D_{ik}^v(x - x')v_k(t, x'). \tag{12}$$

In a functional formulation various stochastic quantities (correlation and structure functions) are calculated as path integrals with weight functional

$$\exp(S_{\text{total}}[\Phi]).$$

The main benefits of such approach are transparency in a perturbation theory and potential use of powerful methods of the quantum field theory, such as Feynman diagrammatic technique and renormalization group procedure [18–20].

3 Renormalization Group Analysis

Ultraviolet renormalizability reveals itself in a presence divergences in Feynman graphs, which are constructed according to simple laws [16, 20] using a graphical notation from Fig. 1. From a practical point of view, an analysis of the 1-particle irreducible Green functions, later referred to as 1-irreducible Green functions following the notation in [16], is of utmost importance. In the case of translationally invariant models [16, 20] two independent scales have to be introduced: the time scale T and the length scale L. Thus the canonical dimension of any quantity F (a field or a parameter) is described by two numbers, the frequency dimension d_F^ω and the momentum dimension d_F^k, defined such that following normalization holds

$$d_k^k = -d_x^k = 1, \quad d_k^\omega = d_x^\omega = 0, \quad d_\omega^\omega = -d_t^\omega = 1, \quad d_\omega^k = d_t^k = 0, \tag{13}$$

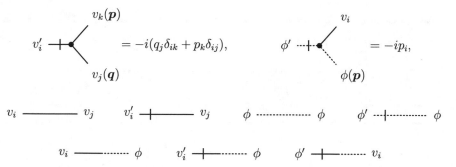

Fig. 1 Graphical representation of elements of the perturbation theory (9)

and the given quantity then scales as

$$[F] \sim [T]^{-d_F^\omega}[L]^{-d_F^k}. \tag{14}$$

The remaining dimensions can be found from the requirement that each term of the action functional (9) be dimensionless, with respect to both the momentum and the frequency dimensions separately.

Based on d_F^k and d_F^ω the total canonical dimension $d_F = d_F^k + 2d_F^\omega$ can be introduced, which in the renormalization theory of dynamic models plays the same role as the conventional (momentum) dimension does in static problems [16]. Setting $\omega \sim k^2$ ensures that all the viscosity and diffusion coefficients in the model are dimensionless. Another option is to set the speed of sound c_0 dimensionless and consequently obtain that $\omega \sim k$, i.e., $d_F = d_F^k + d_F^\omega$. This variant would mean that we are interested in the asymptotic behavior of the Green functions as $\omega \sim k \to 0$, in other words, in sound modes in turbulent medium. Even though this problem is very interesting itself, it is not yet accessible for the RG treatment, so we do not discuss it here. The choice $\omega \sim k^2 \to 0$ is the same as in the models of incompressible fluid, where it is the only possibility because the speed of sound is infinite. A similar alternative in dispersion laws exists, for example, within the so-called model H of equilibrium dynamical critical behavior, see [16, 20].

The canonical dimensions for the model (9) are listed in Table 1. It then directly follows that the model is logarithmic (the coupling constant $g \sim [L]^{-y}$ becomes dimensionless) at $y = 0$. In this work we use the minimal subtraction (MS) scheme for the calculation of renormalization constants. In this scheme the UV divergences in the Green functions manifest themselves as pole in y.

The total canonical dimension of any 1-irreducible Green function Γ is given by the relation

$$\delta_\Gamma = d + 2 - \sum_\Phi N_\Phi d_\Phi, \tag{15}$$

Table 1 Canonical dimensions of the fields and parameters entering velocity part of the total action (9)

F	v_i'	v_i	ϕ'	ϕ	m, μ, Λ	ν_0, ν	c_0, c	g_{10}	u_0, v_0 $w_0, u,$ v, g, α
d_F^k	$d+1$	-1	$d+2$	-2	1	-2	-1	y	0
d_F^ω	-1	1	-2	2	0	1	1	0	0
d_F	$d-1$	1	$d-2$	2	1	0	1	y	0

where N_Φ is the number of the given type of field entering the function Γ, d_Φ is the corresponding total canonical dimension of field Φ, and the summation runs over all types of the fields Φ in function Γ [16, 18, 20].

Superficial UV divergences whose removal requires counterterms can be present only in those functions Γ for which the formal index of divergence δ_Γ is a non-negative integer. A dimensional analysis should be augmented by the several additional considerations. They are all explicitly stated in the previous works [11, 14]. Therefore, we do not repeat them here and continue with a simple conclusion that model with the action (9) is renormalizable.

From a straightforward inspection of RG theory it is clear that for determination of scaling regimes only two Green functions have to be considered. The reason is that we study theory with three charges, g, u and v. Once their fixed values are found, we would be able to study scaling regimes and their stabilities. Thus only graphs that are needed to be calculated are two-point Green functions $\langle vv \rangle_{1PI}$ and $\langle pp \rangle_{1PI}$. In a one-loop approximation [11, 14, 40] the calculation is simple as there are only two Feynman diagrams at this level

$$\text{(16)}$$

For two-loop approximation, following graphs have to be computed for the velocity part

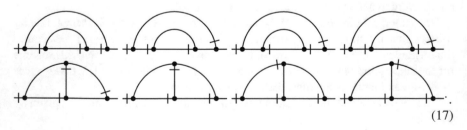

$$\text{(17)}$$

On the other hand, for the pressure part additional eight diagrams are needed

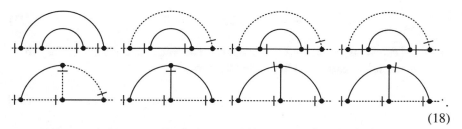

$$(18)$$

The remaining diagrams are needed only for determination of anomalous dimension of fields, which is left for future study.

In contrast to the incompressible case [49] compressible model (9) proved to be much more demanding from technical point of view. This is caused by three reasons. First, in compressible case there are six physical quantities (μ_0, ν_0, v_0, g_0, α, c_0) instead of just two (ν_0 and charge g_0) for incompressible fluid. Second, propagators now contain both transversal and longitudinal parts and last, interaction vertices are not proportional to the momentum of prime field, what implies that the degree of UV divergence could not be lowered.

In evaluation of UV divergent parts of Feynman diagrams we have applied approach suggested in [49]. Using symbolic software [52] we were able to simplify some calculations and to determine divergent parts at least in numerical sense. Because the details of calculation are rather straightforward and proceed in a standard fashion [16–18, 20], we refrain from mentioning them here.

4 Scaling Regimes

The relation between the initial and renormalized action functionals $\mathcal{S}(\varphi, e_0) = \mathcal{S}^R(Z_\varphi \varphi, e, \mu)$ (where e_0 is the complete set of bare parameters and e is the set of their renormalized counterparts) yields the fundamental RG differential equation:

$$\left\{ \mathcal{D}_{RG} + N_\varphi \gamma_\varphi + N_{\varphi'} \gamma_{\varphi'} \right\} G^R(e, \mu, \dots) = 0, \qquad (19)$$

where $G = \langle \varphi \cdots \varphi \rangle$ is a correlation function of the fields φ; N_φ and $N_{\varphi'}$ are the counts of normalization-requiring fields φ and φ', respectively, which are the inputs to G; the ellipsis in expression (19) stands for the other arguments of G (spatial and time variables, etc.). $\mathcal{D}_{RG}$ is the operation $\widetilde{\mathcal{D}}_\mu$ expressed in the renormalized variables and $\widetilde{\mathcal{D}}_\mu$ is the differential operation $\mu \partial_\mu$ for fixed e_0. For the present model it takes the form

$$\mathcal{D}_{RG} = \mathcal{D}_\mu + \beta_g \partial_g + \beta_u \partial_u + \beta_v \partial_v - \gamma_\nu \mathcal{D}_\nu - \gamma_c \mathcal{D}_c. \qquad (20)$$

Here, we have denoted $\mathcal{D}_x \equiv x \partial_x$ for any variable x. The anomalous dimension γ_F of some quantity F (a field or a parameter) is defined as

$$\gamma_F = Z_F^{-1}\widetilde{\mathcal{D}}_\mu Z_F = \widetilde{\mathcal{D}}_\mu \ln Z_F, \tag{21}$$

and the β functions for the four dimensionless coupling constants g, u and v, which we now redefine according to the following rule

$$g \equiv g_1, \quad u \equiv g_2, \quad v \equiv g_3. \tag{22}$$

for convenience. β functions express the flows of parameters under the RG transformation, and are defined through relation $\beta_i = \widetilde{\mathcal{D}}_\mu g_i$. This yields

$$\beta_1 = g_1(-y - \gamma_1), \quad \beta_2 = -g_2\gamma_2, \quad \beta_3 = -3\gamma_3, \tag{23}$$

$$\gamma_1 \equiv \gamma_g, \quad \gamma_2 \equiv \gamma_u, \quad \gamma_3 \equiv \gamma_v. \tag{24}$$

Based on the analysis of the RG equation (19) it follows that the large scale behavior with respect to spatial and time scales is governed by the IR attractive ("stable") fixed points $g^* \equiv \{g_1^*, g_2^*, g_3^*\}$, whose coordinates are found from the conditions [16–18]:

$$\beta_1(g^*) = \beta_2(g^*) = \beta_3(g^*) = 0. \tag{25}$$

Let us consider a set of invariant couplings $\overline{g}_i = \overline{g}_i(s, \{g_i\})$ with the initial data $\overline{g}_i|_{s=1} = g_i$. Here, $s = k/\mu$ and IR asymptotic behavior (i.e., behavior at large distances) corresponds to the limit $s \to 0$. An evolution of invariant couplings is described by the set of flow equations

$$\mathcal{D}_s\overline{g}_i = \beta_i(\overline{g}_j), \tag{26}$$

whose solution as $s \to 0$ behaves approximately like

$$\overline{g}_i(s, g^*) \cong g_i^* + const \times s^{\omega_i}, \tag{27}$$

where $\{\omega_i\}$ is the set of eigenvalues of the matrix

$$\Omega_{ij} = \partial\beta_i/\partial g_j|_{g^*}. \tag{28}$$

The existence of IR attractive solutions of the RG equations leads to the existence of the scaling behavior of Green functions. From (27) it follows that the type of the fixed point is determined by the matrix (28): for the IR attractive fixed points the matrix Ω has to be positive definite.

Altogether two IR attractive fixed points are found, which defines possible scaling regimes of the system. The fixed point FPI (the trivial or Gaussian point) is stable if $y < 0$. This regime is characterized by irrelevance of all his charges, i.e.,

$$g_1^* = g_2^* = g_3^* = 0. \tag{29}$$

On the other hand, the fixed point FPII is fully nontrivial, i.e. all his coordinates attain nonzero value. We have found the following numerical expressions for them

$$g_1^* = 2y + \frac{-2.00625\alpha^2 - 4.8847\alpha + 4.4206}{5\alpha + 12} y^2, \tag{30}$$

$$g_2^* = 1 + \frac{0.125797\alpha^2 - 0.83854\alpha - 0.188233}{5\alpha + 12} y, \tag{31}$$

$$g_3^* = 1 + \frac{0.217295\alpha^3 + 1.7247\alpha^2 - 1.27116\alpha - 6.9228}{(\alpha + 6)(5\alpha + 12)} y. \tag{32}$$

To one-loop order we have thus obtained same results as has been claimed previously [11, 40]. The initial analysis reveals that FPII is nontrivial for $y > 0$ and not very large values of α.

5 Conclusion

In the present paper the compressible fluid governed by the Navier-Stokes velocity ensemble has been examined. The fluid was assumed to be compressible and the space dimension was fixed to $d = 3$. The problem has been investigated by means of renormalization group and expansion scheme in y was constructed.

There are two nontrivial IR stable fixed points in this model and, therefore, the critical behavior in the inertial range demonstrates two different regimes depending on the the scaling exponent y. Coordinates of nontrivial fixed points have been obtained for the first time to two-loop precision. This can be considered as a first step to full two-loop analysis of the model.

Acknowledgements The work was supported by VEGA grant No. 1/0345/17 of the Ministry of Education, Science, Research and Sport of the Slovak Republic, and by the Russian Foundation for Basic Research within the Project No. 16-32-00086. N. M. G. acknowledges the support from the Saint Petersburg Committee of Science and High School.

References

1. U. Frisch, *Turbulence: The Legacy of A. N. Kolmogorov* (Cambridge University Press, Cambridge, 1995)
2. P.A. Davidson, *Turbulence: An Introduction for Scientists and Engineers*, 2nd edn. (Oxford University Press, Oxford, 2015)
3. S.N. Shore, *Astrophysical Hydrodynamics: An Introduction* (Wiley-VCH Verlag GmbH & KGaA, Weinheim, 2007)
4. V. Carbone, R. Marino, L. Sorriso-Valvo, A. Noullez, R. Bruno, Phys. Rev. Lett. **103**, 061102 (2009)
5. F. Sahraoui, M.L. Goldstein, P. Robert, YuV Khotyainstsev, Phys. Rev. Lett. **102**, 231102 (2009)
6. H. Aluie, G.L. Eyink, Phys. Rev. Lett. **104**, 081101 (2010)

7. S. Galtier, S. Banerjee, Phys. Rev. Lett. **107**, 134501 (2011)
8. S. Banerjee, L.Z. Hadid, F. Sahraoui, S. Galtier, Astrophys. J. Lett. **829**, L27 (2016)
9. L.D. Landau, E.M. Lifshitz, *Fluid Mechanics* (Pergamon Press, Oxford, 1959)
10. P. Sagaut, C. Cambon, *Homogeneous Turbulence Dynamics* (Cambridge University Press, Cambridge, 2008)
11. N.V. Antonov, M.M. Kostenko, Phys. Rev. E **90**, 063016 (2014)
12. N.V. Antonov, N.M. Gulitskiy, M.M. Kostenko, T. Lučivjanský, EPJ Web Conf. **125**, 05006 (2016)
13. N.V. Antonov, N.M. Gulitskiy, M.M. Kostenko, T. Lučivjanský, EPJ Web of Conf. **137**, 10003 (2016)
14. N.V. Antonov, N.M. Gulitskiy, M.M. Kostenko, T. Lučivjanský, Phys. Rev. E **95**, 0331200 (2017)
15. G. Falkovich, K. Gawędzki, M. Vergassola, Rev. Mod. Phys. **73**, 913 (2001)
16. A.N. Vasil'ev, *The Field Theoretic Renormalization Group in Critical Behavior Theory and Stochastic Dynamics* (Chapman Hall/CRC, Boca Raton, 2004)
17. D.J. Amit, V. Martín-Mayor, *Field Theory, the Renormalization Group and Critical Phenomena* (World Scientific, Singapore, 2005)
18. J. Zinn-Justin, *Quantum Field Theory and Critical Phenomena*, 4th edn. (Oxford University Press, Oxford, 2002)
19. L.Ts. Adzhemyan, N.V. Antonov, A.N. Vasil'ev: *The Field Theoretic Renormalization Group in Fully Developed Turbulence* (Gordon & Breach, London, 1999)
20. U. Täuber, *Critical Dynamics: A Field Theory Approach to Equilibrium and Non-Equilibrium Scaling Behavior* (Cambridge University Press, New York, 2014)
21. R.H. Kraichnan, Phys. Fluids **11**, 945 (1968)
22. K. Gawędzki, A. Kupiainen, Phys. Rev. Lett. **75**, 3834 (1995), D. Bernard, K. Gawędzki, A. Kupiainen, Phys. Rev. E **54**, 2564 (1996), M. Chertkov, G. Falkovich, Phys. Rev. Lett. **76**, 2706 (1996)
23. L.Ts. Adzhemyan, N.V. Antonov, A.N. Vasil'ev, Phys. Rev. E **58**, 1823 (1998)
24. N.V. Antonov, J. Phys. A: Math. Gen. **39**, 7825 (2006)
25. N.V. Antonov, N.M. Gulitskiy, Lect. Notes Comp. Sci., **7125/2012**, 128 (2012), N.V. Antonov, N.M. Gulitskiy, Phys. Rev. E **85**, 065301(R) (2012), N.V. Antonov, N.M. Gulitskiy, Phys. Rev. E **87**, 039902(E) (2013)
26. E. Jurčišinova, M. Jurčišin, J. Phys. A: Math. Theor., **45**, 485501 (2012), E. Jurčišinova and M. Jurčišin, Phys. Rev. E **88**, 011004 (2013)
27. E. Jurčišinova, M. Jurčišin, Phys. Rev. E **77**, 016306 (2008), E. Jurčišinova, M. Jurčišin, R. Remecky, Phys. Rev. E **80**, 046302 (2009), E. Jurčišinova, M. Jurčišin, R. Remecky, J. Phys. A: Math. Theor. **42**, 275501 (2009)
28. N.V. Antonov, N.M. Gulitskiy, Phys. Rev. E **91**, 013002 (2015), N.V. Antonov, N.M. Gulitskiy, Phys. Rev. E **92**, 043018 (2015), N.V. Antonov, N.M. Gulitskiy, AIP Conf. Proc. **1701**, 100006 (2016), N.V. Antonov, N.M. Gulitskiy, EPJ Web of Conf. **108**, 02008 (2016)
29. E. Jurčišinova, M. Jurčišin, Phys. Rev. E **91**, 063009 (2015)
30. N.V. Antonov, A. Lanotte, A. Mazzino, Phys. Rev. E **61**, 6586 (2000), N.V. Antonov, N.M. Gulitskiy, Theor. Math. Phys., **176**(1), 851 (2013)
31. H. Arponen, Phys. Rev. E **79**, 056303 (2009)
32. M. Hnatič, J. Honkonen, T. Lučivjanský, Acta Phys. Slovaca **66**, 69 (2016)
33. L.Ts. Adzhemyan, N.V. Antonov, J. Honkonen, T.L. Kim, Phys. Rev. E **71**, 016303 (2005)
34. N.V. Antonov, Phys. Rev. Lett. **92**, 161101 (2004)
35. N.V. Antonov, N.M. Gulitskiy, A.V. Malyshev, EPJ Web Conf. **126**, 04019 (2016)
36. E. Jurčišinova, M. Jurčišin, R. Remecky, Phys. Rev. E **93**, 033106 (2016)
37. M. Vergassola, A. Mazzino, Phys. Rev. Lett. **79**, 1849 (1997)
38. A. Celani, A. Lanotte, A. Mazzino, Phys. Rev. E **60**, R1138 (1999)
39. M. Chertkov, I. Kolokolov, M. Vergassola, Phys. Rev. E **56**, 5483 (1997)
40. N.V. Antonov, M. Yu, Nalimov, A.A. Udalov, Theor. Math. Phys. **110**, 305 (1997)
41. N.V. Antonov, M.M. Kostenko, Phys. Rev. E **92**, 053013 (2015)

42. M. Hnatich, E. Jurčišinova, M. Jurčišin, M. Repašan, J. Phys. A: Math. Gen. **39**, 8007 (2006)
43. V.S. L'vov, A.V. Mikhailov, Preprint No. 54, Inst. Avtomat. Electron., Novosibirsk (1977)
44. I. Staroselsky, V. Yakhot, S. Kida, S.A. Orszag, Phys. Rev. Lett. **65**, 171 (1990)
45. S.S. Moiseev, A.V. Tur, V.V. Yanovskii, Sov. Phys. JETP **44**, 556 (1976)
46. A.Z. Patashinskii, V.L. Pokrovskii, *Fluctuation Theory of Phase Transitions* (Pergamon Press, Oxford, 1979)
47. L.Ts. Adzhemyan, N.V. Antonov, A.N. Vasil'ev, Sov. Phys. JETP **68**, 733 (1989)
48. J. Honkonen, M.Yu. Nalimov, Z. Phys, B **99**, 297 (1996)
49. L.Ts. Adzhemyan, J. Honkonen, M.V. Kompaniets, A.N. Vasil'ev, Phys. Rev. E **71**(3), 036305 (2005)
50. L.Ts. Adzhemyan, M. Hnatich, J. Honkonen, Eur. Phys. J B **73**, 275 (2010)
51. D. Yu, Volchenkov, M. Yu. Nalimov, Theor. Math. Phys. **106**(3), 307 (1996)
52. Wolfram Research, *Mathematica, Version 9.0* (Champaign, Illinois, 2012)
53. L.Ts. Adzhemyan, A.N. Vasil'ev, M. Gnatich, Theor. Math. Phys. **64**(2), 777 (1985)
54. N.V. Antonov, A. Lanotte, A. Mazzino, Phys. Rev. E **61**, 6586 (2000)
55. N.V. Antonov, M. Hnatich, J. Honkonen, M. Jurčišin, Phys. Rev. E **68**, 046306 (2003)
56. B. Duplantier, A. Ludwig, Phys. Rev. Lett. **66**, 247 (1991), G.L. Eyink, Phys. Lett. A **172**, 355 (1993)

Stability of a Nonlinear Viscoelastic Problem Governed by Lamé Operator

Meflah Mabrouk and Khoukhi Alae Nore

Abstract In this paper we will investigated the stability of the nonlinear viscoelastic problem governed by Lamé operator. When I' study the existence and uniqueness in (Meflah in Int J Math Arch 2(5):693–697, 2011 [5]). We denote by Ω an open subset of $\mathbb{R}^n$ with regular boundary Γ. Let Q the cylinder $\mathbb{R}^n_x \times \mathbb{R}_t$ with $Q = \Omega \times]0,T[$; T fini, Σ boundary of Q, L designed Lamé system define by $\mu\Delta + (\lambda + \mu)\nabla\mathrm{div}$, f, $u_0(x)$ and $u_1(x)$ are functions. We look for the stabilisation of a function $u = u(x, t)$, $x \in \Omega$, $t \in]0, T[$, solution of the problem (P).

$$(P)\begin{cases} \frac{\partial^2 u}{\partial t^2} - Lu + \int\limits_0^t g(t-s)\Delta u(s)ds + |u|^p u = f & \text{in } \Omega \times]0, T[\\ u = 0 & \text{on } \Sigma \\ u(x, 0) = u_0(x), \frac{\partial u(x,t)}{\partial t}|_{t=0} = u_1(x) & x \in \Omega \end{cases}$$

Keywords Nonlinear · Priori estimate · Stability · Viscoelastic

1 Introduction

Viscoelastic equations arising in many mathematical models, Cavalcanti et al. [1] studies in bounded domain the problem:

$$\frac{\partial^2 u}{\partial t^2} - \Delta u + \int\limits_0^t g(t-\tau)\Delta u(\tau)d\tau + a(x)\frac{\partial u}{\partial t} + |u|^\gamma u = 0;$$

g positive function and $\gamma > 0$.

M. Mabrouk (✉) · K. A. Nore
Faculté des Mathématiques et des Sciences de la Matière, Laboratoire Mathématiques Appliquées, Université Kasdi Merbah Ouargla, 30000 Ouargla, Algeria
e-mail: meflah.mabrouk@univ-ouargla.dz

K. A. Nore
e-mail: khoukhi.alae@univ-ouargla.dz

© Springer Nature Switzerland AG 2019
C. H. Skiadas and I. Lubashevsky (eds.), *11th Chaotic Modeling and Simulation International Conference*, Springer Proceedings in Complexity, https://doi.org/10.1007/978-3-030-15297-0_17

Messaoudi [6] study the system

$$
\begin{cases}
\frac{\partial^2 u}{\partial t^2} - \Delta u + \int_0^t g(t-\tau)\Delta u(\tau)d\tau + a(x)\frac{\partial u}{\partial t} = |u|^\gamma u & t > 0 \\
u(x,t) = 0 & t \geq 0 \\
u(x,0) = u_0(x), \quad u'(x,0) = u_1(x) & x \in \Omega
\end{cases}
$$

When g = o. Lions [2] invented some methods of solving boundary nonlinear problem. The author study [3, 5] a nonlinear problem governed by Lamé operator.

In this work, we needed in the proof of our result, the relaxation function g(t) satisfied with the conditions (G1) and (G2). The first is necessary to guarantee the hyperbolicity and the second using for assure the estimate results.

(G1) g: $\mathbb{R}_+$! $\mathbb{R}_+$ is a bounded C^1 function satisfying

$$
g(0) > 0, \quad \mu - \int_0^{+\infty} g(s)ds = l > 0
$$

(G2) There exists a positive constant ξ such that

$$
g(t) \leq -\xi g^p(t), t \geq 0; \quad 1 \leq p \leq 3/2.
$$

2 The Energy Equation

We multiply the partial diferential equation

$$
\frac{\partial^2 u}{\partial t^2} - Lu + \int_0^{\hat{}} tg(t-s)\Delta u(s)ds + |u|^\hat{}p\, u = f\$\$
$$

by u_t, and integrating over Ω and using green's formula, then we define the energy function related with problem (P) is given

$$
E(t) = \frac{1}{2}\left[\|u_t\|_2^2 + (\mu - \int_0^t g(s)ds)\|\Delta u\|_2^2 + (\lambda+\mu)\|div u\|_2^2 (go\nabla u)(t)\right]
$$
$$
+ \frac{1}{p+1}\|\Delta u\|_2^2 > 0
$$

With

$$
(go\Delta u)(t) = \int_0^t g(t-s)\|u(t)-u(s)\|_2^2(s)ds
$$

We denote by $\| \ \|_k$. The L^K-norm over Ω. In particular, the L^2-norm is denoted $\| \ \|_2$, and throughout this presentation we assume $(u_0, u_1) \in (H_0^1(\Omega) \cap L^{p+2}(\Omega))^n$

We define the function

$$F(t) = E(t) + \varepsilon_1 \varphi(t) + \varepsilon_2 \psi(t)$$

where

$$\varphi(t) = \int_\Omega u.u_t dx$$

And

$$\psi(t) = - \int_\Omega u_t \int_0^t g(t - \tau)(u(t) - u(\tau))d\tau dx$$

Lemma

$$\alpha_1 F(t) \leq E(t) \leq \alpha_2 F(t)$$

Lemma

$$\varphi'(t) = \|u_t(t)\|_2^2 - \frac{l}{2}\|\Delta u\|_2^2 + \frac{1-l}{2l}(go\nabla u)(t)$$

Lemma

$$\psi'(t) = \delta + 2\delta(1 - l)\left(\int_0^t g(s)ds\right)\|\Delta u\|_2^2 + \left(\delta - \int_0^t g(s)ds\|u_t(t)\|_2^2\right)$$

$$+ \frac{1}{4\delta}(go\Delta u)(t) + \left(2 + \frac{1}{4\delta}\right)\int_0^t g(s)ds(go\nabla u)(t)$$

$$+ \frac{1}{4\delta}C_p \cdot (-g'o\nabla u)(t).$$

Stability

We define the Liaponov function

$$F(t) = E(t) + \varepsilon_1 \varphi(t) + \varepsilon_2 \psi(t)$$

$$F'(t) = E'(t) + \varepsilon_1 \varphi'(t) + \varepsilon_2 \psi'(t) \leq \frac{1}{2}(g'o\nabla u)(t)$$

$$-\frac{t}{2}g(t)\|\Delta u\|_2^2 + \varepsilon_1\left[\|u_t(t)\|_2^2 - \frac{l}{2}\|\Delta u\|_2^2 + \frac{1-l}{2l}(go\nabla u)(t)\right]$$

$$+\varepsilon_2\left[\begin{array}{c}\delta + 2\delta(1-l)\int_0^t g(s)ds\,\|\Delta u\|_2^2 + \left(\delta - \int_0^t g(s)ds\right)\|u_t(t)\|_2^2 \\ +\frac{1}{4\delta}(go\Delta u)(t) + \left(2\delta + \frac{1}{4\delta}\right)\int_0^t g(s)ds)(go\nabla u)(t) \\ +\frac{g(0)}{4\delta}C_p\cdot(-g'o\nabla u)(t)\end{array}\right]$$

We have

$$\int_0^t g(s)ds \geq \int_0^{t=0} g(s)ds = g(0)$$

And

$$\int_0^\infty g(s)ds = \frac{1-l}{C_p^2}$$

So

$$F'(t) \leq [\varepsilon_2(g_0 - \delta) - \varepsilon_1]\|u_t(t)\|_2^2 - \left[\frac{\varepsilon_1.l}{2} - \frac{\varepsilon_2\delta}{C_p^2}((1-l) + 2(1-l^2))\right]\|\Delta u\|_2^2$$

$$+\left[\frac{\varepsilon_1.(1-l)}{2l} + \frac{\varepsilon_2}{c_p^2}\left(2\delta + \frac{1}{4\delta}\right)(1-l)\right](go\nabla u)(t) + \frac{1}{4\delta}(go\Delta u)(t)$$

$$+\left[\frac{1}{2} - \frac{g(0)}{4\delta}C_p\right](g'o\nabla u)(t).$$

We choose δ such that $g_0 - \delta > \frac{1}{2}g_0$ And

$$\frac{2\delta}{lC_p^2}((1-l) + 2(1-l)^2) < \frac{1}{4}g(0)$$

We find

$$\varepsilon_2(g_0 - \delta) - \varepsilon_1 > \frac{1}{2}g(0)\varepsilon_2 - \varepsilon_1 > 0 \Rightarrow \varepsilon_1 < \frac{1}{2}g(0)\varepsilon_2$$

And

$$\varepsilon_1 - \frac{2\delta}{lC_p^2}((1-l) + 2(1-l)^2)\varepsilon_2 > \varepsilon_1 - \frac{1}{4}g(0)\varepsilon_2 > 0 \Rightarrow \varepsilon_1 > \frac{1}{4}g(0)\varepsilon_2$$

So

$$\frac{1}{4}g(0)\varepsilon_2 < \varepsilon_1 < \frac{1}{2}g(0)\varepsilon_2$$

Will make

$$K_1 = \varepsilon_2(g_0 - \delta) - \varepsilon_1 > 0$$

$$K_2 = \frac{\varepsilon_{11}}{2} - \frac{2\delta}{lC_p^2}\left((1 - l) + 2(1 - l)^2\right) > 0$$

We then pick ε_1 and ε_2 so small that

$$\alpha_1 F(t) \leq E(t) \leq \alpha_2 F(t)$$

And

$$\frac{1}{4}g(0)\varepsilon_2 < \varepsilon_1 < \frac{1}{2}g(0)\varepsilon_2$$

Remain valid and

$$\frac{1}{2} - \varepsilon_2 C_p \frac{g(0)}{4\delta} > 0$$

Then we find

$$F'(t) \leq -K_1\|u_t(t)\|_2^2 - K_2\|\Delta u(t)\|_2^2 + C[(go\nabla u)(t) + (go\Delta u)(t)];$$
$$F'(t) \leq -\beta E(t) + C[(go\nabla u)(t) + (go\Delta u)(t)]; \quad \forall t > t_0; \quad \forall \beta, C > 0$$

We multiply (*) by $\gamma(t)$, we find

$$\gamma(t)F'(t) \leq \beta\gamma(t)E(t) + C\gamma(t)[(go\nabla u)(t) + (go\Delta u)(t)]$$

Then

$$\beta\int_{t_0}^{t}\gamma(s)E(s)ds \leq \eta$$

We have

$$E(t) \leq E(s), s \leq t \quad \gamma(s) E(t) \leq \gamma(s)E(s)$$

$$\beta E(t) \int_{t_0}^{t} \gamma(s)ds = \int_{t_0}^{t} \beta\gamma(s)E(t)ds \le \int_{t_0}^{t} \beta\gamma(s)E(s)ds \le \eta$$

Then

$$E(t) = \alpha^t / \int_0^t \gamma(s)ds \quad \forall t > t_0.$$

3 Conclusions

In this work we have proof that the energy E(t) decay polynomial with respect to time and it's easier to show that it's also decay exponentially. Which it's give us the stability of the problem.

References

1. M.M. Cavalcanti, V.N. Domingos Cavalcanti, J. Ferreira, Existence and uniform decay for nonlinear viscoelastic equation with strong damping, Math. Meth. Appl. Sci. **24**, 1043–1053 (2001)
2. J.L. Lions, Quelques méthodes de résolution des problèmes aux limites non linéaires. Dunod (1969)
3. M. Meflah, A similar nonlinear telegraph problem governed by Lamé system, Int. J. Nonlinear Sci. ISSN: 2241–0503 (2012)
4. M. Meflah, On a nonlinear viscoelastic problem governed by Lamé operator 8th Chaos conference proceeding, 26–29 Mai 2015, pp. 521–528, Henri Poincaré Institute, Paris France (2015)
5. M. Meflah, Study of non linear elasticity by elliptic regularization with Lamé system. Int. J. Math. Arch. **2**(5), 693–697 (2011)
6. S.A. Messaoudi, General decay of solutions of a viscoelastic equation. J. Math. Anal. Appl. **341**, 1457–1467 (2008)

Renormalization Group Approach of Directed Percolation: Three-Loop Approximation

L. Ts. Adzhemyan, M. Hnatič, M. V. Kompaniets, T. Lučivjanský and L. Mižišin

Abstract Directed percolation near its second-order phase transition is investigated by the means of perturbative renormalization group approach. We study a numerical calculation of the renormalization group functions in the ε-expansion. Within this procedure anomalous dimensions are expressed in terms of irreducible renormalized Feynman diagrams. Numerical calculation of integrals was performed on Hybrilit cluster using Vegas algorithm from CUBA library.

Keywords Directed percolation · Renormalization group · Numerical calculation

1 Introduction

Directed percolation (DP) process is an important model in statistical physics [1–3]. It provides a paramount example of non-equilibrium phase transitions between absorbing (inactive) and active state. In the absorbing state, there are no spreading particles in the system, whereas in the active state the number of active particles fluctuates around a certain mean value. This type of transition can be observed, for instance in model of spreading epidemics [1, 4], forest fires [3], transport in random media [1, 2] and so on. The quantum field theory of DP is used in order to describe cross-section of particles at high energies (Reggeon field theory) [5].

L. Ts. Adzhemyan · M. V. Kompaniets
Department of Physics, St. Petersburg State University,
7/9 Universitetskaya nab., St. Petersburg 199034, Russian Federation

M. Hnatič · T. Lučivjanský
Faculty of Sciences, P.J. Šafárik University, Moyzesova 16, 040 01 Košice, Slovak Republic

M. Hnatič · L. Mižišin (✉)
Bogoliubov Laboratory of Theoretical Physics Joint Institute for Nuclear Research, 141980
Dubna, Russian Federation
e-mail: mizisin@theor.jinr.ru

Institute of Experimental Physics, Slovak Academy of Sciences,
Watsonova 47, 040 01 Košice, Slovak Republic

© Springer Nature Switzerland AG 2019
C. H. Skiadas and I. Lubashevsky (eds.), *11th Chaotic Modeling and Simulation International Conference*, Springer Proceedings in Complexity, https://doi.org/10.1007/978-3-030-15297-0_18

195

An important method for analysis of the second order phase transition is the renormalization group (RG) approach [6] and ε-expansion, where ε is the deviation from the upper critical dimension $d_c = 4$. Within this method, critical exponents of DP are known only to the second-order (two-loop approximation) of the perturbation theory [1, 3]. In this work, the third-order pertubative corrections are calculated numerically by two distinct methods. In the first case, the null-momentum (NM) subtraction scheme is used. In this procedure anomalous dimensions γ are expressed in terms of irreducible renormalized Feynman diagrams and thus the calculation of renormalization constants can be entirely skipped [7]. In the second case, the minimal subtraction (MS) scheme is utilized and contributions to the renormalization constants for each Feynman diagram are determined. Let us note that universal quantities, in the form of the ε-expansion, are independent of the choice of renormalization scheme.

The numerical evaluation of contributions Feynman diagrams plays very important role in renormalization methods. In this work we choose a multidimensional integration algorithm: Vegas [8, 9] is a Monte Carlo algorithm with importance sampling.

2   The Model and Renormalization

A field theoretical formulation of the percolation process [1–3] is based on the following De Dominicis-Janssen action functional

$$S = \psi^\dagger(-\partial_t + D_0\partial^2 - D_0\tau_0)\psi + \frac{D_0\lambda_0}{2}[(\psi^\dagger)^2\psi - \psi^\dagger\psi^2], \qquad (1)$$

where $\partial_t = \partial/\partial t$, ∂^2 is the Laplace operator, ψ is a coarse-grain density of percolating particles, $\psi^\dagger$ is an auxiliary (Martin-Siggia-Rose) response field, D_0 is a diffusion constant, λ_0 is a positive coupling constant and τ_0 is a deviation from the threshold value of injected probability (an analog of critical temperature in static models [6]). The model is studied near its critical dimension $2\varepsilon = 4 - d$ in the region where τ_0 acquires its critical value. The expansion parameter of the perturbation theory is rather λ_0^2 than λ_0 as it could be easily seen by a direct calculation of Feynman diagrams. This motivates introduction of a new charge u as follows $u \equiv \lambda^2$. Further, in Eq. (1) and rest of the paper we employ abbreviated notation, in which integrals over the spatial and time variables are not explicitly written.

The renormalized action functional can be written in the following form

$$S_R = \psi^\dagger(-Z_1\partial_t + Z_2D\partial^2 - Z_3D\tau)\psi + \frac{Z_4D\lambda\mu^\epsilon}{2}[(\psi^\dagger)^2\psi - \psi^\dagger\psi^2], \qquad (2)$$

where μ is renormalization mass [6]. The model (1) exhibits important symmetry with respect to the following replacement

$$\psi(t) \longleftrightarrow -\psi^\dagger(-t). \tag{3}$$

Immediate consequence of this symmetry is that two triple vertice, i.e. $(\psi^\dagger)^2\psi$ and $\psi^\dagger\psi^2$, are renormalized by the same renormalized constant [1].

On the other hand, the action functional S_R can be derived by the standard procedure of multiplicative renormalization of all the fields and parameters

$$\psi_0 = \psi Z_\psi, \qquad \psi_0^\dagger = \psi^\dagger Z_{\psi^\dagger}, \qquad D_0 = DZ_D, \qquad \lambda_0 = \lambda\mu^\epsilon Z_\lambda, \qquad \tau_0 = \tau Z_\tau. \tag{4}$$

and the relations among renormalization constants parameters, fields and Z_i, $i = 1, 2, 3, 4$ can be obtained by a direct comparison.

In this work, we use two methods for calculation of universal quantities. In the first method [10], the renormalization scheme at the normalization point (NP), $p = 0, \omega = 0$ and $\tau = \mu^2$ is considered. The counterterms are then specified at the normalization point, and for numerical calculation, it is advantageous to express renormalization constants in terms of normalized Green functions

$$\bar{\Gamma}_1 = \partial_{i\omega} \Gamma_{\psi^\dagger\psi}\big|_{p=0,\omega=0}, \qquad \bar{\Gamma}_3 = -\frac{\Gamma_{\psi^\dagger\psi} - \Gamma_{\psi^\dagger\psi}\big|_{\tau=0}}{D\tau}\bigg|_{p=0,\omega=0},$$

$$\bar{\Gamma}_2 = -\frac{1}{2D}\partial_p^2 \Gamma_{\psi^\dagger\psi}\big|_{p=0,\omega=0}, \qquad \bar{\Gamma}_4 = \frac{\Gamma_{\psi^\dagger\psi^\dagger\psi} - \Gamma_{\psi^\dagger\psi\psi}}{D\lambda\mu^\epsilon}\bigg|_{p=0,\omega=0}. \tag{5}$$

that satisfy the following conditions

$$\bar{\Gamma}_i\big|_{\tau=\mu^2} = 1, \qquad\qquad i = 1, 2, 3, 4. \tag{6}$$

Further, RG equations are determined from the condition that original (bare) Green function are independent of the momentum scale μ [6], and are the same as in minimal subtraction scheme

$$(\mu\partial_\mu + \beta_u\partial_u - \tau\gamma_\tau\partial_\tau - D\gamma_D\partial_D)\Gamma_i^R = (n_\psi\gamma_\psi + n_{\psi^\dagger}\gamma_{\psi^\dagger})\Gamma_i^R, \tag{7}$$

where μ is a reference mass scale, n_ψ and $n_{\psi^\dagger}$ are the numbers of the corresponding fields entering the Green function under consideration, γ_x are anomalous dimensions defined as follows

$$\gamma_x \equiv \mu\partial_\mu \log Z_x, \tag{8}$$

and $\beta_u = u(-2\epsilon - \gamma_u)$ is a beta function that describes a flow of the charge u under the RG transformation [6]. Using these equations we easily find relations for the normalized functions

$$(\mu\partial_\mu + \beta_u\partial_u - \tau\gamma_\tau\partial_\tau - D\gamma_D\partial_D)\bar{\Gamma}_i^R = \gamma_i\bar{\Gamma}_i^R, \tag{9}$$

which play crucial role in considered procedure. In the renormalized Green function Γ^R, counterterms can be replaced by R operation acting on the Green functions. The main benefit of these procedure is that R operation is taken at normalization point and it can be expressed in terms of subtracting operation $1 - K_i$, which eliminates all divergences from Feynman diagrams [6]

$$R\Gamma = \prod_i (1 - K_i)\Gamma, \tag{10}$$

where product is taken over all relevant subgraphs of concrete Feynman diagram, including given diagram as whole.

Taking into account the renormalization scheme, we can express the anomalous dimension in terms of the renormalized derivatives of the one-particle irreducible Green function $\bar{\Gamma}_i$ at the normalization point [7]

$$f_i \equiv R[-\tilde{\tau}\partial_{\tilde{\tau}}\bar{\Gamma}_i(\tilde{\tau})]\big|_{\tilde{\tau}=1}, \tag{11}$$

where $\tilde{\tau} = \tau/\mu^2$. The redefinition equations for anomalous dimension allows us to express them in terms of the renormalized derivatives of the one-irreducible Green function $\bar{\Gamma}_i$ with respect to the normalized point

$$\gamma_i = \frac{2f_i}{1 + f_2}, \quad i = 1, 2, 4. \tag{12}$$

In the event renormalization scheme are used as subtraction operation for diagrams at the normalization point and then we get a representation for R operation [7] in the null-momentum scheme

$$R\chi = \prod_i \frac{1}{n_i!} \int_0^1 da_i (1 - a_i)^{n_i} \partial_{a_i}^{n_i+1} \chi(\{a\}), \tag{13}$$

where product is taken over all one-irreducible subgraphs χ_i (including diagram χ as a whole) with canonical dimension $n_i \geq 0$ and a_i - parameter that stretches moments flowing into i-th subgraph inside this graph.

In the second method, the minimal subtraction (MS) scheme, which is always used in analytic calculation, it is applied for determination contribution each Feynman diagram to the renormalization constant Z_i. In this case, the divergent part must be identified in graphs and after extraction of the poles in ε only numerical integration of the rest of integral is needed. In the final step the anomalous dimension (8) is calculated.

3 Numerical Calculation

In percolation problem the numerical calculation was performed up to the third-order in perturbation theory (three-loop approximation) basically consists of four main points. They are

(1) Generation of Feynman diagrams without fields – all possible connection between vertices up to third order perturbation theory (three-loop diagrams) with two for $\Gamma_{\psi\psi^\dagger}$ and three for $\Gamma_{\psi\psi\psi^\dagger}$ arguments.
(2) Construction of Feynman diagrams – The fields ψ and $\psi^\dagger$ are attached in diagrams and graphs, which do not obey Feynman rules, are discarded.
(3) Symbolic construction of integrand using procedure on selected Feynman diagram in Python 2.7 (library Graphine and GraphState).
(4) The numerical evaluation of integrand based on the Vegas algorithm [9]. The first method is not suitable for all diagrams (some of them oscillate).

The first step consists in a generation of all relevant one-irreducible Feynman diagrams for perturbative calculation. In the Table 1 final numbers of relevant diagrams are displayed. All of them are needed in order to calculate universal quantities of DP process up to the third order of perturbation theory. The number of Feynman diagram can be reduced by taking advantage of symmetry (3), and identification of pairs of diagrams, which yields same contribution. In graphical representation the transformation is the only reverse field in diagrams and it is easily checked that it obeys Feynman rules Fig. 1. If the new diagram is one on the list of DP diagrams then the number of independent diagrams decrease (Table 1, column – symmetry).

The number of relevant vertex diagrams can also be lowered using self-energy diagram proportional to external frequency (see Table 1, column – relation). The first step in calculation of the contribution Feynman diagram is applied to differentiation with respect to the external frequency ∂_ω. In graphical representation, it is equiva-

Table 1 The number of the Feynman diagrams for DP process up to third order

	1-loop	2-loop	3-loop	$\to$ Symmetry	$\to$ Relation	$\to$ Model A
$\langle \psi^\dagger \psi \rangle$	1	2	17	$\to$ 13	$\to$ 13	$\to$ 13
$\langle \psi^\dagger \psi \psi \rangle$	1	11	150	$\to$ 98	$\to$ 76	$\to$ 74

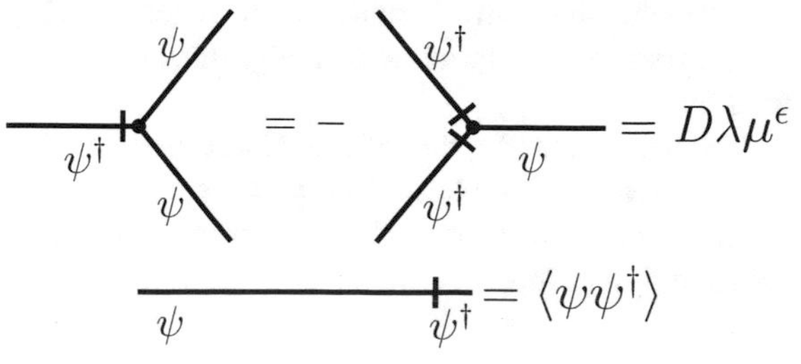

Fig. 1 Interaction vertices and the propagator of DP process

Fig. 2 Two-loop diagram: (left) model A, (middle) model A without external tails (right) vertex diagram DP proces

lent to an inclusion of the unit vertex on every internal line with ω. The self-energy diagram is then replaced by some sum of diagram $\langle \psi^\dagger \psi \psi \rangle$ and contribution to quantities f_1 or Z_1 is replaced by vertex diagram. Further, number of diagrams is reduced by a comparison of relation corresponding different path with external frequency in diagram. The calculation can be also avoided for renormalization constant Z_3 and contribution is determined by sum of vertex diagrams.

The vertex diagrams can be reduced using knowledge of the model A of critical dynamics [6]. The recipe is shown on Fig. 2. The external tails are cut off in diagram for model A and line $\langle \psi \psi \rangle$ is replaced by propagators $\langle \psi^\dagger \psi \rangle \langle \psi \psi^\dagger \rangle$ or $\langle \psi \psi^\dagger \rangle \langle \psi^\dagger \psi \rangle$. New diagrams have the same contribution and this way they can be found groups of diagram with same contribution in DP model (Table 1, column - model A).

In the second step symbolic expressions for all independent Feynman diagrams are constructed. Final expressions are calculated using time-momentum representation for vertex diagram (contribution to f_4, Z_4) and self-energy diagram (contribution only f_2, Z_2). Unfortunately, in first method the quantity f_3 can not be determined, but renormalization constant Z_3 can be calculated in second method.

The last step consists in a numerical calculation of a integrand by modified Monte Carlo methods, which is created on integration of multidimensional space. The method was used on integration function after application $\mathcal{R}$ operation and integrand of schematic form

$$I = S \prod_{\substack{i,j \\ i \neq j}}^{3} \int_0^1 dk_i d\vartheta_i \{da\} J(k_i, \vartheta_{ij}) \chi(k_i, \vartheta_{ij}, \{a\}), \qquad \mathbf{k}_i \cdot \mathbf{k}_j = k_i k_j \cos(\vartheta_{ij}), \quad (14)$$

where k_j is magnitude of the momentum vector, ϑ_{ij} is angle between momentum vectors, $\{a\}$ is a set of parameters that stretches moments flowing into subgraphs, J is the Jacobian determinant and S denotes such parameter, which is not a part of integration. The number of integration variables depends on a given Feynman diagram.

The main idea of quasi-Monte Carlo methods lies in a replacement of a pseudo-random numbers by a low-discrepancy sequences. Integration is the same as in case of Monte Carlo method and that approximation a integrand $f(u)$ by average value function f in points $x_1, \ldots, x_N$

$$\int\limits_{[0,1]^d} du f(u) \approx \frac{1}{N} \sum_{n=1}^{N} f(x_n), \tag{15}$$

where integration is performed in d-dimensional space and every points x_i is d-dimensional vector. In modified Monte Carlo method the low-discrepancy sequence was used a Sobolov sequence [11].

On the numerical calculation was used the Vegas algorithm with new implementation [9]. A detailed description of the Vegas algorithm can be found in paper [8] and shall not be reproduce here.

The integrand is constructed in such a way that not obtain divergent part and discontinuities. On the evaluation accuracy of calculation is used standard error for Monte Carlo integration. In Fig. 3 results of numerical integration for the Feynman diagram proportional to quadratic term by first method in external momentum are displayed. Graphs proportional to p^2 have one of the most complicated structure. As can be seen in Fig. 3, results follow Gauss distribution for 100 different initial condition and 10^{10} samples. Final evaluation of Feynman diagram can be done as average from all results and can be viewed as a numerical calculation for 10^{12} samples. This

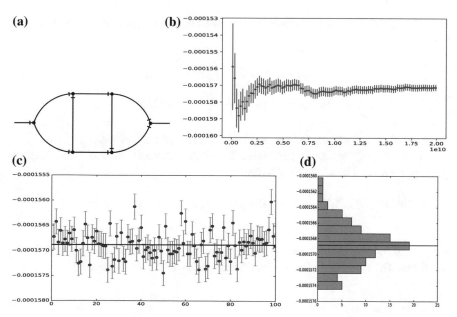

Fig. 3 The numerical evaluation of **a** the Feynman diagram to $\bar{\Gamma}_2$ by the Vegas algorithm in first method. **b** Numerical evaluation value ($\pm$ error) of Feynman diagram with increasing number of sample (10^6 points for each point). **c** Final results of the Feynman diagram generate for 100 different seeds and 10^{10} integrand evaluations in each seed. **d** Histogram of final value. The average value is $-0.000\,156\,882(25)$ for contribution Feynman diagram calculating form these seeds and corresponds solid line

decreases numerical errors[1] and better precision for numerical evaluation of contribution to the quantities f_i. Final result for f_i with included three-loop contributions of diagrams

$$f_2 = u\left(-\frac{1}{32} + \frac{\epsilon}{64} - \frac{\pi^2\epsilon^2}{768}\right) + u^2(0.006280 - 0.009428\epsilon) - 0.00377u^3, \quad (16)$$

$$f_4 = u\left(-\frac{1}{4} + \frac{\epsilon}{8} - \frac{\pi^2\epsilon^2}{96}\right) + u^2(0.11718 - 0.18628\epsilon) - 0.1198u^3, \quad (17)$$

where error is smaller than 10^{-5}. Quantities f_i can not be compared with results from analytic calculation [1]. Then the our attention is focused on the comparison dynamical critical exponents. The critical exponent η, which is related with survival probability of cluster, has following approximate expression in two-loop calculation

$$\text{analytic: } \eta \approx -\frac{\varepsilon}{6} - 0.06805\varepsilon^2,$$

$$\text{1st method: } \eta = -\frac{\varepsilon}{6} - 0.06807\varepsilon^2.$$

The dynamical exponent z, which is associated with mean square radius, have following form

$$\text{analytic } z \approx 2 - \frac{\varepsilon}{12} - 0.02920\varepsilon^2, \quad (18)$$

$$\text{1st method } z = 2 - \frac{\varepsilon}{12} - 0.02921\varepsilon^2. \quad (19)$$

Our numerical calculation is in agreement with analytic calculation.

The final contribution is calculated by second method for the same diagram (Fig. 3a) to Z_2 and has following value

$$-0.005208333373(29)\varepsilon^{-3} - 0.0065160472(23)\varepsilon^{-2} + 0.006946642(19)\varepsilon^{-1}, \quad (20)$$

where it is not needed to accede to evaluate the diagram in 100 different initial condition. The final value for renormalization constants have following form

---

[1]The number in brackets correspond a numerical error on last digits.

$$\text{analytic } Z_2 \to \sum^{2loop} = -0.025390625\varepsilon^{-2} + 0.0011110791\varepsilon^{-1},$$

$$\text{2nd method } Z_2 \to \sum^{2loop} = -0.02539062498(7)\varepsilon^{-2} + 0.0011110783(12)\varepsilon^{-1},$$

$$\text{analytic } Z_4 \to \sum^{2loop} = -0.3125\varepsilon^{-2} - 0.125\varepsilon^{-1},$$

$$\text{2nd method } Z_4 \to \sum^{2loop} = -0.3124999998(3)\varepsilon^{-2} - 0.124999972(47)\varepsilon^{-1},$$

where analytic results are taken from article [1]. The accuracy is better then first method.

4 Summary

The numerical calculation was carried out for the DP process up to the third order in perturbation theory by two different methods. The first method yields good agreement with analytic calculation in two-loop approximation, but in third order we have encountered problems with several diagrams: oscillation of numerical result, what brings about smaller precision and problems with extraction of divergence. The second method achieves better agreement with analytic calculation than previous method, and full three-loop calculation is still in progress.

Acknowledgements We would like to thank M. Val'a, whose expertize helped us with implementation of numerical calculation on Hybrilit cluster. The work was supported by VEGA Grant 1/0345/17 of the Ministry of Education, Science, Research and Sport of the Slovak Republic.

References

1. H.K. Janssen, U.C. Täuber, Ann. Phys. **315**, 147–192 (2004)
2. J.L. Cardy, R.L. Sugar, J. Phys. A **13**, L423–L427 (1980)
3. H. Hinrichsen, Adv. Phys. **49**, 815–958 (2000)
4. D. Mollison, Math. Biosci. **107**(2), 255–287 (1991)
5. M. Moshe, Phys. Rep. **37**(3), 255–354 (1978)
6. A.N. Vasilev, *The Field Theoretic Renormalization Group in Critical Behavior Theory and Stochastic Dynamics*, (in Russian), PIYaF, St. Petersburg (1998); English trans., (Chapman and Hall/CRC, Boca Raton, Fla, 2004)
7. L.Ts. Adzhemyan, M.V. Kompaniets, *Theor. Math. Phys.* **169**, (2011), 1450–1459, L.Ts. Adzhemyan, M.V. Kompaniets, S.V. Novikov, V.K. Sazonov, *Theor. Math. Phys.* 175(3), (2013) 719–728, L.Ts. Adzhemyan, M.V. Kompaniets, *Journal of Physics Conference Series* **523**, 012049, (2014)
8. G.P. Lepage, J. Comput. Phys. **27**(2), 192–203 (1978)

9. T. Hahn, Comput. Phys. Commun. **168**(2), 78–95 (2005)
10. L.Ts. Adzhemyan et al., *EPJ Web of Conferences*, vol. **108**, EDP Sciences, (2016)., L.Ts Adzhemyan et al., *EPJ Web of Conferences*, vol. **173**, EDP Sciences, (2018)
11. I.M. Sobol, Zh. Vychislitel'noi Matematiki i Matematicheskoi Fiz. **7**(4), 784–802 (1967)

Reduced-Order Modeling
of the Fluidic Pinball

Luc R. Pastur, Nan Deng, Marek Morzyński and Bernd R. Noack

Abstract The fluidic pinball is a geometrically simple wake flow configuration with three rotating cylinders on the vertex of an equilateral triangle. Yet, it remains physically rich enough to host a range of interacting frequencies and to allow testing of control laws within minutes on a laptop. The system has multiple inputs (the three cylinders can independently rotate around their axis) and multiple outputs (downstream velocity sensors). Investigating the natural flow dynamics, we found that the first unsteady transition undergone by the wake flow, when increasing the Reynolds number, is a Hopf bifurcation leading to the usual time-periodic vortex shedding phenomenon, typical of cylinder wake flows, in which the mean flow field preserves axial symmetry. We extract dynamically consistent modes from the flow data in order to built a reduced-order model (ROM) of this flow regime. We show that the main dynamical features of the primary Hopf bifurcation can be described by a non-trivial lowest-order model made of three degrees of freedom.

Keywords Fluid mechanics · Flow control · Reduced-order modeling · Transition to chaos

L. R. Pastur (✉) · N. Deng
IMSIA — ENSTA ParisTech, 828 Bd des Maréchaux, 91120 Palaiseau, France
e-mail: luc.pastur@ensta-paristech.fr

N. Deng · B. R. Noack
LIMSI-CNRS, Université Paris Sud, Université Paris-Saclay, 91405 Orsay, France
e-mail: nan.deng@ensta-paristech.fr

M. Morzyński
Institute of Combustion Engines and Basics of Machine Design, Poznań University
of Technology, 60-965 Poznań, Poland
e-mail: marek.morzynski@put.poznan.pl

B. R. Noack
Institute for Turbulence-Noise-Vibration Interaction and Control, Harbin Institute of Technology,
Shenzhen, People's Republic of China

Institut für Strömungsmechanik und Technische Akustik (ISTA),
Technische Universität Berlin, 10623 Berlin, Germany
e-mail: noack@limsi.fr

© Springer Nature Switzerland AG 2019
C. H. Skiadas and I. Lubashevsky (eds.), *11th Chaotic Modeling
and Simulation International Conference*, Springer Proceedings
in Complexity, https://doi.org/10.1007/978-3-030-15297-0_19

1   Introduction

Machine learning control (MLC) has been recently successfully applied to closed-loop turbulence control experiments for mixing enhancement [10], reduction of circulation zones [4], separation mitigation of turbulent boundary layers [5, 6], force control of a car model [7] and strongly nonlinear dynamical systems featuring aspects of turbulence control [2, 3]. In all cases, a simple genetic programming algorithm has learned the optimal control for the given cost function and out-performed existing open- and closed-loop approaches after few hundreds to few thousands test runs. Yet, there are numerous opportunities to reduce the learning time by avoiding the testing of similar control laws and to improve the performance measure by generalizing the considered control laws. In addition, running thousands of tests before converging to the optimal control law can be out-of-reach when dealing with heavy numerical simulations of the Navier-Stokes equations.

In order to further improve MLC strategies, it is therefore of the utmost importance to handle numerical simulations of the Navier-Stokes equations in flow configurations that are geometrically simple enough to allow testing of control laws within minutes on a Laptop, while being physically rich enough to host a range of complex dynamical flow regimes. With that aim in mind, Noack and Morzynski [8] proposed as an attractive flow configuration the uniform flow around 3 cylinders which can be rotated around their axis (3 control inputs), with multiple downstream velocity sensors as multiple outputs. As a standard objective, the control goal could be to stabilize the wake or reduce the drag. This configuration, proposed as a new benchmark for multiple inputs-multiple outputs (MIMO) nonlinear flow control, was named as the *fluidic pinball* as the rotation speeds allow to change the paths of the incoming fluid particles like flippers manipulate the ball of a real pinball.

With non-rotating cylinders, the steady base flow looses stability, beyond a critical value of the Reynolds number, with respect to an oscillatory vortex-shedding instability. In this flow regime, we show that the reduced order model (ROM) of lowest dimension, though still able to reproduce the dynamical features of the flow regime, has three degrees of freedom. Designing a relevant ROM to describe a complex system is a first step toward the design of winning control strategies, as ROMs both allow testing hundreds to thousands of controllers within a minute and are predictive over a finite time horizon.

2   The Fluidic Pinball

The *fluidic pinball* is made of three equal circular cylinders of radius R that are placed in parallel in a viscous incompressible uniform flow with speed U_∞. The centers of the cylinders form an equilateral triangle with side-length $3R$, symmetrically positioned with respect to the flow (see Fig. 1). The leftmost triangle vertex points upstream, while the rightmost side is orthogonal to the oncoming flow. Thus, the transverse extend of the three cylinder configuration is given by $L = 5R$. This flow is described in a Cartesian coordinate system where the x-axis points in the direction of the flow, the z-axis is aligned with the cylinder axes, and the y-axis is orthogonal to both. The origin of this coordinate system coincides with the mid-point of the rightmost bottom and top cylinder. The location is described by $\mathbf{x} = (x; y; z) = x\,\mathbf{e}_x + y\,\mathbf{e}_y + z\,\mathbf{e}_z$, where $\mathbf{e}_{x;y;z}$ are unit vectors pointing in the direction of the corresponding axes. Analogously, the velocity reads $\mathbf{u} = (u; v; w) = u\,\mathbf{e}_x + v\,\mathbf{e}_y + w\,\mathbf{e}_z$. The pressure is denoted by p and time by t. In the following, we assume a two-dimensional flow, i.e. no dependency of any flow quantity on z and vanishing spanwise velocity $w \equiv 0$. The Newtonian fluid is characterized by a constant density ρ and kinematic viscosity ν. In the following, all quantities are assumed to be non-dimensionalized with the cylinder diameter $D = 2R$, velocity U_∞ and fluid density ρ. The corresponding Reynolds number is defined as $Re_D = U_\infty D/\nu$. The Reynolds number based on the transverse length $L = 5D$ is 2.5 times larger. The computational domain extends from $x = -6$ up to $x = 20$ in the streamwise direction, and from $y = -6$ up to $y = 6$ in the crosswise direction. In these units, the cylinder axes are located at

$$
\begin{aligned}
x_F &= -\sqrt{3/2}\cos 30, & y_F &= 0, \\
x_B &= 0, & y_B &= -\sqrt{3/4}, \\
x_T &= 0, & y_T &= +\sqrt{3/4}.
\end{aligned}
$$

Here, and in the following, the subscripts 'F', 'B' and 'T' refer to the front, bottom and top cylinder.

The dynamics of the flow is governed by the incompressible Navier-Stokes equations:

$$
\frac{\partial \mathbf{u}}{\partial t} + \mathbf{u} \cdot \nabla \mathbf{u} = -\nabla p + \frac{1}{Re_D}\Delta \mathbf{u}, \tag{1}
$$

$$
\nabla \cdot \mathbf{u} = 0, \tag{2}
$$

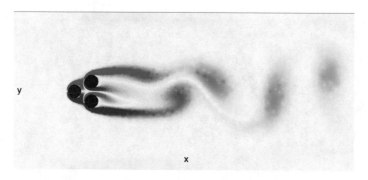

Fig. 1 Configuration of the fluidic pinball: the three cylinders are in black, the flow is coming from the left. The colormap encodes the vorticity field (arbitrary units)

where ∇ represents the Nabla operator, ∂_t and Δ denote the partial derivative and the Laplace operator. Without forcing, the boundary conditions comprise a no slip-condition ($\mathbf{u} = 0$) on the cylinder and a free-stream condition ($\mathbf{u} = \mathbf{e}_x$) in the far field. The flow can be forced by rotating the cylinders. In the forthcoming part of the paper, however, the cylinders are kept fixed.

For more details about the numerical setup and the Navier-Stokes solver, the interested reader can refer to the technical report and user manual by Noack and Morzynski [8].

3   Reduced-Order Model of the Vortex Shedding Flow Regime

The steady solution, shown in Fig. 2 for $Re_D = 10$, is stable up to the critical value $Re_c \simeq 18$ of the Reynolds number (the critical value would be about 45 in units of L). Beyond this value, the system undergoes a supercritical Hopf bifurcation characterized in the flow field by the usual vortex shedding phenomenon and generation of the von Kármán vortex street. The associated mean flow field is shown in Fig. 3 for $Re_D = 30$.

Low-dimensional and yet relevant ROMs must rely on the identification of the manifold on which the dynamics takes place. As an illustration, we consider the oscillatory flow regime observed at $Re_D = 30$. The inertial manifold hosts both the final oscillatory state and the transient dynamics to the final state. Following [9], we apply a proper orthogonal decomposition (POD) to the data set made of the fluctuating velocity field, $\mathbf{u}'(x, y; t) = \mathbf{u}(x, y; t) - \bar{\mathbf{u}}(x, y)$, where $\mathbf{u}(x, y; t)$ is the velocity flow field and $\bar{\mathbf{u}}(x, y) = \lim_{T \to \infty} 1/T \int_0^T \mathbf{u}(x, y; t) \, dt$ is the time-averaged mean flow field. The POD modes $\mathbf{u}_k(x, y)$, $k = 1, \ldots, N - 1$ with N the number of snapshots $\mathbf{u}(x, y; t)$ in the data set, provide a complete basis of orthogonal modes

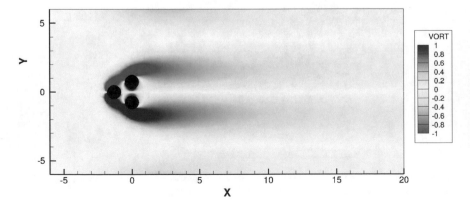

Fig. 2 Steady base flow at $Re_D = 10$. The colormap encodes the vorticity field

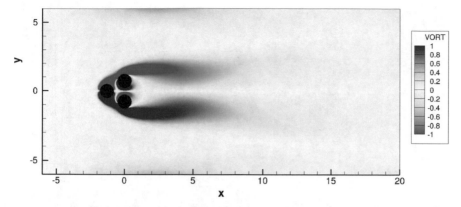

Fig. 3 Mean flow field at $Re_D = 30$. The colormap encodes the vorticity field

for the decomposition of any flow field in the data set [1]:

$$\mathbf{u}(x, y; t) = \bar{\mathbf{u}}(x, y) + \underbrace{\sum_{k=1}^{N-1} a_k(t)\mathbf{u}_k(x, y)}_{\mathbf{u}'(x, y; t)}, \tag{3}$$

where the a_k's are the mode amplitudes of the decomposition. The two leading POD modes $\mathbf{u}_{1,2}(x, y)$ are associated with the vortex shedding phenomenon, as shown in Fig. 4a, b, together with the power spectral densities of their associated time coefficients $a_1(t)$ and $a_2(t)$ in Fig. 5a, b, where a dominant peak is found at $St_D = fD/U \simeq 9 \times 10^{-2}$ ($St_L = fL/U \simeq 0.22$). Modes $\mathbf{u}_{1,2}(x, y)$, however, are associated with the final oscillatory state around the mean flow field $\bar{\mathbf{u}}(x, y)$. In order to describe the transient dynamics from the (unstable) steady solution $\mathbf{u}_s(x, y)$ to the final state, it is necessary to introduce as an additional degree of freedom the so-called shift mode $\mathbf{u}_\Delta(x, y)$ defined as $\mathbf{u}_\Delta(x, y) = \bar{\mathbf{u}}(x, y) - \mathbf{u}_s(x, y)$ and orthonormalized with respect to the leading POD modes [9]. The steady solution $\mathbf{u}_s(x, y)$ is obtained by a Netwon method and the shift mode $\mathbf{u}_\Delta(x, y)$ is shown in Fig. 4c. Following [9], let us consider the following truncated flow field:

$$\tilde{\mathbf{u}}(x, y; t) = \mathbf{u}_s(x, y) + a_\Delta(t)\mathbf{u}_\Delta(x, y) + a_1(t)\mathbf{u}_1(x, y) + a_2(t)\mathbf{u}_2(x, y), \tag{4}$$

The dynamics of a_1, a_2, a_Δ should write:

$$\begin{aligned}
\dot{a}_1 &= (\sigma - \kappa_r a_\Delta)a_1 - (\omega + \kappa_i a_\Delta)a_2, \\
\dot{a}_2 &= (\sigma - \kappa_r a_\Delta)a_2 + (\omega + \kappa_i a_\Delta)a_1, \\
\dot{a}_\Delta &= -\lambda\left(a_\Delta - \kappa_\Delta(a_1^2 + a_2^2)\right),
\end{aligned} \tag{5}$$

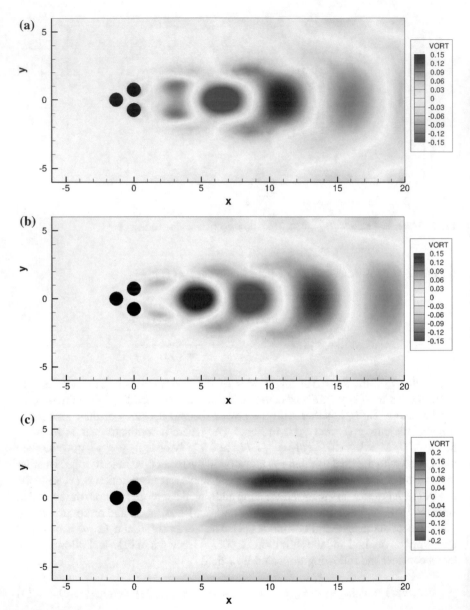

Fig. 4 First two leading POD modes **a** $\mathbf{u}_1(x, y)$, **b** $\mathbf{u}_2(x, y)$ and **c** shift mode $\mathbf{u}_\Delta(x, y)$, at $Re_D = 30$. The colormap encodes the vorticity field

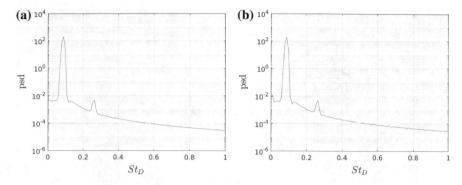

Fig. 5 Power spectral densities of **a** $a_1(t)$ and **b** $a_2(t)$

in order to account for the Hopf bifurcation normal form with triadic interaction-s between the individual modes, as imposed by the quadratic nonlinearities of the underlying Navier-Stokes equations. Identifying the coefficients of the dynamical system (5) from the transient and final flow regimes, one gets $\sigma = 4.2 \times 10^{-2}$, $\omega = 0.5$, $\kappa_r = 1.5 \times 10^{-2}$, $\kappa_i = 2.2 \times 10^{-2}$, $\kappa_\Delta = 0.2$ and $\lambda \gg 1$, slaving a_Δ to $(a_1^2 + a_2^2)$.

The dynamics of the ROM from some arbitrary initial condition to the final oscillatory state, integrated with a Runge-Kutta 4.5 numerical scheme, is compared to the dynamics of the fluidic pinball from the same initial condition, see Fig. 6. In both cases, the final oscillatory states in the phase portraits spanned by (a_1, a_2) are two limit cycles of identical amplitude. In the phase portraits spanned by (a_1, a_Δ), the parabolic shape of the manifold is identical in the two cases. This means that the inertial manifold on which the dynamics takes place is correctly identified at the leading order by our ROM. This also means that the lowest-order model able to reproduce the dynamics of the fluidic pinball, at $Re_D = 30$, has at least three degrees of freedom, namely a_1, a_2, a_Δ, the latter being slaved to the two former. Only the time scales of the transient are not perfectly reproduced, but this should be improved by introducing for instance few additional degrees of freedom or better calibrating λ.

4 Conclusions

We have considered the fluidic pinball, a newly introduced benchmark configuration for MIMO nonlinear fluid flow control, beyond its primary instability towards the vortex-shedding flow regime. We could propose a reduced-order model based on POD of at least three degrees of freedom which is able to catch the main features of the manifold on which the dynamics takes place. The degrees of freedom are the two leading POD modes, associated with the vortex shedding in the final oscillatory state, and, slaved to them, the shift mode that account for the steady solution deformation towards the mean flow field in the final state.

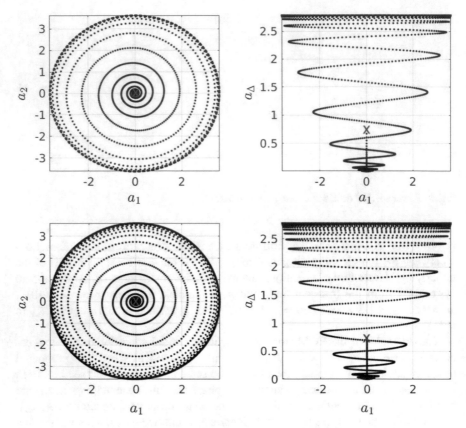

Fig. 6 Phase portraits of the ROM (top) and the fluidic pinball (bottom) from the initial condition (red cross in the figures) to the final oscillatory state (larger limit cycle)

The ROM was derived for non-rotating cylinders and for a given Reynolds number. Yet, the fluidic pinball can display a much richer spectrum of dynamical behaviors using the three cylinder rotations as free constant parameters.

Acknowledgements This work is part of a larger project involving S. Brunton, G. Cornejo Maceda, J.C. Loiseau, F. Lusseyran, R. Martinuzzi, C. Raibaudo, R. Ishar and many others.

This work is supported by the ANR-ASTRID project "FlowCon" (ANR-17-ASTR0022), by a public grant overseen by the French National Research Agency (ANR) as part of the "Investissement d'Avenir" program, ANR-11-IDEX-0003-02, and by the Polish National Science Center (NCN) under the Grant No.: DEC-2011/01/B/ST8/07264 and by the Polish National Center for Research and Development under the Grant No. PBS3/B9/34/2015.

References

1. G. Berkooz, P. Holmes, J.L. Lumley, The proper orthogonal decomposition in the analysis of turbulent flows. Ann. Rev. Fluid Mech. **25**(1), 539–575 (1993)
2. S.L. Brunton, B.R. Noack, Closed-loop turbulence control: progress and challenges. Appl. Mech. Rev. **67**(5), 050801 (2015)
3. T. Duriez, S.L. Brunton, B.R. Noack, *Machine Learning Control-Taming Nonlinear Dynamics and Turbulence* (Springer, Berlin, 2017)
4. N. Gautier, J.-L. Aider, T. Duriez, B.R. Noack, M. Segond, M. Abel, Closed-loop separation control using machine learning. J. Fluid Mech. **770**, 442–457 (2015)
5. J. Hu, Y. Zhou, Flow structure behind two staggered circular cylinders. Part 1. Downstream evolution and classification. J. Fluid Mech. **607**,51–80 (2008)
6. J. Hu, Y. Zhou, Flow structure behind two staggered circular cylinders. Part 2. Heat and momentum transport. J. Fluid Mech. **607**, 81–107 (2008)
7. R. Li, D. Barros, J. Borée, O. Cadot, B.R. Noack, L. Cordier, Feedback control of bimodal wake dynamics. Exp. Fluids **57**(10), 158 (2016)
8. B.R. Noack, M. Morzyński, The fluidic pinball—a toolkit for multiple-input multiple-output flow control (version 1.0). Technical report, Institute of Combustion Engines and Transport (Poznań University of Technology, 2017)
9. B.R. Noack, K. Afanasiev, M. Morzyński, G. Tadmor, F. Thiele, A hierarchy of low-dimensional models for the transient and post-transient cylinder wake. J. Fluid Mech. **497**, 335–363 (2003)
10. V. Parezanović, L. Cordier, A. Spohn, T. Duriez, B.R. Noack, J.-P. Bonnet, M. Segond, M. Abel, S.L. Brunton, Frequency selection by feedback control in a turbulent shear flow. J. Fluid Mech. **797**, 247–283 (2016)

Bifurcations of One-Dimensional One-Parametric Maps Revisited

Lenka Přibylová

Abstract A parameter dependent family of maps $z \mapsto f(z, \alpha)$ with a real or complex variable and parameter is studied. We deal with dynamics and bifurcations of iterates of this map in dependence on the parameter α and real bifurcations are analysed in a section of the phase-parameter complex hyperplane. Structure of bifurcation polynomials of polynomial maps will be presented on a logistic map.

Keywords Bifurcations of maps · Period-doubling · Chaos · Complex dynamics · Logistic map

1 Introduction

We consider a family of polynomial one-parameter dependent real maps

$$x \mapsto f(x, \alpha), \tag{1}$$

with parameter $\alpha \in \mathbb{R}$ and phase variable $x \in \mathbb{R}$. In particular, the logistic map

$$f(x, \alpha) = \alpha x(1 - x) \tag{2}$$

or topologically equivalent Mandelbrot map $f(x, \alpha) = x^2 + \alpha$ as well as other maps (1) generally exhibit a well-known phenomenon of a period doubling route to chaos.

L. Přibylová (✉)
Department of Mathematics and Statistics, Faculty of Science,
Masaryk University, Kotlářská 2, 611 37 Brno, Czech Republic
e-mail: pribylova@math.muni.cz

© Springer Nature Switzerland AG 2019
C. H. Skiadas and I. Lubashevsky (eds.), *11th Chaotic Modeling and Simulation International Conference*, Springer Proceedings in Complexity, https://doi.org/10.1007/978-3-030-15297-0_20

Fold or flip bifurcation points of a cycle $x_1, x_2, \ldots, x_n$ of the map (1) can be found as solutions $\alpha = \alpha^*$ of a set of equations

$$f(x_1, \alpha) = x_2,$$
$$f(x_2, \alpha) = x_3,$$
$$\vdots \qquad\qquad (3)$$
$$f(x_n, \alpha) = x_1,$$
$$f'(x_n, \alpha) \cdot \ldots \cdot f'(x_1, \alpha) = \pm 1,$$

or equivalently

$$f^{(n)}(x_i, \alpha) = x_i,$$
$$(f^{(n)})'(x_i, \alpha) = \pm 1$$

for any $i \in \{1, \ldots, n\}$, with $+1$ in the fold case and -1 in the flip case. The state variables x_i can be eliminated using for example the Gröbner basis method and the fold bifurcation points can be expressed as roots of a polynomial $P_n(\alpha)$. The polynomial $P_n(\alpha)$ is usually denoted as the fold bifurcation polynomial of an n-cycle in case of $+1$, since fold bifurcation points of any n-cycle are its roots. In case that the last equation determines that the n-cycle eigenvalue is -1 instead of $+1$, we usually talk about flip bifurcation polynomial of an n-cycle. Not all zeros of $P_n(\alpha)$ are fold (or flip) bifurcation points of an n-cycle, but later we will see that all the roots are some bifurcation points, so we will use the notion bifurcation polynomial for $P_n(\alpha)$. We generally cannot specify the type of bifurcation nor the corresponding cycle period for a chosen specific root of $P_n(\alpha)$.

It is practically impossible to compute the fold (or flip) bifurcation polynomials of higher period cycles, since the polynomial degrees grow exponentially with a cycle period. The last explicitly known bifurcation polynomial for the logistic map was computed for the 8-cycle flip bifurcation by Kotsireas and Karamanos [7]. First positive points of the flip bifurcations of a 2^{k-1}-cycle of a real logistic map are usually denoted as B_k, since exactly kth flip bifurcation occurs in the row of the period doubling cascade. The mentioned point B_4 of the logistic map (8-cycle flip) is an algebraic number of degree 240. The degree of the polynomial was conjectured already in Bailey and Broadhurst [1], where authors conjectured also that $B_4(B_4 - 2)$ would be a root of a polynomial of degree 120. This was prooved using Gröbner basis method and long time-consuming computations with results presented in Kotsireas and Karamanos [7]. Later, Kotsireas and Karamanos presented more rapid computation method based on transformation to more natural variables according to the Gröbner basis elimination method for equations (3). This method implemented in program Magma computed the 120° polynomial approximately 38 min. For the next bifurcation point B_5, they conjectured that $B_5(B_5 - 2)$ is an algebraic integer of degree 32640 (see Kotsireas and Karamanos [6]). Other interesting method to deal with (3) was also suggested in Zhang [8] using cyclic polynomial basis. It is obvious that this computational problem is still opened and interesting (see Bailey et al. [2]).

Moreover, the structure of the bifurcation polynomials is related to many parts of mathematics as complex dynamics or number theory, so still it seems that deeper insight to this problem is needed and important.

This paper reveals the structure of the polynomial map bifurcation polynomials $P_n(\alpha)$ together with the particular case of the logistic map bifurcation polynomials structure. We show possibility to compute partial factors of the fold bifurcation polynomials and possibility to recognize the type of bifurcation for a chosen root. We rediscover paper Guckenheimer and McGehee [5] that presents a normal form of the n-fold bifurcation in complex domain. The flip bifurcation in $\mathbb{R}$ is the 2-fold bifurcation in complex domain. This contribution unifies notions of fold and flip bifurcations in $\mathbb{R}$ (or generally n-fold complex bifurcation referred as Guckenheimer's bifurcation) through the concept of transcritical bifurcation in $\mathbb{C}$. Generic transcritical bifurcation points of one-dimensional one-parameter holomorphic maps are given by intersections of two transversal equilibrium branches in $\mathbb{C}$. Consequently the maximum modulus principle implies the period doubling cascade in real domain and tripling or n-tupling in complex domain, so the concept of transcritical bifurcation in $\mathbb{C}$ is an illustrative tool that explains existence of the period doubling route to chaos and its universality. The stable gaps in the chaotic area are born as intersections of two n-cycle branches of f in complex domain and on the other hand qualitatively different case of the n-fold Guckenheimer's bifurcation is the transcritical bifurcation that arise as the intersection of an n-cycle branch and a fixed point branch in complex domain. Together with Sharkovsky ordering it explains the complexity of dynamics in the chaotic domain that is present behind the accumulation point of the first Feigenbaum period doubling sequence of a fixed point. Moreover, full fractal structure of the Mandelbrot hyperbolic components distribution is beautifully visible, since the order of stable bulbs and its roots is given by primitive roots of unity sequence on all boundaries of hyperbolic components.

2 Transcritical Bifurcation in $\mathbb{C}$

Let's remind the normal form of the fold bifurcation: $\varphi_{fold}(x, \alpha) = \alpha + x - x^2$, $x, \alpha \in \mathbb{R}$. The assumption $\frac{\partial \varphi_{fold}}{\partial x}(0, 0) = 1$ of unit eigenvalue is accompanied by condition $\frac{\partial \varphi_{fold}}{\partial \alpha}(0, 0) \neq 0$ that implies existence of an implicitly defined unique equilibrium branch near the bifurcation point (see Fig. 1). Transcritical bifurcation with normal form $\varphi_{trans}(x, \alpha) = x(1 + \alpha - x)$, $x, \alpha \in \mathbb{R}$, satisfies the same condition of having unit eigenvalue $\frac{\partial \varphi_{trans}}{\partial x}(0, 0) = 1$, but $\frac{\partial \varphi_{trans}}{\partial \alpha}(0, 0) = 0$ implies multiplicity of the equilibria near the bifurcation point (see Fig. 2).

The logistic map (2) evince both cases. Transcritical bifurcation of equilibrium can be found at $\alpha = 1$ since two equilibria branches $x = 0$ and $x = \frac{\alpha - 1}{\alpha}$ intersect transversally at this bifurcation point (see Fig. 3). Fold bifurcation of a 3-cycle can be checked at $\alpha^* = 1 + \sqrt{8}$ (beginning of the stable 3-cycle gap in the chaotic area, see Fig. 4), since the map $f^{(3)}(x, \alpha) = \alpha^3 x (1 - x) (\alpha x^2 - \alpha x + 1) (\alpha^3 x^4 - 2 \alpha^3 x^3 + \alpha^3 x^2 +$

Fig. 1 Folding equilibrium branch $\varphi_{fold}(x, \alpha) - x = 0$

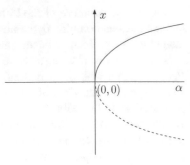

Fig. 2 Transcritical bifurcation - two branches intersection

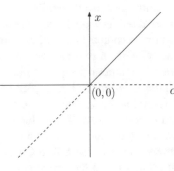

Fig. 3 Logistic map - transcritical bifurcation at $\alpha = 1$

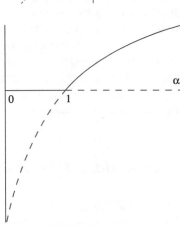

$\alpha^2 x^2 - \alpha^2 x + 1$) has one folded equilibrium branch with limit multiple fixed point (3-cycles of f) at the bifurcation value α^*. The limit point 3-cycle values are zeros of

$$343\,x^3 + \left(-49\sqrt{2} - 490\right)x^2 + \left(112\sqrt{2} + 91\right)x - 41\sqrt{2} + 31.$$

In the complex plane there is no difference between generic fold and transcritical bifurcations, since there are two complex branches of equilibria for every α in the complex neighbourhood of the bifurcation point α^*. Visualisation of this complex transcritical bifurcation is computed and plotted for the above mentioned 3-cycle

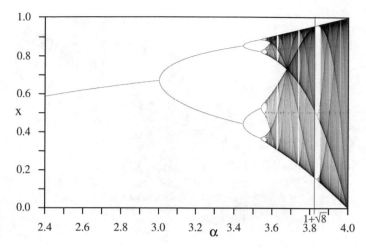

Fig. 4 Logistic map - fold bifurcation of a 3-cycle

fold of the logistic map at Fig. 5, where the two 3-cycle branches near $\alpha^* = 1 + \sqrt{8}$ are figured in 3D section of the 4D phase-parametric space. The gray real stable gap in the chaotic area is zoomed. Every real fold of an n-cycle is a transcritical bifurcation in $\mathbb{C}$, intersection of two different branches of equilibria of $f^{(n)}$ that are two different n-cycle branches of f. But intersections of equilibria of $f^{(n)}$ can happen to be intersections of an n-cycle and a fixed point (see Fig. 6). This case is known as n-fold bifurcation. Its complex normal form was derived and published by Guckenheimer and McGehee in [5] that has few citations till now. The fixed point of $f^{(n)}$ can be any d-cycle for $d|n$ (n/d-fold Guckenheimer bifurcation). For $n = 2d$ it is flip bifurcation of a d-cycle. Any period doubling cascade in real domain is a sequence of transcritical bifurcations in $\mathbb{C}$, that is a sequence of 2-fold Guckenheimer bifurcations. They are intersections of two branches of a d-cycle and a $2d$-cycle at bifurcation point α^* that belong to second primitive root of unity, that is -1 (see normal form at Guckenheimer and McGehee [5]). Due to the fact that the only real nth primitive root of unity for $n \geq 2$ is -1, only period doubling (flip) bifurcation appears in real systems (higher order Guckenheimer's bifurcations happen only in complex domain) and births of stable real cycles not connected by period doubling are exclusively real folds with eigenvalue 1 (that is births of stable gaps in the chaotic area).

Since the d-cycle branch is given by

$$f^d(z, \alpha) - z = 0,$$

implicit function theorem guaranties local function representation of the d-cycle branch and its eigenvalue $\lambda(\alpha)$ near the Guckenheimer's bifurcation value $\alpha = \alpha^*$, since the eigenvalue $(f^d)'(z(\alpha^*), \alpha^*) \neq 1$. This is true for any n-fold Guckenheimer's bifurcation, $n \geq 2$. The function $\lambda(\alpha)$ is holomorphic near $\alpha = \alpha^*$ and $|\lambda(\alpha^*)| = 1$.

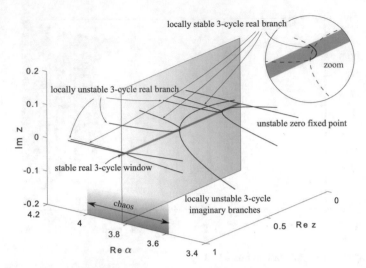

Fig. 5 Fold bifurcation of a 3-cycle of the logistic map (2) as transcritical bifurcation in $\mathbb{C}$. The section of the 4D space is at Im $\alpha = 0$

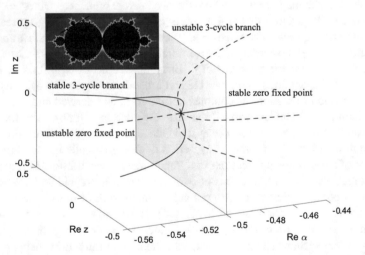

Fig. 6 3-fold Guckenheimer's bifurcation of the zero equilibrium of the logistic map (2). The section of the 4D space is at $\alpha^* = -\frac{1}{2} + \frac{\sqrt{3}}{2}i$ orthogonally to the stable bulb boundary (under the angle Arg $\alpha = \frac{2\pi}{3}$)

According to the maximum modulus principle, there have to be nearby values α such that $|\lambda(\alpha)| > 1$ and also nearby values α such that $|\lambda(\alpha)| < 1$ in each neighbourhood of α^*. Due to this and due to the proof of the Mandelbrot conjecture in Guckenheimer and McGehee [5] relating the eigenvalue derivatives on the two branches, this bifurcation point give birth to a stable d-cycle bulb that is tangent to a stable n-cycle bulb (hyperbolic component of the Mandelbrot set of the map f) in case of n/d-fold

Fig. 7 Period doubling in $\mathbb{C}$

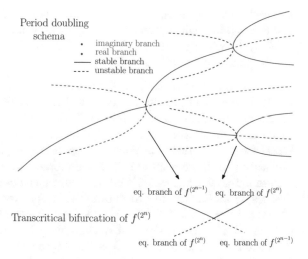

Period doubling schema

· imaginary branch
· real branch
—— stable branch
---- unstable branch

eq. branch of $f^{(2^{n-1})}$ eq. branch of $f^{(2^n)}$

Transcritical bifurcation of $f^{(2^n)}$

eq. branch of $f^{(2^n)}$ eq. branch of $f^{(2^{n-1})}$

bifurcation (where $d\,|\,n$). The ratio between moduli of eigenvalue derivatives in the n/d-fold bifurcation point is $(n/d)^2$ (see Guckenheimer and McGehee [5]), so it is 4 in case of period doubling. The 2-fold bifurcation points (see Fig. 7) give birth to a cascade of stable bulbs that are chained along the real axes, so the real flip bifurcation of any stable real n-cycle branch generally continues in period doubling cascade into its accumulation point. Intersections of $f^{(k)}$ and $f^{(l)}$ branches in other cases would give a birth to an isolated n-cycle, $n = LCM\,(k, l)$. According to provided computations, there are no such intersections in case of the logistic map, but the proof is not evident.

3   Structure

Solutions of

$$f(z_1, \alpha) = z_2,$$
$$f(z_2, \alpha) = z_3,$$
$$\vdots$$
$$f(z_n, \alpha) = z_1,$$
$$f'(z_n, \alpha) \cdot \ldots \cdot f'(z_1, \alpha) = 1$$

are n-tupples $z_1, \ldots z_n$ that are

- n-cycles with eigenvalue 1 (generic fold of an n-cycle)
- k-cycles ($k|n$) with eigenvalue 1 (fold of a k-cycle)
- n/d-cycles with eigenvalue λ: $\lambda^d = 1$ (Guckenheimer's d-fold bifurcation), where λ is a primitive dth root of unity.

Let's remind that elimination of the phase value gives the bifurcation polynomial $P_n(\alpha)$. It is evident that all the roots of the bifurcation polynomial $P_n(\alpha)$ satisfy necessary conditions for some fold or Guckenheimer's bifurcation. Moreover, for Mandelbrot or logistic map, Douady-Hubbard-Sullivan theorem implies that all Guckenheimer's bifurcation points are generic (for proof see Guckenheimer and McGehee [5]). If we want to restrict the problem to find the bifurcation points of a given d-fold Guckenheimer's bifurcation, we can use the cyclotomic polynomial that belong to dth primitive roots of unity. So for example 3-fold bifurcation point of the zero fixed point from Fig. 6 of the logistic map (2) is a root of the bifurcation polynomial $a^2 + a + 1$, since the cyclotomic polynomial that belongs to the third primitive roots of unity is $\lambda^2 + \lambda + 1$ and eigenvalue of the trivial equilibrium is $\lambda = \alpha$. Specific bifurcation points that belong to d-fold bifurcation of an n-cycle are given by a set of equations

$$f(z_1, \alpha) = z_2,$$
$$f(z_2, \alpha) = z_3,$$
$$\vdots$$
$$f(z_n, \alpha) = z_1,$$
$$C_d(f'(z_n, \alpha) \cdot \ldots \cdot f'(z_1, \alpha)) = 0,$$

or equivalently

$$f^{(n)}(z, \alpha) = z,$$
$$C_d((f^{(n)})'(z, \alpha)) = 0$$

where C_d is the dth cyclotomic polynomial. Elimination of z is possible (hypothetically always, but the computation is time-consuming for big n) and we get the bifurcation polynomial in α.

Due to Douady-Hubbard-Sullivan theorem (see for example Carleson [3] Theorem 2.1 page 134), all primitive roots of unity are mapped in order to the boundary of each hyperbolic component of the Mandelbrot set as a given dense sequence of bifurcation points (see Fig. 8), so the self-similarity of the Mandelbrot fractal is beautifully visible (see Figs. 9 and 10).

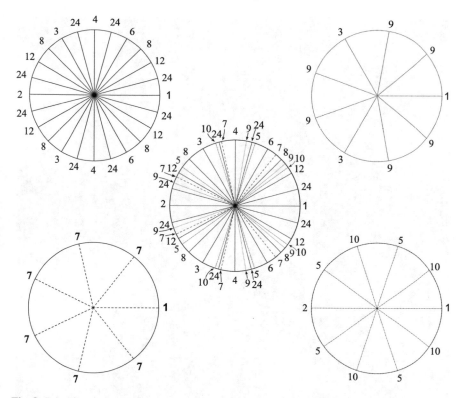

Fig. 8 Primitive roots of unity order sequence of Guckenheimer's bifurcations

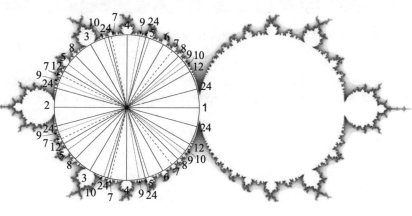

Fig. 9 Mandelbrot set of the logistic map

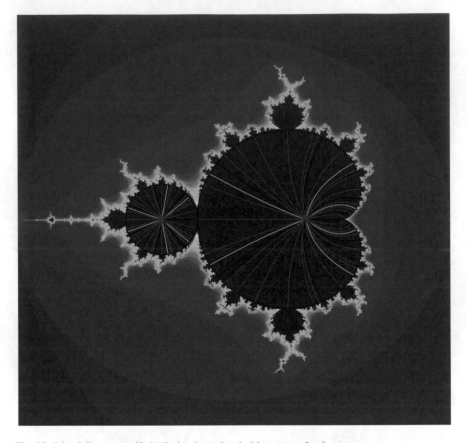

Fig. 10 Mandelbrot set self-similarity through primitive roots of unity sequence

Moreover, using Möbius inversion formula, we can find bifurcation points, that are births exactly of an n-cycle. Let's introduce the set of equations

$$f(z_1, \alpha) = z_2,$$
$$f(z_2, \alpha) = z_3,$$
$$\vdots$$
$$f(z_n, \alpha) = z_1,$$
$$(f'(z_n, \alpha) \cdot \ldots \cdot f'(z_1, \alpha)) = \lambda,$$
$$\lambda^d = 1,$$

or equivalently

$$f^{(n)}(z, \alpha) = z,$$
$$(f^{(n)})'(z, \alpha) = \lambda,$$
$$\lambda^d = 1,$$

(4)

or equivalently

$$f^{(n)}(z, \alpha) = z,$$
$$((f^{(n)})'(z, \alpha))^d = 1.$$

Since z can be eliminated from the last set of equations by Gröbner basis method, we can eliminate z from the set of equations (4). Let's denote the eliminated polynomial (monic in λ) as $\chi_n(\alpha, \lambda)$. In case of the logistic map and Mandelbrot map, χ_n are polynomials in both variables.

Theorem *Births of n-cycles of* (1) *are exactly at points of vanishing*

$$M_n(\alpha) = \prod_{\substack{dk=n \\ d|n}} \left(\prod_{j \in R(d)} \chi_k(\alpha, j) \right)^{\mu(d)},$$

where μ is Möbius function, $R(d)$ is set of all dth root of unity and χ_k are monic in λ bifurcation polynomials eliminated from (4).

As an example we reveal the structure of the 6-cycle bifurcation points for Mandelbrot map $Q_c(z) = z^2 + c$. The characteristic polynomials are as follows:

$$P_1(c, \lambda) = \lambda^2 - 2\lambda + 4c$$
$$P_2(c, \lambda) = (\lambda - 4c - 4)\left(\lambda^2 + 8c\lambda - 4\lambda + 16c^2\right)$$
$$P_3(c, \lambda) = \left(\lambda^2 - 8c\lambda - 16\lambda + 64c^3 + 128c^2 + 64c + 64\right)$$
$$\left(\lambda^2 + 24c\lambda - 8\lambda + +64c^3\right)$$
$$P_6(c, 1) = (4c + 7)(4c - 1)\left(1099511627776\,c^{20} + 10445360463872\,c^{19} + \right.$$
$$+ 44873818308608\,c^{18} + 121736553037824\,c^{17} + 245929827368960\,c^{16}$$
$$+ 399107688497152\,c^{15} + 535883874828288\,c^{14} + 617743938224128\,c^{13}$$
$$+ 631168647036928\,c^{12} + 576952972869632\,c^{11} + 484537901514752\,c^{10}$$
$$+ 376633058918400\,c^9 + 263974525796352\,c^8 + 173544017002496\,c^7$$
$$+ 104985522188288\,c^6 + 58905085704192\,c^5 + 33837528259584\,c^4$$
$$+ 15555915962496\,c^3 + 8558772746832\,c^2 + 1167105374568\,c$$
$$+ 3063651608241)(4c + 3)^2\left(16c^2 - 12c + 3\right)^2\left(16c^2 + 36c + 21\right)^2$$
$$\cdot \left(16c^2 + 4c + 7\right)^2\left(64c^3 + 128c^2 + 72c + 81\right)^2$$

and it yields that

$$\prod_{j\in R(1)} P_6(c,j) = P_6(c,1),$$

$$\prod_{j\in R(2)} P_3(c,j) = P_3(c,1) \cdot P_3(c,-1) = (4c+7)(4c+3)(4c-1)$$
$$\cdot \left(16c^2 - 12c + 3\right)\left(64c^3 + 128c^2 + 72c + 81\right)\left(16c^2 + 4c + 7\right)^2,$$

$$\prod_{j\in R(3)} P_2(c,j) = P_2(c,1) \cdot P_2(c,-1/2 + i\sqrt{3}/2) \cdot P_2(c,-1/2 - i\sqrt{3}/2)$$
$$= -(4c-1)\left(16c^2 - 12c + 3\right)\left(16c^2 + 36c + 21\right)\left(16c^2 + 4c + 7\right)(4c+3)^2,$$

$$\prod_{j\in R(6)} P_1(c,j) = (4c+3)(4c-1)\left(16c^2 + 4c + 7\right)\left(16c^2 - 12c + 3\right),$$

so the bifurcation polynomial of the birth of a 6-cycle is

$$M_6(c) = \frac{\prod_{j\in R(1)} P_6(c,j) \prod_{j\in R(6)} P_1(c,j)}{\prod_{j\in R(2)} P_3(c,j) \prod_{j\in R(3)} P_2(c,j)} = \left(16c^2 - 12c + 3\right)$$
$$\cdot \left(16c^2 + 36c + 21\right)\left(64c^3 + 128c^2 + 72c + 81\right)\left(1099511627776\, c^{20}\right.$$
$$+ 10445360463872\, c^{19} + 44873818308608\, c^{18} + 121736553037824\, c^{17}$$
$$+ 245929827368960\, c^{16} + 399107688497152\, c^{15} + 535883874828288\, c^{14}$$
$$+ 617743938224128\, c^{13} + 631168647036928\, c^{12} + 576952972869632\, c^{11}$$
$$+ 484537901514752\, c^{10} + 376633058918400\, c^9 + 263974525796352\, c^8$$
$$+ 173544017002496\, c^7 + 104985522188288\, c^6 + 58905085704192\, c^5$$
$$+ 33837528259584\, c^4 + 15555915962496\, c^3 + 8558772746832\, c^2$$
$$\left. + 1167105374568\, c + 3063651608241\right),$$

where roots are 6-folds of a fixed point, 3-folds of a 2-cycle, 2-fold (flip) of a 3-cycle and fold of a 6-cycle in the order of factors. The unique homeomorphism $H_c(z) = \frac{1}{2} - \frac{z}{\alpha}$ for $c = \frac{1}{4}\alpha(2-\alpha)$ between the Mandelbrot map and the logistic map give structured bifurcation polynomials for $f(z,\alpha) = \alpha z(1-z)$. For example, the flip of a 3-cycle gives the end point of the 3-cycle stable gap at $\alpha \doteq 3.841499008$ as one of the two real roots of

$$\alpha^6 - 6\alpha^5 + 4\alpha^4 + 24\alpha^3 - 14\alpha^2 - 36\alpha - 81 = -\left(64c^3 + 128c^2 + 72c + 81\right)$$

in agreement with Gordon [4].

References

1. D.H. Bailey, D. Broadhurst, Parallel integer relation detection: techniques and applications. Math. Comput. **70**(236), 1719–1736 (2001)
2. D.H. Bailey, J.M. Borwein, V. Kapoor, E.W. Weisstein, Ten problems in experimental mathematics. Am. Math. Mon. **113**(6), 481–509 (2006)
3. L. Carleson, T. Gamelin, *Complex Dynamics* (Springer Science & Business Media, 2013)
4. W.B. Gordon, Period three trajectories of the logistic map. Math. Mag. **69**(2), 118–120 (1996)
5. J. Guckenheimer, R. McGehee. *A proof of the Mandelbrot n^2 conjecture* (Institut Mittag-Leffler, 1984)
6. I.S. Kotsireas, K. Karamanos, Computing the bifurcation points and superstable orbits of the logistic map, www.researchgate.net https://www.researchgate.net/publication/265005019
7. I.S. Kotsireas, K. Karamanos, Exact computation of the bifurcation point B4 of the logistic map and the bailey-broadhurst conjectures. Int. J. Bifurc. Chaos **14**(7), 2417–2423 (2004)
8. Ch. Zhang, Cycles of the logistic map. Int. J. Bifurc. Chaos **24**(1), 1450005 (2014)

Memristor: Modeling and Research of Information Properties

Volodymyr Rusyn and Sviatoslav Khrapko

Abstract This paper describe about general information and properties of memristor based on Chua's scheme. The circuit of connection of the memristor to obtain I–V characteristics is presented. Practical realization and research of information properties are also presented. Memristor scheme is one of the main part in modern telecommunication systems of transmitting and receiving signals and experimental results can be used for masking and decrypt of information carrier.

Keywords Memristor · Chaos · Chua's scheme

1 Introduction

Chaos theory has been established since the 1970s due to its applications in many different research areas, such as electronic circuits [1, 2], secure communication systems [3–6], magnetism [7], economy [8] etc.

It is known that the memristor-based circuits show different types of oscillations [9–15]. The properties of the main electric circuits, constructed of three ideal elements, a resistor, a capacitor, an inductor and an ideal voltage source $v(t)$ are standard basic radio elements. These circles show a variety of phenomena such as exponential charge and discharge of the resistor-capacitor (RC) circuit with time constant $\tau_{RC} = RC$, exponential growth and current decay in the circuit of a resistor-inductor (RL) with a constant time $\tau_{RL} = L/R$—not dissipative oscillations in the circuit of inductance-capacitor (LC) with frequency $= 1/\sqrt{LC}$, as well as resonant oscillations in the circuit resistor-capacitor-inductor (RCL) induced by an alternating current source with frequency $\omega \sim \omega_{LC}$ [16].

A memristor, a capacitive memristor and a memristive inductor are all passive memory devices that can store information without a power supply. Thanks to their

V. Rusyn (✉) · S. Khrapko
Department of Radio Engineering and Information Security, Yuriy Fedkovych Chernivtsi National University, Chernivtsi, Ukraine
e-mail: rusyn_v@ukr.net

© Springer Nature Switzerland AG 2019
C. H. Skiadas and I. Lubashevsky (eds.), *11th Chaotic Modeling and Simulation International Conference*, Springer Proceedings in Complexity, https://doi.org/10.1007/978-3-030-15297-0_21

unique memory property and dynamic storage capabilities, they can be used in areas such as non-volatile storage and adaptive and spontaneous behavioral modeling [17, 18]. Much attention was paid to the use of these devices in memory in nonlinear circuits. Now, as a memory device, the memristor is widely studied in nonvolatile memory, artificial neural networks and nonlinear circuits, such as chaotic oscillators [19].

Memristors are passive two-component circuit elements that combine resistance and memory. Although in the theory of memristors, they are very promising approach to the manufacture of hardware with adaptive properties, there are only a few implementations that can reflect its basic properties [20].

Discovered that chaos can be useful in many areas; in particular, the strengthening of the mechanism of reaction kinetics in the transport of heat /mass transfer is one of the advantages. Because of the unpredictability resulting from purely deterministic systems and the combination of synchronization, chaos leads to some interesting communication applications, in particular, in the encoding of private electronic communications). Nevertheless, it has also been established that chaos may be useless in some cases where regular oscillations are required, such as metal cutting processes and power electronics, where chaos can not be avoided to improve system performance [21].

Successful manufacturing of the memristor by the Hewlett Packard (HP) [22, 23] greatly increased the interest of researchers to the memristor. Many scientists began to operate a memristor in analog systems and in digital media. Others used the memristors as resistive memory modules and logical applications. The potential of using a memristor in neuromorphic applications has also been studied by many researchers [22].

Modern computational evolution is closely linked to the development of the latest high-performance and energy-efficient devices of nanoelectronics, such as resistive random-access memory (RRAM) based on the structures of the memristor [24].

2 Memristor as Nonlinear Element

The concept of a memristor was introduced for the first time in [25] as a two-terminal element of the circuit, which connects a residual missing pair of four main variables, namely, flow and charge. It was formally defined as the fourth element of the chain [12] and named as a memristor in 1971 [10].

The memristor differs from most types of modern semiconductor memory elements because its properties are not stored as a charge. This is his main advantage, since he is not afraid of the leakage charge. Another remarkable property of the memristor is that it can accept not two positions of memory—0 and 1, but any other in the intervals between zero and one, so the commutator can work both in analog and in digital (discrete) mode. Also, the advantage of the memristor is its energy independence. This property ensures that the data stored on the memristor is as much time as there are materials used in its manufacture.

We focus our attention in this article on the practical reception of a hysteresis loop, since the hysteresis loop captures the essence of all mathematical models, as an ideal memristor as well real memristor.

One of the representative features of the memristor is the programmability of the resistance with the help of external voltage or current.

In our study, we used the KNOWM memristor BS-AF-W 16DIP series based on the tungsten substrate.

The connection of the memristor occurred according to the diagram in Fig. 1.

To achieve phase-shift operation, the device is operated under high voltage and under specified conditions. In order to obtain a device from the phase transition mode, pulse, higher return voltage and pulse melting are used. Figure 2 shows the response of the phase change to a higher applied voltages in the abrasion region.

Figure 3 shows the experimental result of phase change from the value of the applied voltage.

Practically the following characteristics are obtained, according to the connection circuit shown in Fig. 1, the memristor on Fig. 4 at the voltage of the source E at the level 1 V and frequency 30 Hz (Figs. 5, 6, 7, 8 and 9).

3   Chua's Sceme Based on the Memristor

It is known that a memristor can be used as a nonlinear element in the circuits of oscillators of chaotic signals. In our research, we used a memristor as a non-linear element in Chua's scheme.

Components with the following denominations were used in the scheme of the oscillator of chaotic oscillations based on Chua circuit:

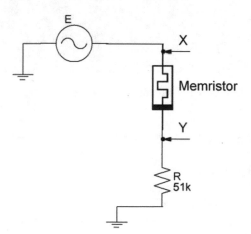

Fig. 1 The circuit of connection of the memristor to obtain I–V characteristics

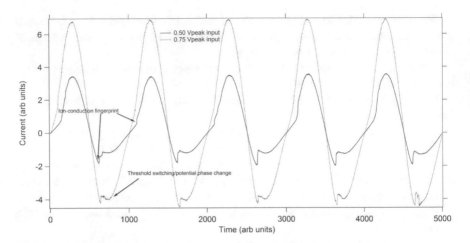

Fig. 2 Reaction of phase change from the value of the applied voltage

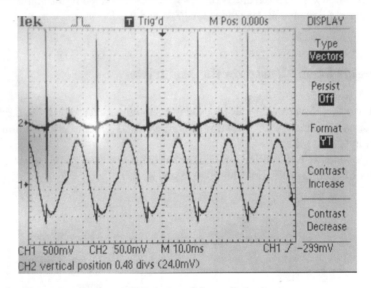

Fig. 3 Experimental phase change of the value of the applied voltage

$$L = 30\,\text{mH},\ C1 = 100\,\text{nF},\ C2 = 10\,\text{nF},\ R1 = 1,5\,\text{k}\Omega,\ R2 = 51\,\text{k}\Omega.$$

The results of the experiment:

Figure 10 shows a chaotic attractor and temporal dependence of the signal at a voltage of 0.3 V and a frequency of 20 Hz.

We also received spectral characteristics, which are presented in Fig. 11.

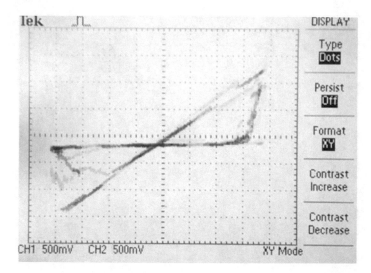

Fig. 4 I–V characteristics of the memristor at a voltage of 1 V and a frequency of 30 Hz

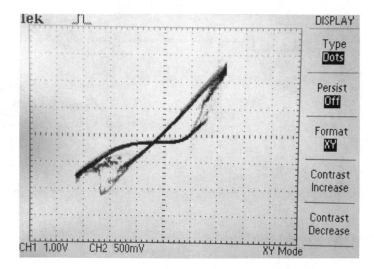

Fig. 5 I–V characteristics of the memristor at a voltage of 1 V and a frequency of 50 Hz

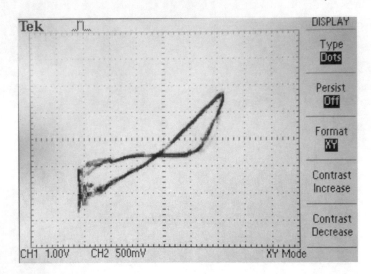

Fig. 6 I–V characteristics of the memristor at a voltage of 1 V And a frequency of 80 Hz

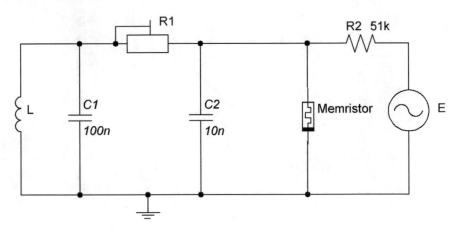

Fig. 7 Chua's scheme based on the memristor

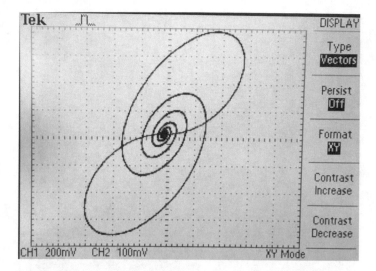

Fig. 8 Attractor appearance in saturation mode

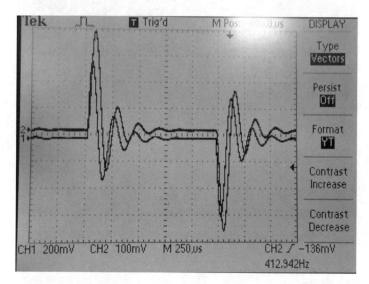

Fig. 9 The appearance of time charts in saturation mode

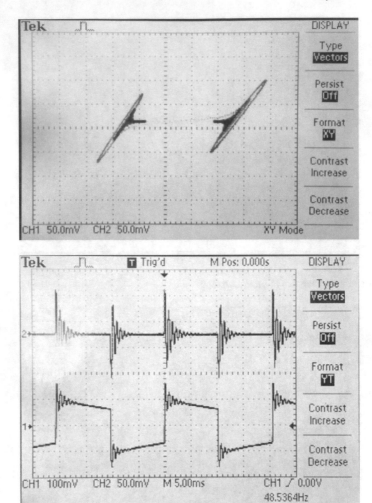

Fig. 10 Chaotic attractor and temporal dependence at $U = 0.3$ V; $f = 20$ Hz

4 Conclusions

General information and properties of memristor based on Chua's scheme, the circuit of connection of the memristor to obtain I–V characteristics is presented. Practical realization and research of information properties are also presented. Memristor scheme is one of the main part in modern telecommunication systems of transmitting and receiving signals and experimental results can be used for masking and decrypt of information carrier.

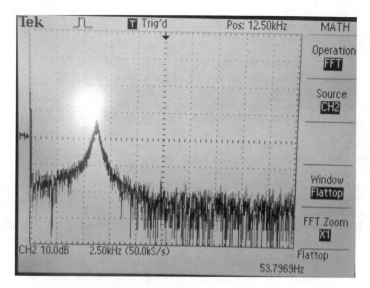

Fig. 11 Spectral characteristic of the non-autonomous oscillator of chaotic signals at the level of 15 kHz

References

1. M. Sanjaya WS, D.S Maulana, M. Mamat, Z. Salleh, Nonlinear dynamics of chaotic attractor of chua circuit and its application for secure communication. J. Oto. Ktrl. Inst (J. Auto. Ctrl. Inst), **3**(1), 1–16 (2011)
2. V. Rusyn, M. Kushnir, O. Galameiko, Hyperchaotic control by thresholding method. TCSET'2012, February 21–24, 2012, Lviv-Slavske, Ukraine, p. 67
3. A. Sambas, M. Sanjaya WS, H. Diyah, Unidirectional chaotic synchronization of rossler circuit and its application for secure communication. WSEAS Trans. Syst. **9**(11), 506–515 (2012)
4. A. Sambas, M. Sanjaya WS, M. Mamat, H. Diyah, Design and analysis bidirectional chaotic synchronization of rossler circuit and its application for secure communication. Appl. Math. Sci. **7**(1), 11–21 (2013)
5. A. Sambas, M. Sanjaya WS, M. Mamat, NV Karadimas, O. Tacha, Numerical simulations in jerk circuit and it's application in a secure communication system. Recent Advances in Telecommunications and Circuit Design. *WSEAS 17th International Conference on Communications Rhodes Island, Greece July* (2013), pp. 16–19 190–196, ISBN: 978-960-474-310-0
6. A. Sambas, M. Sanjaya WS, M. Mamat, O. Tacha, Design and numerical simulation of unidirectional chaotic synchronization and its application in secure communication system. Recent advances in nonlinear circuits: theory and applications. J. Eng. Sci. Technol. Review. **6**(4), 66–73 (2013)
7. P.P. Horley, M. Ya, Kushnir, Mishel Morales-Meza, Alexander Sukhov, Volodymyr Rusyn. Period-doubling bifurcation cascade observed in a ferromagnetic nanoparticle under the action of a spin-polarized current. Phys. B: Condens. Matter **486**, 60–63 (2016)
8. V. Rusyn, O. Savko, Modeling of chaotic behavior in the economic model. Chaotic modeling and simulation. An Int. J. Nonlinear Sci., pp. 291–298 (2016)
9. W. Marszalek, H. Podhaisky, 2D bifurcation and Newtonian properties of memristive Chua's circuits, *2016 EPL* **113**, 10002 (2016)
10. L.O. Chua, IEEE Trans. Circuit Theory **18**, 507 (1971)

11. M. Itoh, L.O. Chua, Int. J. Bifurcat. Chaos **18**, 3183 (2008)
12. L.O. Chua, Proc. IEEE **100**, 1920 (2012)
13. B.C. Bao, Z. Liu, J.P. Xu, Electron. Lett. **46**, 237 (2010)
14. Y.V. Pershin, M. Di Ventra, Adv. Phys. **60**, 145 (2011)
15. W. Marszalek, Z. Trzaska, Circuits Syst. Signal Process. **29**, 1075 (2010)
16. Yogesh N. Joglekar, Stephen J. Wolf, Eur. J. Phys. **30**, 661 (2009)
17. C.M. Jung, J.M. Choi, K.S. Min, IEEE Trans. Nanotechnol. 11611 (2012)
18. K. Eshraghian, K.R. Cho, O. Kavehei, S.K. Kang, D. Abbott, S.M. Steve, IEEE Trans. Vlsi Systa. **19**, 1407 (2011)
19. Guang-Yi Wang et al., Chinese Phys. B **25**, 090502 (2016)
20. R. Sigala et al., Modelling Simul. Mater. Sci. Eng. **21**, 075007 (2013)
21. E.B. Megam Ngouonkadi et al., Phys. Scr. **89**, 035201 (2014)
22. Changju Yang et al., Semicond. Sci. Technol. **30**, 015007 (2015)
23. D.B. Strukov, G.S. Snider, D.R. Stewart, R.S. Williams, Nature **453**, 80 (2008)
24. V.I. Avilov et al., IOP Conf. Ser.: Mater. Sci. Eng. **256**, 012001 (2017)
25. L.O. Chua, *Introduction to Nonlinear Network Theory* (McGraw-Hill, New York, 1969)

Coexistence of Chaotic and Non-chaotic Orbits in a New Nine-Dimensional Lorenz Model

B.-W. Shen, T. Reyes and S. Faghih-Naini

Abstract In this study, we present a new nine-dimensional Lorenz model (9DLM) that requires a larger critical value for the Rayleigh parameter (a rc of 679.8) for the onset of chaos, as compared to a rc of 24.74 for the 3DLM, a rc of 42.9 for the 5DLM, and a rc 116.9 for the 7DLM. Major features within the 9DLM include: (1) the coexistence of chaotic and non-chaotic orbits with moderate Rayleigh parameters, and (2) the coexistence of limit cycle/torus orbits and spiral sinks with large Rayleigh parameters. Version 2 of the 9DLM, referred to as the 9DLM-V2, is derived to show that: (i) based on a linear stability analysis, two non-trivial critical points are stable for all Rayleigh parameters greater than one; (ii) under non-dissipative and linear conditions, the extended nonlinear feedback loop produces four incommensurate frequencies; and (iii) for a stable orbit, small deviations away from equilibrium (e.g., the stable critical point) do not have a significant impact on orbital stability. Based on our results, we suggest that the entirety of weather is a superset that consists of both chaotic and non-chaotic processes.

Keywords Lorenz model · Limit cycle · Nonlinear feedback loop · Coexistence · Incommensurate frequencies · Aggregated negative feedback

1 Introduction

A view that weather is chaotic was proposed and is recognized based on the pioneering work of Prof. Lorenz [6] who first introduced the concept of deterministic chaos. Chaos is defined as the sensitive dependence of solutions on initial conditions, also known as the butterfly effect. The appearance of deterministic chaos suggests finite predictability, in contrast to the Laplacian view of deterministic predictabil-

B.-W. Shen (✉) · T. Reyes · S. Faghih-Naini
Department of Mathematics and Statistics,
San Diego State University, 5500 Campanile Drive, San Diego, CA 92182, USA
e-mail: bshen@mail.sdsu.edu
URL: http://bwshen.sdsu.edu

© Springer Nature Switzerland AG 2019
C. H. Skiadas and I. Lubashevsky (eds.), *11th Chaotic Modeling and Simulation International Conference*, Springer Proceedings in Complexity, https://doi.org/10.1007/978-3-030-15297-0_22

ity. After a follow-up study in 1972 [8], the butterfly effect has come to be known as a metaphor for indicating that a tiny perturbation such as a butterfly's flap may ultimately cause a large impact, such as the creation of a tornado. The two studies discussed above, as well as Lorenz's 1969 study [7], laid the foundation for chaos theory that is viewed as one of the three scientific achievements of the 20th century (e.g., [1, 4]), inspiring numerous studies in multiple fields. Using a comprehensive literature review (e.g., [16]), we suggested that two kinds of butterfly effects can be derived from Lorenz's studies [6–8]. The butterfly effect of the first kind (BE1) is defined as the sensitive dependence of solutions on initial conditions, while the butterfly effect of the second kind (BE2) indicates the hypothesized enabling role of tiny perturbations in producing an organized large-scale system (e.g., a tornado).

The 3DLM and high-dimensional LMs have been extensively studied over the past fifty years (e.g., [2, 9]). While some high-dimensional LMs have led to a deceasing degree of chaotic responses with better predictability (e.g., [10] and references therein), others have produced an increasing degree of chaotic responses (e.g., [11] and references therein). Within the 3DLM or a high-dimensional LM, chaotic solutions display two major features of trajectory divergence and boundedness. While the linear uncoupled geometric model (e.g., [5]) displays the role of a saddle point in producing the important feature of divergence, the limiting equations [18] and the non-dissipative Lorenz model [14] reveal the role of nonlinearity (or the nonlinear feedback loop, NFL) in producing recurrence that is responsible for the boundedness of solutions. Recurrent solutions in non-dissipative higher-dimensional LMs may be more complicated (than in lower-dimensional LMs) as a result of successive extension of the nonlinear feedback loop that produces more incommensurate frequencies (e.g., [3, 17]). On the other hand, in dissipative versions, smaller scale modes that have larger coefficients for the dissipative terms possess stronger dissipations, reducing the complexities of solutions. For example, as a result of the negative nonlinear feedback associated with the collective impact of dissipations and nonlinearity, the critical values (rc) for the onset of chaos for the 5DLM and 7DLM are 42.9 and 116.9, respectively, as compared to a rc of 24.74 for the 3DLM (e.g., [10, 12]). While negative nonlinear feedback associated with small scale processes can suppress chaotic responses (e.g., within the 5DLM of [10]), positive feedback associated with an additional heating process may appear to destabilize the system (e.g., within the 6DLM of [11]). The result suggests the importance of proper selection of new modes for improving predictability.

The 3DLM produces various types of solutions, including steady-state, chaotic, and nonlinear oscillatory solutions (i.e., limit cycle solutions) at small, moderate, and large Rayleigh parameters, respectively (e.g., [17] and references therein). More importantly, the 3DLM allows the coexistence of chaotic and non-chaotic solutions over a small range of Rayleigh parameters (i.e., $24.06 < r < 24.74$). When two types of solutions coexist within a system with the same model parameters, their appearance is solely determined by the corresponding initial conditions. Under the condition of $r < 24.74$, a linear stability analysis indicates that two non-trivial critical points are stable. Therefore, the result may indicate a relationship between the appearance of stable critical points and the coexistence of chaotic and non-chaotic solutions.

High-dimensional LMs also produce various types of solutions that appear within various ranges of parameters in different LMs. However, whether the coexistence of chaotic and non-chaotic solutions appears within high-dimensional LMs has not been investigated.

Based on the above, here, we investigate the role of the extended NFL and its collective impact with dissipations and heating using a nine-dimensional Lorenz model (9DLM). We examine: (1) the extent to which successive extension of the NFL within the 9DLM may qualitatively change the stability of non-trivial critical points as compared to those within the low-dimensional LMs, and (2) under which conditions chaotic and non-chaotic solutions may coexist. The paper is organized as follows. In Sect. 2, we present a new 9DLM, versions 1 and 2. In Sect. 3.1, we discuss the role of the NFL in producing incommensurate frequencies and providing a mechanism for energy transferring across scales. We additionally provide a mathematical analogy between the non-dissipative linearized 9DLM and a coupled system with four masses and four springs. In Sect. 3.2, we discuss stability and aggregated negative feedback within the 9DLM. In Sect. 3.3, we present numerical results for the coexistence of chaotic and non-chaotic solutions and the coexistence of limit cycle and steady-state solutions. In Sect. 3.4, we discuss the dependence of various types of solutions on the initial conditions. Concluding remarks are provided at the end. Appendix presents solutions for non-trivial critical points within the 9DLM.

2 A New Nine-Dimensional Lorenz Model (9DLM)

2.1 Governing Equations

Following derivations, the nine equations for the 9DLM are obtained:

$$\frac{dX}{d\tau} = -\sigma X + \sigma Y, \tag{1}$$

$$\frac{dY}{d\tau} = -XZ + rX - Y, \tag{2}$$

$$\frac{dZ}{d\tau} = XY - XY_1 - bZ, \tag{3}$$

$$\frac{dY_1}{d\tau} = XZ - 2XZ_1 - d_o Y_1, \tag{4}$$

$$\frac{dZ_1}{d\tau} = 2XY_1 - 2XY_2 - 4bZ_1, \tag{5}$$

$$\frac{dY_2}{d\tau} = 2XZ_1 - 3XZ_2 - d_1Y_2, \tag{6}$$

$$\frac{dZ_2}{d\tau} = 3XY_2 - 3XY_3 - 9bZ_2, \tag{7}$$

$$\frac{dY_3}{d\tau} = 3XZ_2 - 4XZ_3 - d_2Y_3, \tag{8}$$

$$\frac{dZ_3}{d\tau} = 4XY_3 - 16bZ_3. \tag{9}$$

Here, τ, σ, and r represent dimensionless time, the Prandtl number, and the normalized Rayleigh number (or the heating parameter), respectively. More detailed information may be found in [10]. $b = 4/(1 + a^2)$, $d_o = (9 + a^2)/(1 + a^2)$, $d_1 = (25 + a^2)/(1 + a^2)$, and $d_2 = (49 + a^2)/(1 + a^2)$. We referr to (Y, Z), (Y_1, Z_1), (Y_2, Z_2), and (Y_3, Z_3) as the primary, secondary, tertiary, and quaternary modes, respectively. The 9DLM is reduced to become the 7DLM when (Y_3, Z_3) are neglected, the 5DLM when (Y_2, Z_2, Y_3, Z_3) are neglected, and the 3DLM when $(Y_1, Z_1, Y_2, Z_2, Y_3, Z_3)$ are ignored. As discussed below, the 9DLM is also called the 9DLM V1 in comparison with V2. In the following, unless stated otherwise, while various values of r may be used, the remaining parameters are kept constant, including $\sigma = 10$, $a = 1/\sqrt{2}$, and $b = 8/3$.

2.2 The 9DLM Version 2 (9DLM-V2)

Using the critical point solution as the basic state, the total field (A) can be decomposed into the basic state (A_c) and perturbations (A'), i.e., $A = A_c + A'$. Here, A represents each of $(X, Y, Z, Y_1, Z_1, Y_2, Z_2, Y_3, Z_3)$. By applying this perturbation method to Eqs. (1)–(9), we may obtain the following equations:

$$\frac{dX'}{d\tau} = -\sigma X' + \sigma Y', \tag{10}$$

$$\frac{dY'}{d\tau} = (r - Z_c)X' - Y' - X_cZ' - FN(X'Z'), \tag{11}$$

$$\frac{dZ'}{d\tau} = (Y_c - Y_{1c})X' + X_cY' - bZ' - X_cY_1' + FN(X'Y' - X'Y_1'), \tag{12}$$

$$\frac{dY_1'}{d\tau} = (Z_c - 2Z_{1c})X' + X_cZ' - dY_1' - 2X_cZ_1' + FN(X'Z' - 2X'Z_1'), \tag{13}$$

$$\frac{dZ_1'}{d\tau} = (2Y_{1c} - 2Y_{2c})X' + 2X_c Y_1' - 4bZ_1' - 2X_c Y_2' + FN(2X'Y_1' - 2X'Y_2'), \quad (14)$$

$$\frac{dY_2'}{d\tau} = (2Z_{1c} - 3Z_{2c})X' + 2X_c Z_1' - d_1 Y_2' - 3X_c Z_2' + FN(2X'Z_1' - 3X'Z_2'), \quad (15)$$

$$\frac{dZ_2'}{d\tau} = (3Y_{2c} - 3Y_{3c})X' + 3X_c Y_2' - 9bZ_2' - 3X_c Y_3' + FN(3X'Y_2' - 3X'Y_3'), \quad (16)$$

$$\frac{dY_3'}{d\tau} = (3Z_{2c} - 4Z_{3c})X' + 3X_c Z_2' - d_2 Y_3' - 4X_c Z_3' + FN(3X'Z_2' - 4X'Z_3'), \quad (17)$$

$$\frac{dZ_3'}{d\tau} = 4Y_{3c}X' + 4X_c Y_3' - 16bZ_3' + FN(4X'Y_3'). \quad (18)$$

Equations (10)–(18) are referred to as the 9DLM V2. A system with $FN = 0$, which is linear with respect to the critical point, is referred to as the linearized 9DLM or the locally linear 9DLM. A system with $FN = 1$ is fully nonlinear. V2 systems with $FN = 0$ (or $FN = 1$) can describe the linear (or nonlinear) evolution of "perturbations" that depart from a non-trivial critical point. As discussed in Sects. 3.3 and 3.4, the nonlinear V2 is also capable of producing chaotic solutions. As a result, V2 systems may be used for: (1) performing a linear stability analysis with $FN = 0$, (2) revealing incommensurate frequencies of oscillatory components under non-dissipative and linear conditions, (3) illustrating the time evolution of local (oscillatory) solutions and their transition to another type of solution (e.g., chaotic or limit cycle solutions), (4) examining the impact of nonlinearity by comparing linear simulations with nonlinear simulations, and (5) tracing the "destination" of various orbits (i.e., a strange attractor or point attractor) beginning with initial conditions distributed over a hypersphere with a center at the non-trivial critical point and various radii. In this study, using the 9DLM V2, we illustrate (1–2) and (5).

3 Discussion

3.1 Incommensurate Frequencies and Scale Interaction

Previously, the appearance of oscillatory components with commensurate or incommensurate frequencies was illustrated using the 3D and 5D non-dissipative Lorenz models (NLMs, [3, 14, 17]). These studies indicated that the 3D-NLM with $r = 0$ is identical to the simplified set of the Lorenz model in [18], who applied the simplified model in order to reveal the nature of the limit cycle (i.e., oscillatory) solution. To reveal the role of the extended NFL within the 9DLM in producing additional incommensurate frequencies and transferring energy across scales, we analyze the

V2 system (i.e., Eqs. (10)–(18)) under a non-dissipative condition and provide a mathematical analogy of a coupled system containing four masses and four springs.

The non-dissipative version of the 9DLM V2 possesses the following critical point solution: $Y_c = 0$, $Y_{1c} = 0$, $Y_{2c} = 0$, $Z_c = r$, $Z_{1c} = \frac{r}{2}$, $Z_{2c} = \frac{r}{3}$, $Z_{2c} = \frac{r}{4}$. Applying the above critical point solution as the basic state to Eqs. (10)–(18) with $FN = 0$ leads to a linearized 9D non-dissipative LM (9D-NLM). From Eqs. (11)–(12) of the linearized 9D-NLM, we can obtain:

$$\frac{d^2 Y'}{d\tau^2} = -X_c \frac{dZ'}{d\tau} = -X_c^2(Y' - Y_1'). \tag{19}$$

Eqs. (12)–(14) lead to:

$$\frac{d^2 Y_1'}{d\tau^2} = X_c \frac{dZ'}{d\tau} - 2X_c \frac{dZ_1'}{d\tau} = X_c^2(Y' - 5Y_1' + 4Y_2'). \tag{20}$$

Eqs. (14)–(16) yield:

$$\frac{d^2 Y_2'}{d\tau^2} = 2X_c \frac{dZ_1'}{d\tau} - 3X_c \frac{dZ_2'}{d\tau} = X_c^2(4Y_1' - 13Y_2' + 9Y_3'). \tag{21}$$

and Eqs. (16)–(18) lead to:

$$\frac{d^2 Y_3'}{d\tau^2} = 3X_c \frac{dZ_2'}{d\tau} - 4X_c \frac{dZ_3'}{d\tau} = X_c^2(9Y_2' - 25Y_3'). \tag{22}$$

As discussed below, the above system is mathematically identical to a system containing four springs and four masses of the same weight, vertically connected. For a coupled system with spring constants $k_0 - k_3$ and masses $m_0 - m_3$, the uppermost spring that has a constant k_3 is attached to the ceiling on one end and to mass m_3 on the other end. Then, the pair of (spring, mass), (k_{2-i}, m_{2-i}) and $i \in \mathbb{Z} : i \in [0, 2]$, are sequentially attached. The lowest spring whose low end connects mass m_0 has a spring constant k_0. In Table 1, where x_j are displacements of the centers of masses

Table 1 A comparison between the non-dissipative linearized 9DLM and a coupled system containing four masses and four springs. Two systems are mathematically identical when $Y_j' = x_j$ and $k_j = (j+1)^2 X_c^2$ and $j \in \mathbb{Z} : j \in [0, 3]$, here $Y_0' = Y'$

The non-dissipative linearized 9DLM	The coupled spring-mass system
$\frac{d^2 Y'}{d\tau^2} = -X_c^2(Y' - Y_1')$	$\frac{d^2 x_0}{d\tau^2} = -k_0(x_0 - x_1)$
$\frac{d^2 Y_1'}{d\tau^2} = -4X_c^2(Y_1' - Y_2') - X_c^2(Y_1' - Y')$	$\frac{d^2 x_1}{d\tau^2} = -k_1(x_1 - x_2) - k_0(x_1 - x_0)$
$\frac{d^2 Y_2'}{d\tau^2} = -9X_c^2(Y_2' - Y_3') - 4X_c^2(Y_2' - Y_1')$	$\frac{d^2 x_2}{d\tau^2} = -k_2(x_2 - x_3) - k_1(x_2 - x_1)$
$\frac{d^2 Y_3'}{d\tau^2} = -16X_c^2 Y_3' - 9X_c^2(Y_3' - Y_2')$	$\frac{d^2 x_3}{d\tau^2} = -k_3 x_3 - k_2(x_3 - x_2)$

from equilibrium, proper selection of the spring constants leads to a coupled spring-mass system that is identical to the linearized 9D-NLM.

The above analogy indicates the role of the extended NFL in producing incommensurate frequencies and transferring energy across scales. The extended NFL increases the complexity of solutions through spatial interactions associated with the inclusion of small-scale modes. By comparison, the linearized system, which still reserves scale coupling processes associated with spatial basic state-perturbation interactions, does decrease the complexity of solutions in the temporal space. Such a linearized system, that is mathematically simpler (than its nonlinear version), is effective for revealing the role of the NFL in transferring energy across scales and in creating incommensurate frequencies. However, as discussed below, the linearized system is not suitable for studying chaotic processes that only appear in the full system with the collective impact of the extended NFL, dissipative and heating terms.

3.2 Stability and Aggregated Feedback

Here, we briefly present a linear stability analysis near a non-trivial critical point. As discussed in [10, 12], and Appendix for the time-independent 5DLM, 7DLM, and 9DLM, respectively, we can obtain the following single equation for the critical point solution X:

$$X^2 - XY_1 - b(r - 1) = 0, \tag{23a}$$

$$Y_1 = \frac{bXZT_1}{X^2 + T_2}. \tag{23b}$$

As shown in Table 2, the above three systems share the same equations as Eqs. (23a) and (23b). The only difference among the three systems is the T_2 term. The increasing complexity of T_2 in higher-dimensional LMs indicates an aggregated feedback from smaller-scale modes (see [15] for details). Using Eq. (23), we obtain the critical solution for X and then the remaining variables, denoted as X_c, Y_c, Z_c, etc., found in Appendix. Plugging the critical point solution, as the basic state, into the V2 system with $FN = 0$ (e.g., Eqs. (10)–(18)), we perform a linear stability analysis in order to

Table 2 Negative nonlinear feedback within the 5D, 7D, and 9D Lorenz models. The nonlinear feedback term is represented by the term XY_1, which is a function of T_1 and T_2

	3DLM	5DLM	7DLM	9DLM
Y_1	N/A	$\dfrac{bXZT_1}{X^2 + T_2}$	as in the 5DLM	as in the 5DLM
T_1	N/A	1	$\dfrac{2X^2 + bd_1}{X^2 + bd_1}$	$\dfrac{3X^4 + 2bX^2(d_1 + d_2) + b^2 d_1 d_2}{X^4 + (bd_2 + 2bd_1)X^2 + b^2 d_1 d_2}$
T_2	N/A	$bd_o T_1$	as in the 5DLM	as in the 5DLM

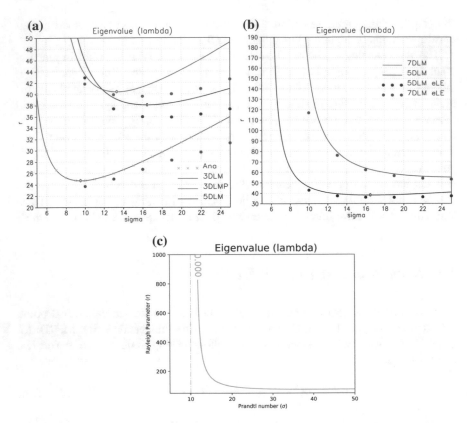

Fig. 1 Linear stability analysis near the non-trivial critical points within the 3DLM, 5DLM, 7DLM, and 9DLM, showing the leading eigenvalue $Re(\lambda)$ as a function of σ and r. The pink, black, blue, and orange solid lines indicate a constant contour of $Re(\lambda) = 0$ for the linearized 3DLM, 5DLM, 7DLM, and 9DLM, respectively. Panels (**a-b**) are from [10] and [12], respectively

reveal the stability of the non-trivial critical points. In Fig. 1, the pink, black, blue, and orange solid lines indicate a constant contour of $Re(\lambda) = 0$ for the linearized 3DLM, 5DLM, 7DLM, and 9DLM, respectively. Higher-dimensional LMs require large values of r for instability (i.e., $Re(\lambda) > 0$). More importantly, for a typical value of $\sigma = 10$, the non-trivial critical points of the linearized 3D, 5D, and 7D LMs are unstable when the Rayleigh parameters exceed critical values of 24.74, 45.94, and 160.3, respectively (e.g., [12]). The non-trivial critical points of the 9DLM are stable for all Rayleigh parameters greater than one.

The origin within the above model systems is a saddle point. Therefore, the 9DLM possesses one unstable critical point (at the origin) and two stable non-trivial critical points. The feature is similar to that of the 3DLM for $24.06 < r < 24.74$. Therefore, as compared to the 3DLM, the 9DLM also produces chaotic solutions that coexist with non-chaotic solutions. As shown in Table 3, the critical value for the onset of chaos is 679.8 within the 9DLM. The unique feature for the 9DLM is the coexistence

Table 3 The characteristics of various Lorenz models. Values for r_c are determined based on analyses of the ensemble Lyapunov exponents. The "Heating terms" column indicates heating terms within the corresponding LM

Model	r_c	Heating terms	References
3DLM	23.7	rX	[6]
5DLM	42.9	rX	[10]
6DLM	41.1	rX, rX_1	[11]
7DLM	116.9	rX	[12]
8DLM	103.4	rX, rX_1	[13]
9DLM	102.9	rX, rX_1, rX_2	[13]
9DLM (new)	679.8	rX	This study

of chaotic and non-chaotic orbits over a wide range of Rayleigh parameters (e.g., [15]). Additionally, the 9DLM displays the coexistence of limit cycle and steady state solutions. In the following, we discuss the two kinds of attractor coexistence using numerical solutions.

3.3 Numerical Results for the Coexistence of Two Types of Orbits

In this section, we present numerical solutions using both the 9DLM V1 and V2 for the coexistence of two types of solutions. Results obtained from the V1 system are provided for "verification" of results obtained from the V2 system that is then used to perform ensemble runs, as discussed in Sect. 3.4. With the exception of the Rayleigh parameter, the other parameters are constants, including $\sigma = 10$ and $a = 1/\sqrt{2}$ and $b = 8/3$. While perturbations are computed using the V2 system, the total fields (i.e., the sum of the basic state and the perturbation) are plotted.

In Fig. 2, the left and right panels display the coexistence of non-chaotic and chaotic solutions, respectively, that were obtained using models with $r = 680$ but different initial conditions (ICs). The ICs are $(X', Y', Z', Y'_1, Z'_1, Y'_2, Z'_2, Y'_3, Z'_3) = (0, 1, 0, 0, 0, 0, 0, 0, 0) - \overrightarrow{X_c}$ and $(100, 100, 100, 100, 100, 100, 100, 100, 100) - \overrightarrow{X_c}$, respectively. Here, $\overrightarrow{X_c}$ represents a vector at the non-trivial critical point (i.e., $(X_c, Y_c, Z_c, Y_{1c}, Z_{1c}, Y_{2c}, Z_{2c}, Y_{3c}, Z_{3c})$). The first set of ICs is near the origin while the second set of ICs is away from the origin. While $(X, Y, Z) = (0, 1, 0)$ that appears in the first set of ICs has been used to generate a chaotic orbit within the 3DLM, the first set of ICs leads to a steady solution that moves toward the non-trivial critical point at $(X_c, Y_c) = (84.489, 84.489)$ within the 9DLM, as shown in the left panels. In the bottom panels, we display the time evolution of solutions from both the 9DLM V1 and V2. For the steady state solutions, both V1 and V2 produce the same results. For the chaotic solutions, both produce the same initial results (e.g., $\tau \in [0, 3]$) and

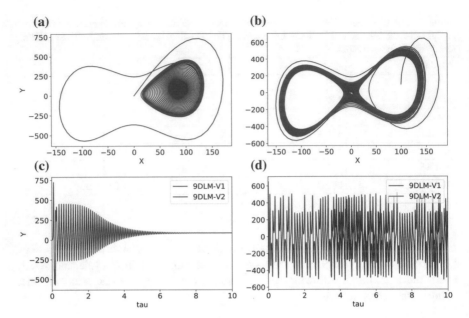

Fig. 2 Solutions of the 9DLM V2 for coexistence of non-chaotic (left panels) and chaotic (right panels) solutions. The two runs use the same model parameters (e.g., $r = 680$) but different initial conditions, $(X', Y', Z', Y_1', Z_1', Y_2', Z_2', Y_3', Z_3') = (0, 1, 0, 0, 0, 0, 0, 0, 0) - \overrightarrow{X_c}$ and $(100, 100, 100, 100, 100, 100, 100, 100, 100) - \overrightarrow{X_c}$, respectively. Here, $\overrightarrow{X_c}$ represents the non-trivial critical point (i.e., $(X_c, Y_c, Z_c, Y_{1c}, Z_{1c}, Y_{2c}, Z_{2c}, Y_{3c}, Z_{3c})$). Bottom panels compare time-varying solutions of the 9DLM V1 and V2

display diverged solutions (e.g., $\tau \in [3.8, 6.5]$), indicating the sensitive dependence of solutions on "round-off" differences in results between the V1 and V2 systems. Interestingly, both "chaotic" trajectories become overlapped again at a later time (e.g., $\tau \in [7, 8]$). Based on the above discussions, we suggest that the approach with V2 is robust for producing the coexistence of chaotic and non-chaotic solutions. As discussed in [15], such a coexistence appears over a wide range of Rayleigh parameters within the 9DLM, as compared to a small range of Rayleigh parameters within the 3DLM.

By comparison, the 9DLM possesses a second kind of coexistence for two types of solutions that has not been previously documented using Lorenz-like systems. Within the 9DLM V2 with $r = 1600$, Fig. 3 displays numerical solutions for the coexistence of limit cycle (left panels) and steady-state (right panels) solutions. Both runs have the same model parameters but different ICs, $(X', Y', Z', Y_1', Z_1', Y_2', Z_2', Y_3', Z_3')$ $= (0, 1, 0, 0, 0, 0, 0, 0, 0) - \overrightarrow{X_c}$ and $(100, 100, 1600, 100, 800, 100, 530, 100, 400)$ $- \overrightarrow{X_c}$, respectively. The first set of IC, close to the origin, produces a limit cycle solution, while the second set of IC, away from the origin, leads to a steady state solution that moves toward the non-trivial critical point at $(130.196, 130.196)$, as shown in the right panels. Since both types of solutions are stable, no visual differ-

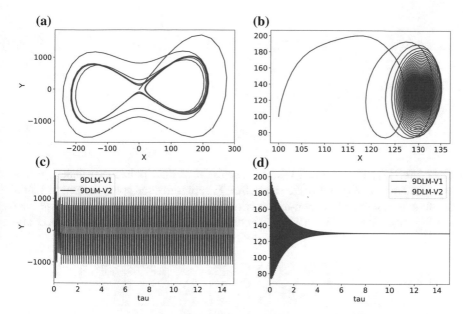

Fig. 3 Numerical solutions of the 9DLM V2 for coexistence of the limit cycle (left panels) and steady-state (right panels) solutions. The two runs use the same model parameters (e.g., $r = 1, 600$) but different initial conditions, $(X', Y', Z', Y'_1, Z'_1, Y'_2, Z'_2, Y'_3, Z'_3) = (0, 1, 0, 0, 0, 0, 0, 0, 0) - \overrightarrow{X_c}$ and $(100, 100, 1600, 100, 800, 100, 530, 100, 400) - \overrightarrow{X_c}$, respectively. $\overrightarrow{X_c}$ represents the non-trivial critical point (i.e., $(X_c, Y_c, Z_c, Y_{1c}, Z_{1c}, Y_{2c}, Z_{2c}, Y_{3c}, Z_{3c})$). Bottom panels compare time-varying solutions of the 9DLM V2 and V1

ence appears in the solutions that were obtained from the 9DLM V1 and V2 systems (e.g., the bottom panels). Once again, the above results indicate the robustness of the V2 system.

3.4 Ensemble Runs

Previously, we illustrated the dependence of two types of solutions on the initial conditions. Since non-trivial critical points are stable, it is reasonable to hypothesize that orbits beginning "near" the non-trivial critical points may move toward the critical point, at least from a statistical perspective. To verify this hypothesis, a large number of runs with different initial conditions are performed. We first apply a Gaussian random generator to produce N data points, V_{ij}, within the 9D space. Here, $i \in \mathbb{Z} : i \in [1, 9]$ indicates each of the nine variables within the phase space, while $j \in \mathbb{Z} : j \in [1, N]$ is located within the ensemble space. We then distribute these points over a hypersphere centered at a non-trivial critical point, as follows:

$$W_{ij} = R\frac{V_{ij}}{norm_j}, \quad norm_j = \sqrt{\sum_{i=1}^{9} V_{ij}^2}, \quad (24a)$$

which yields:

$$\sum_{i=1}^{9} W_{ij}^2 = R^2. \quad (24b)$$

Therefore, $W_{1j}, W_{2j}, W_{3j} \ldots, W_{8j}, W_{9j}$ represent N data points distributed over a hypersphere with a radius of R. The radius, R, represents the spatial extent of initial conditions centered at the non-trivial critical point within the phase space. We then use $W_{1j}, W_{2j}, W_{3j} \ldots, W_{8j}, W_{9j}$ as a set of ICs for $X, Y, Z, \ldots, Y_3, Z_3$, respectively.

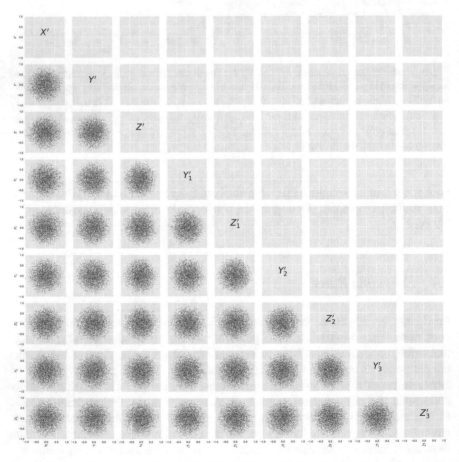

Fig. 4 A matrix of scatter plots for 4096 initial conditions distributed over the hypersphere within the 9D phase space

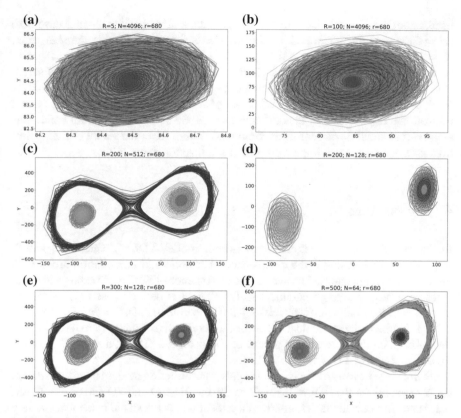

Fig. 5 Dependence of numerical simulations for chaotic and non-chaotic orbits on initial conditions (ICs). **a** 4096 ICs distributed over a hypersphere with a radius of $R = 5$; **b** 4096 ICs with a $R = 100$; **c** 512 ICs with a $R = 200$; **d** 128 ICs with a $R = 200$; **e** 128 ICs with a $R = 300$; and **f** 64 ICs with a $R = 500$. τ in [1.25, 10]

Figure 4 displays a matrix of scatter plots for 4096 initial conditions distributed over the hypersphere with $R = 1$ within the 9D phase space. In the following, we present numerical solutions for the initial conditions using various radii (i.e., R), as well as various numbers of ensemble members (i.e., N).

We begin with a large number of ensemble runs (i.e., a large N) with various ICs on a small hypersphere (i.e., a small R) to illustrate steady-state solutions. We then gradually increase the value of R and decrease the number of ensemble runs to show the coexistence of steady state and chaotic solutions. Figure 5 displays six cases for various values of (R, N) using the same Rayleigh parameter, $r = 680$, and the same period, $\tau \in [1.25, 10]$. In Fig. 5a, all of the orbits obtained from 4096 ICs distributed over a hypersphere with a radius of $R = 5$ move toward the non-trivial critical point at $(X_c, Y_c) = (84.489, 84.489)$. As shown in Fig. 5b, similar results can be found in the case with $R = 100$ where the spatial scale of 4096 ICs is larger. Therefore, the domain of the plot is larger. As shown in Fig. 5c, when a larger R ($R = 200$) is used, chaotic orbits coexist with the steady state solutions when smaller ensemble ($N = 512$) runs are performed. By further decreasing the ensemble members (i.e., $N = 128$), the case with $R = 200$ only produces steady state solutions (e.g., Fig. 5d), while the case containing a larger R ($R = 300$) possesses both chaotic and stead-state solutions (e.g., Fig. 5e). As shown in Fig. 5f, when a very large R ($R = 500$) is used, 64 ensemble runs contain both types of solutions. Ensemble simulations that indicate the dependence of chaotic and non-chaotic solutions on initial conditions suggest that small deviations away from equilibrium (e.g., the stable critical point) do not have a significant impact on orbital stability for stable orbits.

4   Conclusions

In this study, a new nine-dimensional Lorenz model (9DLM), versions 1 and 2, is presented to reveal: (1) the coexistence of chaotic and non-chaotic orbits with moderate Rayleigh parameters and (2) the coexistence of limit cycle orbits and spiral sinks with large Rayleigh parameters. The 9DLM is derived based on successive extension of the nonlinear feedback loop (NFL) within the 3D, 5D, and 7DLMs. The 9DLM requires a larger critical value for the Rayleigh parameter (a rc of 679.8) for the onset of chaos as compared to a rc of 24.74 for the 3DLM, a rc of 42.9 for the 5DLM, and a rc of 116.9 for the 7DLM.

To illustrate the role of the NFL, we apply a perturbation method in order to derive the V2 system for the 9DLM, as well as for the 3D, 5D and 7D LMs. The 9DLM V2 is used to indicate that: (i) non-trivial critical points are stable in a linear stability analysis, and (ii) four incommensurate frequencies appear under non-dissipative and linear conditions. The non-dissipative linearized V2 system is mathematically identical to a system with four masses and four springs. Within the dissipative 9DLM, since the negative nonlinear feedback associated with increasing small-scale dissipative processes can be aggregated to stabilize the system, non-trivial critical points of the 9DLM become stable for Rayleigh parameters greater than one. As compared to the 3DLM, the appearance of stable critical points is important so that the 9DLM can possess the coexistence of chaotic and non-chaotic orbits over a wider range of Rayleigh parameters. The 9DLM additionally displays the coexistence of limit cycle and steady-state solutions, a phenomenon not previously documented in Lorenz-like systems.

Applying the full version of the 9DLM V2, we additionally examine the dependence of solutions (i.e., chaotic or non-chaotic) on the spatial extent (R) and the ensemble member (N) of initial conditions distributed over a hypersphere with a center at the non-trivial critical point. As a result of the stable non-trivial critical point, 4,096 orbits beginning near the critical point (e.g., $R = 5$ or $R = 100$) move toward the critical point, being classified as non-chaotic solutions. By comparison, the probability for the appearance of chaotic solutions becomes high when initial locations are placed "away" from the critical point. For example, chaotic orbits appear for the cases with $(R, N) = (300, 128)$ and $(500, 64)$. The former displays a smaller spatial extent but a larger number of ensemble members for the initial conditions.

As discussed above, the coexistence of chaotic and non-chaotic solutions (as well as the appearance of various types of solutions) indicates that the entirety of weather is a superset that consists of both chaotic and non-chaotic processes. The revised view is contrary to the traditional view that weather is chaotic. Ensemble simulations suggest that for a stable orbit, small deviations away from equilibrium (e.g., the stable critical point) do not have a significant impact on orbit stability.

Acknowledgements We are grateful for support from the College of Science at San Diego State University.

Appendix: Solutions of the Non-trivial Critical Points

The time-independent 9DLM can be reduced as follows:

$$X^2 - XY_1 - bZ = 0, \tag{25}$$

where,

$$Z = r - 1, \tag{26}$$

$$Y_1 = \frac{bXZT_1}{X^2 + T_2}, \tag{27}$$

$$T_1 = \frac{3X^4 + 2bX^2(d_1 + d_2) + b^2 d_1 d_2}{X^4 + (bd_2 + 2bd_1)X^2 + b^2 d_1 d_2}, \tag{28}$$

and

$$T_2 = bd_0 T_1. \tag{29}$$

Equation (25) along with Eqs. (26)–(29) represents an equation for a single variable X. Therefore, we can solve Eqs. (25)–(29) for the solution of X and for the solutions of the remaining variables since they can be expressed in terms of X, as follows:

$$Y = X, \tag{30}$$

$$Z_1 = Y_1 \left(\frac{X^5 + bX^3(d_2 + 2d_1) + b^2 X d_2 d_1}{6bX^4 + 4b^2 X^2(d_2 + d_1) + 2b^3 d_2 d_1} \right), \tag{31}$$

$$Y_2 = 2XZ_1 \left(\frac{2bX^2 + b^2 d_2}{X^4 + X^2 b(d_2 + 2d_1) + b^2 d_2 d_1} \right), \tag{32}$$

$$Z_2 = 3XY_2 \left(\frac{X^2 + bd_2}{18bX^2 + 9b^2 d_2} \right), \tag{33}$$

$$Y_3 = \frac{3bXZ_2}{X^2 + bd_2},\tag{34}$$

$$Z_3 = \frac{3Z_2}{4} - \frac{d_2 Y_3}{4X}.\tag{35}$$

References

1. R. Anthes, Turning the tables on chaos: is the atmosphere more predictable than we assume?, UCAR Magazine, spring/summer, available at: https://news.ucar.edu/4505/turning-tables-chaos-atmosphere-more-predictable-we-assume (Last access: 2 April 2019), 2011
2. J.H. Curry, J.R. Herring, J. Loncaric, S.A. Orszag, Order and disorder in two- and three-dimensional Benard convection. J. Fluid. Mech. **147**, 1–38 (1984)
3. S. Faghih-Naini, B.-W. Shen, Quasi-periodic in the five-dimensional non-dissipative lorenz model: the role of the extended nonlinear feedback loop. Int. J. Bifurc. Chaos **28**(6), 1850072 (20 pages) (2018). https://doi.org/10.1142/S0218127418500724
4. J. Gleick, *Chaos: Making a New Science* (Penguin, New York, 1987), p. 360
5. J. Guckenheimer, R.F. Williams, Structural stability of lorenz attractors. Publ. Math. IHES. **50**, 59 (1979)
6. E.N. Lorenz, Deterministic nonperiodic flow. J. Atmos. Sci. **20**, 130–141 (1963)
7. E.N. Lorenz, The predictability of a flow which possesses many scales of motion. Tellus, **21**, 289–307 (1969)
8. E.N. Lorenz, Predictability: does the flap of a butterfly's wings in Brazil set off a tornado in Texas?, in *American Association for the Advancement of Science, 139th Meeting*, 29 December 1972, Boston, Mass., AAAS Section on Environmental Sciences, New Approaches to Global Weather, GARP. Available at: http://eaps4.mit.edu/research/Lorenz/Butterfly_1972.pdf, Last access: 14 Dec 2015 (1972)
9. D. Roy and Z.E. Musielak.: Generalized Lorenz models and their routes to chaos, I. Energy-conserving vertical mode truncations, Chaos Soliton. Fract., **32**, 1038–1052 (2007)
10. B.-W. Shen, Nonlinear feedback in a five-dimensional Lorenz model. J. Atmos. Sci. **71**, 1701–1723 (2014). https://doi.org/10.1175/JAS-D-13-0223.1
11. B.-W. Shen, Nonlinear feedback in a six dimensional Lorenz model: impact of an additional heating term. Nonlin. Processes Geophys. **22**, 749–764 (2015). https://doi.org/10.5194/npg-22-749-2015
12. B.-W. Shen, Hierarchical scale dependence associated with the extension of the nonlinear feedback loop in a seven-dimensional Lorenz model. Nonlin. Processes Geophys. **23**, 189–203 (2016). https://doi.org/10.5194/npg-23-189-2016
13. B.-W. Shen, On an extension of the nonlinear feedback loop in a nine-dimensional Lorenz model. Chaotic Model. Simul. (CMSIM) **2**, 147–157 (2017)
14. B.-W. Shen, On periodic solutions in the non-dissipative lorenz model: the role of the nonlinear feedback loop. Tellus A **70**, 1471912 (2018). https://doi.org/10.1080/16000870.2018.1471912
15. B.-W. Shen, Aggregated Negative Feedback in a Generalized Lorenz Model. Int. J. Bifurc. Chaos **29**(3), 1950037, (20 pages) (2019). https://doi.org/10.1142/S0218127419500378
16. B.-W. Shen, R.A. Pielke Sr., X. Zeng, S. Faghih-Naini, C.-L. Shie, R. Atlas, J.-J. Baik, T. A. L. Reyes: Butterfly effects of the first and second kinds: new insights revealed by high-dimensional lorenz models, in *The 11th Chaos International Conference (CHAOS2018)*, Rome, Italy, June 5–8, 2018

17. B.-W. Shen, S. Faghih-Naini, On recurrent solutions within high-dimensional non-dissipative Lorenz models: the role of the nonlinear feedback loop, in *The 10th Chaos Modeling and Simulation International Conference (CHAOS2017)*, Barcelona, Spain, 30 May–2 June, 2017
18. C. Sparrow, *The Lorenz Equations: Bifurcations, Chaos, and Strange Attractors* (Springer, New York. Appl. Math. Sci., 41, 1982)

Transition to Deterministic Chaos in Some Electroelastic Systems

Aleksandr Shvets and Serhii Donetskyi

Abstract The electroelastic system "generator-piezoceramic transducer" is considered, which is nonideal in the sense of Sommerfeld-Kononenko. The presence in the system of various types of regular and chaotic attractors, namely, equilibrium positions, limit cycles, invariant tori, chaotic and hyperchaotic attractors, is revealed. An atypical alternation of cascades of bifurcations of period doubling and intermittency at transitions to chaos is revealed. The so-called rare attractors are found, which are located both inside invariant tori and inside chaotic attractors.

Keywords Invariant torus · Chaotic attractor · Rare attractor · Scenario of transition to chaos

1 Introduction

One of the most important elements of modern navigation equipment are piezoceramic transducers. Different types of such transducers are widely used in depthmeters, range finders, devices for scanning underwater space, systems of transmission and reception of information under water. At recently, as a device for exciting oscillations piezoceramic transducer was again used electrolamp $LC-$ generators. This is due to the renaissance of the analog lamp generators that allow significantly higher metrological characteristics of the output signals in comparison with digital devices.

Any oscillating dynamical system, despite a huge variety of such systems, in fact consists of two basic elements. The first element is oscillating system itself, which we will name oscillating loading and second element - the any source of excitation of oscillations. All variety of existing oscillating dynamical systems can be divided

A. Shvets (✉) · S. Donetskyi
National Technical University of Ukraine Kyiv Polytechnic Institute,
37, Prospect Peremohy, Kiev 03056, Ukraine
e-mail: alex.shvets@bigmir.net

S. Donetskyi
e-mail: dsvshka@gmail.com

© Springer Nature Switzerland AG 2019
C. H. Skiadas and I. Lubashevsky (eds.), *11th Chaotic Modeling and Simulation International Conference*, Springer Proceedings in Complexity, https://doi.org/10.1007/978-3-030-15297-0_23

into two class. Ideal oscillating dynamical systems are understood as such systems, at which the energy source of oscillations has a power considerably exceeding power consumed oscillating loading. In turn systems, at which a power consumed by the oscillating loading is comparable on value with a power of the energy source of oscillations, are called as nonideal or systems with limited power-supply [1]. At mathematical modelling of nonideal systems taking into account interactions between oscillating loading and the energy source of oscillations is obligatory.

Problems of global energy-saving force to minimise maximum power of those or others sources of excitation of oscillating systems. Therefore now majority of real oscillating systems in essence should be treated as nonideal.

2 Mathematical Model and Map of Dynamic Regimes

Let us consider a piezoceramic rod transducer, which is loaded on the acoustic medium and to which electrodes the electrical voltage is affixed, raised by the LC–generator. Let's denote through $\phi(t)$ the value proportional to electric voltage of grid lamp of generator and through $V(t)$ – electric potential difference on electrodes of transducer. As shown in works [2, 3], the mathematical model of system "piezoceramic transducer – generator", can be describes by following nonlinear system of differential equations:

$$
\begin{aligned}
\ddot{\phi} + \omega_0^2 \phi &= a_1 \dot{\phi} + a_2 \dot{\phi}^2 - a_3 \dot{\phi}^3 - a_4 V(t), \\
\ddot{V}(t) + \omega_1^2 V(t) &= a_5 \phi + a_6 \dot{\phi} - a_7 \dot{V}(t).
\end{aligned}
\tag{1}
$$

Here $a_1, \ldots, a_7, \omega_0, \omega_1$ – the parameters determined through electromagnetic, geometrical and deformation properties of generator and transducer [2].

Let's introduce dimensionless variables:

$$
\xi = \frac{\phi \omega_0}{E_g}, \quad \frac{d\xi}{d\tau} = \zeta, \quad \beta = \frac{V}{E_g}, \quad \frac{d\beta}{d\tau} = \gamma, \quad \tau = \omega_0 t,
\tag{2}
$$

After that the system (1) will have the following form:

$$
\begin{aligned}
\frac{d\xi}{d\tau} &= \zeta, \\
\frac{d\zeta}{d\tau} &= -\xi + \alpha_1 \zeta + \alpha_2 \zeta^2 - \alpha_3 \zeta^3 - \alpha_4 \beta, \\
\frac{d\beta}{d\tau} &= \gamma, \\
\frac{d\gamma}{d\tau} &= -\alpha_0 \beta + \alpha_5 \xi + \alpha_6 \zeta - \alpha_7 \gamma,
\end{aligned}
\tag{3}
$$

where

$$\alpha_0 = \frac{\omega_1^2}{\omega_0^2}, \quad \alpha_1 = \frac{a_0}{\omega_0}, \quad \alpha_2 = \frac{a_2 E_g}{\omega_0}, \quad \alpha_3 = \frac{a_3 E_g^2}{\omega_0},$$

$$\alpha_4 = \frac{a_4}{\omega_0}, \quad \alpha_5 = \frac{a_5}{\omega_0^3}, \quad \alpha_6 = \frac{a_6}{\omega_0^2}, \quad \alpha_7 = \frac{a_7}{\omega_0}.$$

As the system (3) is a nonlinear system of differential equations of fourth order all its researches were carried out by means of various numerical methods. The technique for such calculations was developed and described in detail in [4–6].

A very clear presentation of the dynamic behavior of the system is given by a map of dynamic regimes. On such a flat map, the coordinate axes correspond to those or other parameters of the system and areas corresponding to the different types of steady-state dynamic regimes, which are plotted on the map by the various colors. For construction a dynamic regimes map, the plane of any selected parameters of the system is split using a vertical-horizontal grid into a points with a small grid step. At each grid point of the map is numerically calculated the spectrum of the Lyapunov's characteristic exponents. The type of the dynamic regime of the system is identified on the basis of the study the signature of the spectrum of Lyapunov's characteristic exponents. After identifying the type of dynamic regime in all grid points of the plane parameters, the corresponding pixel of the computer screen is assigned a certain color code. As a result, we get a multicolor map on the computer screen, which gives a visual representation about the type of steady-state dynamic regimes at changing the parameters of system.

Let the paramemeters of system (3) be equal to $\alpha_0 = 0.995$, $\alpha_1 = 0.04$, $\alpha_4 = 0.103$, $\alpha_5 = -0.0604$, $\alpha_6 = -0.12$, $\alpha_7 = 0.01$. In Fig. 1, a map of the dynamic regimes of the system for the parameters α_3 and α_2 is constructed. The regions of the limit cycles are plotted in black, the invariant toruses are red, the chaotic attractors are violet, the hyperchaotic attractors are white, and the equilibrium positions are blue. Signatures of the spectrum of the Lyapunov's characteristic exponents have the form $\langle 0, -, -, - \rangle$ for limit cycles, $\langle 0, 0, -, - \rangle$ – for invariant toruses, $\langle +, 0, -, - \rangle$ for chaotic attractors, and $\langle +, +, 0, - \rangle$ for hyperchaotic attractors. We emphasize that hyperchaotic attractors have two positive Lyapunov's exponents. The signature of the spectrum for equilibrium positions has the form $\langle -, -, -, - \rangle$. Especially carefully the identification of the types of the steady-state regimes should be carried out in the neighborhood of the boundaries of the existence of regimes of various types. It is necessary in addition to analyze phase portraits, Poincare sections and Fourier–spectrum of the attractors of the system. As can be seen from this map in the system "generator–piezoceramic transducer" there are dynamic steady-state regimes of absolutely all types. Particularly we emphasize that rather large areas on this map are occupied by chaotic and, in particular, hyperchaotic regimes.

Fig. 1 Map of dynamic
regimes

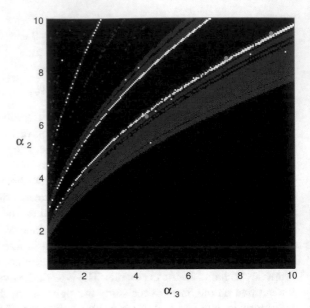

3 Atypical Transition to Chaos and Rare Attractors

As is known [7], the main scenarios for the transition to chaos in dynamic systems are the Feigenbaum's scenario and the Pomeau–Manneville's scenario. In the first scenario, the transition to chaos occurs through a cascade of bifurcations of doubling the period of limit cycles, and in the second it occurs through intermittency. Moreover, the most typical is situation when at increasing (decreasing) some of the system parameters, a number of consecutive transitions to chaos are observed through a cascade of bifurcations of periods doubling of limit cycles. Conversely, with decreasing (increasing) the parameter in this case, a transition to chaos through intermittency is observed.

Figure 2 shows a fragment of the phase-parametric characteristics of the system, the so-called bifurcation tree. In this figure, splitting of individual "branches" of a tree is clearly seen when going over to chaos through the cascade of bifurcations of period doubling. Also we can see a hard transition to chaos after one bifurcation through intermittency. The study of this tree reveals some atypical situation for dynamic systems. As can be seen from Fig. 2, there is a violation of the strict sequence of transitions to chaos with increasing (decreasing) bifurcation parameter. There is a certain symmetry of scenarios of transitions to chaos. In relatively small regions of variation of the parameter α_2, transitions to chaos are observed according to the Feigenbaum's scenario both with increasing and decreasing bifurcation parameter. The same feature is inherent in the transitions to chaos in the Pomeau–Manneville's scenario.

Fig. 2 Fragment of
phase-parametric
characteristic of system

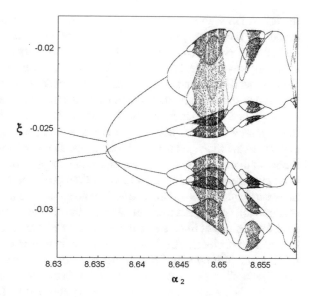

Another interesting feature of the dynamic behavior of system (3) is the presence
in it of so-called rare attractors. Recently, interest in the study of rare, hidden and
self-excited attractors has increased significantly [8, 9].

Under a rare attractor we mean an attractor that:

1. Is located inside the localization area, some other attractor;

2. The phase volume of the basin of attraction of such attractor is much smaller
than the phase volume of basin of attraction of the attractor, in which the rare attractor
is located.

For construct the rare attractors of the system (3), following algorithm can be
proposed. As is known that system (3) has a zero equilibrium position

$$\xi = 0, \ \zeta = 0, \ \beta = 0, \ \gamma = 0, \tag{4}$$

the stability conditions of which are obtained in [2, 3]. In the neighborhood of the
unstable zero equilibrium position, the initial conditions for the trajectories of system
are chosen. Next, for chosen trajectory, the spectrum of Lyapunov's characteristic
exponents (LCE) is calculated and the signature of this spectrum is determined. On the
basis of the signature of the spectrum, identification of the type of a possible attractor
is made. However, in the system (3) a paradoxical situation is possible, when there
are no zero exponents in the calculated spectrum of the LCE, for trajectories that are
not equilibrium positions. Such situation contradicts the general theory of continuous
dynamical systems [7]. Such a contradiction may indicate the appearance of a rare
attractor, the construction of which requires significantly decreasing the error of the
applied computing scheme. In our calculations we used the Runge–Kutta method of

the fifth order with a variable step of numerical integration. For selection the step of numerical integration, the corrective procedure of Prince–Dormand was applied.

In the overwhelming majority of cases, the stability of the computation of the LCE spectrum was achieved for a sufficiently large local error of order $O(10^{-4})$–$O(10^{-5})$. When revealing a paradoxical situation (absent of zero characteristic exponents), it is required to substantially increase the local accuracy of the computations to $O(10^{-7})$–$O(10^{-8})$, and also to increase the duration of numerical integration. After increasing the accuracy of numerical computations, at least one zero characteristic exponent appears in the LCE spectrum of trajectories. Thus, the contradiction with the general theory of dynamical systems disappears. Such a paradox can be explained as follows. Rare attractors have very small basins of attraction in the phase space. Therefore, in numerical calculations with insufficient accuracy, the trajectory can leave the basins of attraction of one attractor and enter the basin of attraction of another attractor. And such transutions of trajectories through the boundary of different basins of attraction can occur repeatedly. This leads to incorrect results in calculation the LCE spectrum of attractor. With a significant increase in the accuracy of numerical calculations, such "jumps" from one basin of attraction to another become impossible.

Let's illustrate the rare attractors of system (3) in few examples. In Fig. 3 the projections of phase portraits of two pairs of attractors that exist in the system (3) at $\alpha_2 = 7.847$ (a) and at $\alpha_2 = 7.95$ (b) are shown. One of the attractor of these pairs is the invariant torus (black), and the second (red) is the limit cycle. In this case, both limit cycles are rare attractors. We note, that one limit cycle is located strictly inside the invariant torus.

The next pairs of attractor that exist in the system are shown in Fig. 4. Here in Fig. 4a the projections of phase portraits of limit cycles constructed at $\alpha_2 = 8.625$. One of the limit cycles (black) is a rare attractor. It should be emphasized that a rare limit cycle has significantly smaller amplitudes of oscillations of time realizations of phase variables. Accordingly, in Fig. 4b the projections of the phase portraits of

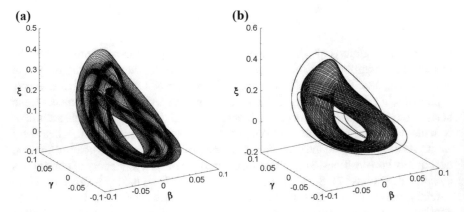

Fig. 3 Projections of phase portrait of invariant torus (black) and rare limit cycle (red) at $\alpha_2 = 7.847$ (**a**) and $\alpha_2 = 7.95$ (**b**)

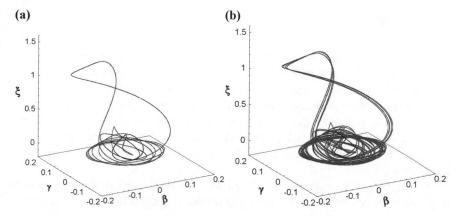

Fig. 4 Projections of phase portrait of limit cycle (red) and rare limit cycle (black) at $\alpha_2 = 8.625$ (**a**); chaotic attractor (red) and rare limit cycle (black) at $\alpha_2 = 8.647$ (**b**)

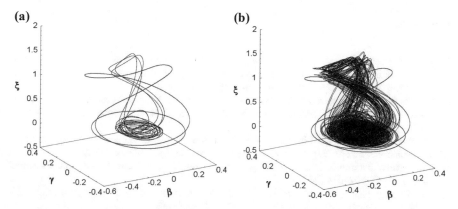

Fig. 5 Projections of phase portrait of limit cycle (black) and rare limit cycle (red) at $\alpha_2 = 8.959$ (**a**); chaotic attractor (black) and rare limit cycle (red) at $\alpha_2 = 8.96$ (**b**)

the chaotic attractor (red) and the rare limit cycle (black) at $\alpha_2 = 8.647$ are shown. A rare limit cycle lies strictly inside the chaotic attractor and has a much smaller volume of the localization region in the phase space than the chaotic attractor.

Finally, in Fig. 5 two more pairs of attractors existing in the system (3) are shown. One pair (Fig. 5a) is two limit cycles at $\alpha_2 = 8.959$, one of which (red) is a rare attractor. The second pair (Fig. 5b) is a chaotic attractor (black) and a rare limit cycle (red) at $\alpha_2 = 8.96$. In both cases, the localization region in the phase space of the rare limit cycles is comparable with the corresponding regions of localization of "not rare" attractors. However, rare attractors have significantly smaller basins of attraction. We also emphasize that the chaotic attractor shown in (Fig. 5b) has a much more complex structure than the chaotic attractor shown in (Fig. 4b).

4 Conclusion

Thus, for a deterministic dynamical system "generator-piezoceramic transducer", in some areas of the space of its parameters, the bistability is typical situation. The limit sets of trajectories of the system are pairs of attractors. One of these attractors is a rare attractor, which has a very small basin of attraction. The second of the attractors is an "ordinary" attractor, which has a large basin of attraction. There are a variety of combinations of such pairs of attractors, one of which is rare attractor. For example, "invariant torus, limit cycle", "limit cycle, limit cycle", "chaotic attractor, limit cycle".

Also in the system, a symmetry is found in the alternation of the cascade of period-doubling bifurcations and intermittency at the transitions from regular attractors to chaotic ones.

References

1. V.O. Kononenko, *Vibrating Systems with a Limited Power Supply* (Iliffe Books, London, 1969)
2. T.S. Krasnopolskaya, A.Y. Shvets, Deterministic chaos in a system generator—piezoceramic transducer. Nonlinear Dyn. Syst. Theor. **6**(4), 367–387 (2006)
3. J.M. Balthazar, J.L. Palacios Felix et al., Nonlinear interactions in a piezoceramic bar transducer powered by vacuum tube generated by a nonideal source. J. Comput. Nonlinear Dyn. **4**(1–7), 011013 (2009)
4. T.S. Krasnopolskaya, A.Y. Shvets, Chaotic oscillations of a spherical pendulum as an example of interaction with an energy source. Int. Appl. Mech. **28**(10), 669–674 (1992)
5. A.Y. Shvets, Deterministic chaos of a spherical pendulum under limited excitation. Ukrainian Math. J. **59**(4), 602–614 (2007)
6. T.S. Krasnopolskaya, A.Y. Shvets, Dynamical chaos for a limited power supply for fluid oscillations in cylindrical tanks. J. Sound Vibr. **322**(3), 532–553 (2009)
7. S.P. Kouznetsov, *Dynamic Chaos* (Physmatlit, Moscow, 2006)
8. G. Leonov, N. Kuznetsov, Nonlinear mathematical models of phase-locked loops, in *Stability and Oscillations* (Cambridge Scientific Publisher, Cambridge, 2014)
9. N. Kuznetsov, Hidden attractors in fundamental problems and engineering models. Lect. Notes Electr. Eng. **4**, 13–25 (2016)

Hyperchaos in Oscillating Systems with Limited Excitation

Aleksandr Shvets and Vasiliy Sirenko

Abstract The oscillating systems of a pendulum type, nonideal in the sense of Sommerfeld-Kononenko are considered. Such systems are used for modeling oscillations in hydrodynamics, shell theory and other applications. The complex scenario of transition to the hyperchaos is revealed and described in details. Revealed scenario begins with symmetric limit cycles and ends with a transition to hyperchaos through generalized intermittency with two coarse grained laminar phases. This scenario is illustrated in detail by projections of phase portraits, Poincaré sections and other characteristics of attractors of the system.

Keywords Limited excitation · Chaotic and hyperchaotic attractor · Poincaré section · Scenario of transition to chaos

1 Introduction

When studying the occurrence of deterministic chaos in dynamic systems, much attention is paid to describing scenarios of transitions from regular regimes to chaotic ones. Despite a huge number of mathematical models of dynamic systems, scenarios of transition to chaos in such systems can be divided into three groups. To the first group belongs the Feigenbaum scenario, in which a transition to chaos occurs through an infinite cascade of bifurcations of doubling the periods of limit cycles [1, 2]. The second group includes scenarios of transitions to chaos through intermittency by Pomeau–Manneville [3, 4]. Finally, scenarios describing the transition to chaos through the destruction of invariant tori [5–7] belong to the third group.

A. Shvets (✉) · V. Sirenko
National Technical University of Ukraine Kyiv Polytechnic Institute, 37,
Prospect Peremohy, Kyiv 03056, Ukraine
e-mail: alex.shvets@bigmir.net

V. Sirenko
e-mail: sirenko.vasiliy@ukr.net

© Springer Nature Switzerland AG 2019
C. H. Skiadas and I. Lubashevsky (eds.), *11th Chaotic Modeling
and Simulation International Conference*, Springer Proceedings
in Complexity, https://doi.org/10.1007/978-3-030-15297-0_24

265

Recently scenarios of transitions to chaos have been described, which represent various generalizations of the Pomeau-Manneville's scenarios [8–10], as well as scenarios that includes are both cascades of period doubling bifurcations and different types of intermittency [11, 12]. However, many questions concerning possible scenarios of transitions to chaos remain unclear.

2 Statement of the Problem and Mathematical Model

Any oscillatory system consists of two main subsystems, a source of excitation of oscillations and the actual vibrational load. If the power of the excitation source is comparable with the power consumed by the vibrational load, then such a system is called nonideal by Sommerfeld - Kononenko [13]. If the power of the excitation source is much higher than the power consumed by the vibrational load, then such system is called ideal.

The tasks of global energy saving make it necessary to minimize the power of the applied oscillation sources as much as possible. In connection with this, most of the real modern oscillatory systems are imperfect. At mathematical modeling of nonideal systems, it is necessary to take into account the inverse influence of the vibrational load on the functioning of the excitation source of oscillations. This leads to the fact that the differential equations of motion of a nonideal system contain additional equations in comparison with the ideal case. Note that neglecting the influence of the vibrational load on the source of excitation can lead to a loss of information about the real, both regular and chaotic, steady-state oscillation regimes [8, 9].

Consider the following system of differential equations:

$$
\begin{aligned}
\frac{dp_1}{d\tau} &= \alpha p_1 - [\beta + \frac{A}{2}(p_1^2 + q_1^2 + p_2^2 + q_2^2)]q_1 + B(p_1 q_2 - p_2 q_1)p_2; \\
\frac{dq_1}{d\tau} &= \alpha q_1 + [\beta + \frac{A}{2}(p_1^2 + q_1^2 + p_2^2 + q_2^2)]p_1 + B(p_1 q_2 - p_2 q_1)q_2 + 1; \\
\frac{d\beta}{d\tau} &= N_3 + N_1 \beta - \mu_1 q_1; \\
\frac{dp_2}{d\tau} &= \alpha p_2 - [\beta + \frac{A}{2}(p_1^2 + q_1^2 + p_2^2 + q_2^2)]q_2 - B(p_1 q_2 - p_2 q_1)p_1; \\
\frac{dq_2}{d\tau} &= \alpha q_2 + [\beta + \frac{A}{2}(p_1^2 + q_1^2 + p_2^2 + q_2^2)]p_2 - B(p_1 q_2 - p_2 q_1)q_1,
\end{aligned}
\tag{1}
$$

here $p_1, q_1, \beta, p_2, q_2$ are phase coordinates, τ is a time, $A, B, \alpha, N_1, N_3, \mu_1$ are some parameters.

As it is established in the works [8, 9, 14–17], the system of equations (1) is used to describe fluid oscillations in cylindrical tanks, for modeling oscillations of thin-walled shells, to study pendulum systems with a vibrating suspension point and a number of other topical problems of nonlinear dynamics. Depending on the applied application, the parameters $A, B, \alpha, N_1, N_3, \mu_1$ have different physical or geometric meaning. The phase variables p_1, q_1, p_2, q_2 are generalized coordinates of the vibrational subsystem, the phase variable β describes the operation of the oscillation source [8, 9].

3 Symmetric Scenario of Transition to Hyperchaos

Since the system of equations (1) is nonlinear, in the general case, a detailed and comprehensive study of its dynamics can be carried out only using various numerical and computer methods [6]. The techniques of carrying out such calculations is developed and described in [6, 7, 9, 18, 19].

Let the parameters of the system (1) have the following values:

$$A = 1.12, B = -1.531, N_3 = -1, N_1 = -1, \mu_1 = 4.024.$$

As a result of carrying out a large number of numerical computer experiments, it was possible to detect and describe a rather complex chain of scenarios for transition from regular attractors to chaotic ones.

It is established that for $-0.026 < \alpha < -0.01$ in the system there exists single-cycle limit cycles symmetric with respect to the axis $q_1 = 0$. The projections of phase portraits of limit cycles of this type are shown in Fig. 1a, b. Each of the limit cycles exists separately and has its own basin of attraction. At decreasing the value of the parameter α at $\alpha = -0.027$, the existing limit cycles lose stability and symmetric invariant tori arise in the system. In Fig. 1c, d it is shown the projections of phase portraits of symmetric invariant tori constructed at $\alpha = -0.034$. Each of the constructed invariant tori exists separately, has its own basin of attraction and arises in the neighborhood of the corresponding limit cycle, the "upper" or "lower" as a result of the Neimark's bifurcation [6].

As the parameter decreases further, at $\alpha = -0.035$, each of the symmetric tori destroys and symmetric resonance cycles appear on the tori (Fig. 2a, b). Each of the appearing resonant cycles arises on the corresponding toroidal surface and is characterized by its own basin of attraction. Although already at the value $\alpha = -0.0365$, the resonance symmetric limit cycles disappear and symmetric chaotic attractors appear in the system (Fig. 2c, d). In the spectrum of Lyapunov's characteristic exponents (LCE) of each of the obtained chaotic attractors, there is one positive exponent which indicates their chaotic nature. The emergence of each of the chaotic attractors occurs according to the classical Pomeau-Manneville intermittency scenario [3, 4]. Chaotic attractors have different basins of attraction, which have no common points.

Chaotic attractors (Fig. 2c, d) exist in a sufficiently small interval of variation of the parameter α, and already at $\alpha = -0.0405$ in the system occur the bifurcation "chaos–hyperchaos", as a result of which each chaotic attractor is destroyed and the corresponding hyperchaotic attractors appear (Fig. 3a, b). The signature of the LCE spectrum of each hyperchaotic attractor has the form $\langle +, +, 0, -, - \rangle$. The

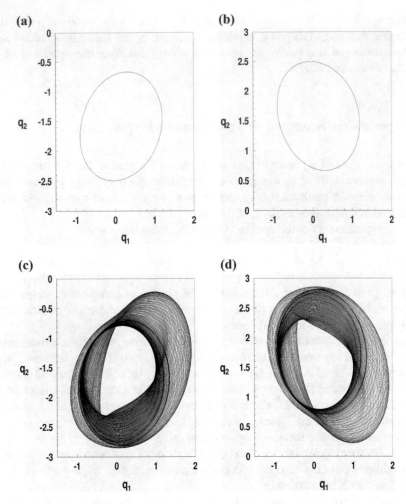

Fig. 1 Projections of phase portraits of symmetric limit cycles at $\alpha = -0.02$ (**a**, **b**) and invariant tori at $\alpha = -0.027$ (**c**, **d**)

projections of the phase portraits (Fig. 3a, b) of the constructed hyperchaotic attractors visually slightly different from the corresponding chaotic attractors (Fig. 2c, d). But the essential difference between them is that the LCE spectrum of the hyperchaotic attractor has two positive Lyapunov exponents, while the LCE spectrum of the chaotic attractor has only one positive Lyapunov exponent.

The arising symmetric hyperchaotic attractors also exist in a small interval of variation of the parameter α and already at the value $\alpha = -0.0407$ the bifurcation "hyperchaos–hyperchaos" occurs in the system, as a result of which hyperchaotic attractors (Fig. 3a, b) disappear and a new type of hyperchaotic attractor appears in the system. The projection of the phase portrait of such hyperchaotic attractor is shown

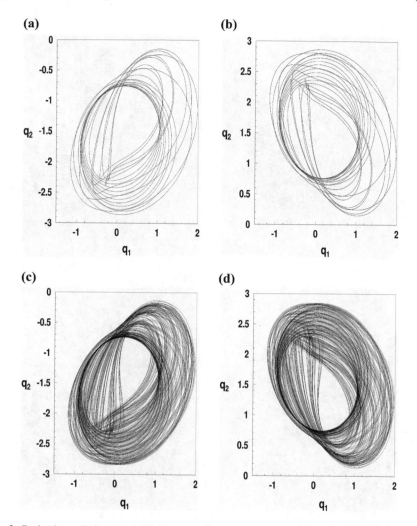

Fig. 2 Projections of phase portraits of symmetric limit cycles on tori for $\alpha = -0.035$ (**a**, **b**) and chaotic attractors for $\alpha = -0.0365$ (**c**, **d**)

in Fig. 3c. This attractor is a "glueing" of two symmetric hyperchaotic attractors that disappear after passing the bifurcation point. In Fig. 3d it is shown a fragment of a central (darker) part of the hyperchaotic attractor (Fig. 3c). This darkened part corresponds to the regions of localization of the vanished hyperchaotic attractors.

The transition "hyperhaos–hyperhaos" here occurs according to the modified scenario of generalized intermittency. Such intermittency has significant differences from the generalized intermittency described in [9, 10]. The motion of a typical trajectory of an attractor consists of three phases. Two of these phases represent chaotic motions of the trajectory in one of the darkened (upper or lower) regions

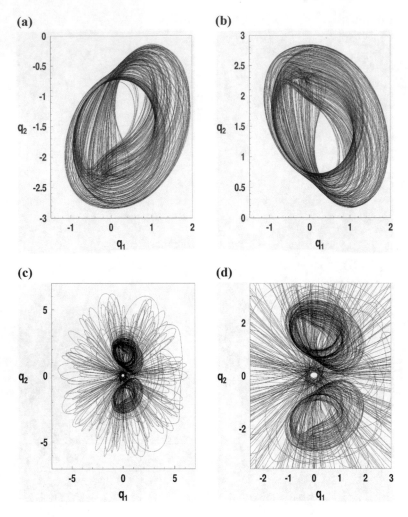

Fig. 3 Projections of phase portraits of symmetric hyperchaotic attractors at $\alpha = -0.0405$ (**a**), **b** and hyperchaotic attractor at $\alpha = -0.0407$ (**c, d**)

of Fig. 3c, d. That is, in the regions of localization of the vanishing hyper-chaotic attractors (Fig. 3a, b). Such phases of motion called coarse grained laminar. The third phase of the motion is the departure of the trajectory to the remote regions of the phase space. This phase of motion is called turbulent. Thus, the trajectory, having started motion for example in one of the coarse grained laminar phases, at an unanticipated instant of time transits into the turbulent phase. Then, again at an unpredictable moment of time it returns to one of the coarse grained laminar phases. And starting the motion in the upper darkened area of Fig. 3a, b, the trajectory, after a turbulent spike, can return both to the upper darkened area and go to the lower one. Note that the "switching" of trajectories between coarse grained laminar phases is

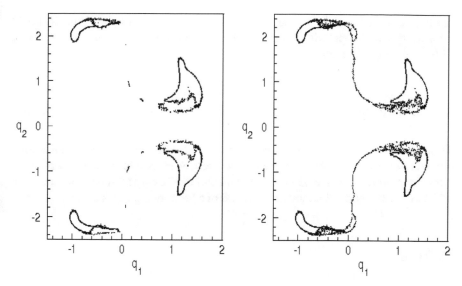

Fig. 4 Projections of Poincaré section of symmetric "small" hyperchaotic attractors at $\alpha = -0.0405$ (**a**) and "large" hyperchaotic attractor at $\alpha = -0.0407$ (**b**)

unpredictable. Such complex character of the change in the phases of the motion of the trajectory is repeated infinite number of times. Unlike the scenario of generalized intermittency described in the [9, 10] scenario, in this case there are two coarse grained laminar phases of motion. We also note that the hyperchaotic attractor that is shown in Fig. 3c has amplitudes of oscillations of phase variables that significantly exceed the amplitudes of oscillations of phase variables of hyperchaotic attractors from Fig. 3a, b. Next, for brevity, the hyperchaotic attractors shown in Fig. 3a, b will be called "small", and the hyperchaotic attractor from Fig. 3c will be called "large".

Similar peculiar properties, that is typical for intermittency, are also observed in the analysis of the Poincaré sections. In Fig. 4a the projections of Poincaré sections of symmetric "small" hyperchaotic attractors is shown. Accordingly in Fig. 4b is shown the projection of Poincaré sections of "large" hyperchaotic attractor. The plane $\beta = -1.7$ was chosen as the secant. The constructed sections are developed chaotic sets of points. The Poincare section of each of the "small" hyperchaotic attractors consists of two fragments located respectively in the upper and lower parts of Fig. 4a. All these four fragments are components of the Poincare section of the "large" hyperchaotic attractor (Fig. 4b). The two upper fragments correspond to one of the coarse grained laminar phases. Accordingly, the two lower fragments correspond to another coarse grained laminar phase. The turbulent phase of the motion of the trajectories along the attractor corresponds to points that "unite" these fragments. At the given moment of time, it is impossible to predict the location of the Poincare section point of "large" attractor in this or that its part. True, it should be noted that the trajectories of the attractor are, in average, most of the time in one of the coarse grained laminar phases.

4 Conclusion

Thus, in a deterministic pendulum system with limited excitation, hyper-chaotic attractors of various types have been discovered. A transition from a hyperchaotic attractor of one type to a hyperchaotic attractor of another type through intermittency with two coarse grained laminar phases and one turbulent phase is described.

References

1. M.J. Feigenbaum, Quantative universality for a class of nonlinear transformations. J. Stat. Phys. **19**(1), 25–52 (1978)
2. M.J. Feigenbaum, The universal metric properties of nonlinear transformations. J. Stat. Phys. **21**(6), 669–706 (1979)
3. P. Manneville, Y. Pomeau, Different ways to turbulence in dissipative dynamical systems. Physica D. Nonlinear Phenom. **1**(2), 219–226 (1980)
4. Y. Pomeau, P. Manneville, Intermittent transition to turbulence in dissipative dynamical systems. Comm. Math. Phys. **74**(2), 189–197 (1980)
5. V. Afraimovich, S.-B. Hsu, *Lectures on Chaotic Dynamical Systems* (International Press, Sommerville, 2003)
6. S.P. Kouznetsov, *Dynamic Chaos* (Physmatlit, Moscow, 2006)
7. V.S. Anishchenko, T.E. Vadivasova, *Lectures on Nonlinear Dynamics* (R&C Dynamics, Moscow-Izhevsk, 2011)
8. T.S. Krasnopolskaya, A.Y. Shvets, Chaotic surface waves in limited power-supply cylindrical tank vibrations. J. Fluids Struct. **8**(1), 1–18 (1994)
9. T.S. Krasnopolskaya, A.Y. Shvets, Dynamical chaos for a limited power supply for fluid oscillations in cylindrical tanks. J. Sound Vibr. **322**(3), 532–553 (2009)
10. A.Yu. Shvets, V.O. Sirenko, Peculiarities of transition to chaos in nonideal hydrodynamics systems. CMSIM J. **2**, 303–310 (2012)
11. A.Yu. Shvets, V.O. Sirenko, New ways of transitions to deterministic chaos in non ideal oscillating systems, in *Research Bulletin of National Technical University of Ukraine "Kyiv Polytechnic Institute"*, vol. 1, no. 99 (2015), pp. 45–51
12. A.Yu. Shvets, V. Sirenko, Complicated scenarios of transitions to deterministic chaos in non-ideal dynamic systems, in *Nonlinear Dynamics-2016 (ND-KhPI2016): Proceedings of 5th International Conference* (2016), pp. 222–229
13. V.O. Kononenko, *Vibrating Systems with a Limited Power Supply* (Iliffe Books, London, 1969)
14. R.A. Ibrahim, *Liquid Sloshing Dynamics: Theory and Applications* (Cambridge University Press, Cambridge, MA, 2005)
15. I.A. Lukovskyi, *Mathematical Models of Nonlinear Dynamics of Solids with Liquid* (Nauk. dumka, Kyiv, 2010)
16. H.A. Navarro, J.M. Balthzar et al., On nonlinear dynamics in a free fluid surface of a tank excited by non-ideal power source, in *Proceedings of the ASME International Mechanical Engineering Congress & Exposition* IMECE2011-65404 (2011). Denver, Colorado, USA

17. T.S. Krasnopol'skaya, A.Yu. Shvets, Properties of chaotic fluid oscillations in cylindrical basins. Int. Appl. Mech. **28**(6), 386–394 (1992)
18. T.S. Krasnopolskaya, A.Yu. Shvets, Chaotic oscillations of a spherical pendulum as an example of interaction with an energy source. Int. Appl. Mech. **28**(10), 669–674 (1992)
19. A.Yu. Shvets, Deterministic chaos of a spherical pendulum under limited excitation. Ukrainian Math. J. **59**(4), 602–614 (2007)

Non-linear Stability Observation Using Magneto-Controlled Diffraction with Opto-Fluidics

Adriana P. B. Tufaile, Michael Snyder, Timm A. Vanderelli and Alberto Tufaile

Abstract We have developed a magneto-optical system which simulates the stability of fixed points and the trajectories of orbits present in dynamical systems. The question of stability is significant because a real-world system is constantly subject to small perturbations, and these orbits can be observed with a Ferrocell, a device using ferrofluid, which is a superparamagnetic fluid obtained with a kind of colloid containing surfactant coated nanometer ferromagnetic particles dispersed in a carrier liquid, and this device can be used in applications of optical effects. Our magneto-optical system is based in a Hele-Shaw cell containing ferrofluid, illuminated with an external light source, such as LED. By injecting a light propagating along the in-plane direction of the liquid film, the orbits can be observed, in a way that we can bend the light. The trajectories of the orbits are obtained by the diffracted light, which consists of light patterns, and these light patterns are related to Faraday effect, linear dichroism, and linear birefringence. The diffraction pattern is different from that produced by a wire because there are no fringes in these light patterns, and the absence of well-defined spacing between the fringes indicates the existence of multiple diffraction. Under certain circumstances, these light patterns can have the same properties of the force lines of magnetic fields. The main idea of this work is to propose a device applied to non-linear systems, based on magneto-photonics. We present the patterns obtained for different magnetic fields simulating dynamical systems.

Keywords Non-linear system analysis · Ferrofluids · Multiple diffraction

A. P. B. Tufaile (✉) · A. Tufaile
Soft Matter Laboratory, Escola de Artes, Ciencias e Humanidades, Universidade de Sao Paulo, Sao Paulo, SP, Brazil
e-mail: atufaile@usp.br

M. Snyder
Technical Space Science Center, Morehead State University, 235, Martindale Drive, Morehead, KY 40531, USA

T. A. Vanderelli
Ferrocell USA, 739 Route 259, Ligonier, PA 15658, USA

© Springer Nature Switzerland AG 2019
C. H. Skiadas and I. Lubashevsky (eds.), *11th Chaotic Modeling and Simulation International Conference*, Springer Proceedings in Complexity, https://doi.org/10.1007/978-3-030-15297-0_25

Fig. 1 Ferrocell using illumination of green LEDs with three magnets. The line pattern is formed due to the presence of magnetic field

1 Introduction

Dynamical systems can be described in the abstract space known as phase space, which represents the evolution of one system from quantities as a function of time. In such space, we can explore the aspects of the geometry of the solutions of this dynamical system and the changes observed when some parameters controlling the dynamical system are changed. In this space, we can find different types of singular points, such as node, saddle point, focus, vortex, and so on. The solutions of the dynamical systems form trajectories around these singular points. For example, the solutions of a pendulum with two modes of oscillation can be represented in the surface of a torus, and each one of these solutions is called attractor. The stability of these attractors is related to the distribution of the singular points around them. Mechanical stability can be represented by potential energy of one object disturbed by small displacements, with stable equilibrium, unstable equilibrium, and indifferent equilibrium, so that we can describe the evolution of this object (Fig. 1).

Inspired by the observation of light patterns in a Ferrocell [1, 2], a Hele-Shaw cell containing ferrofluid controlled by magnetic fields, we explored some properties of the stability of dynamical systems. The aim of this work is to develop new insights for this area of dynamical systems, developing new structures for chaotic attractors using the Ferrocell device.

Ferrofluid is a colloidal suspension of nanoparticles in a liquid carrier, and this magnetic fluid has interesting optical properties.

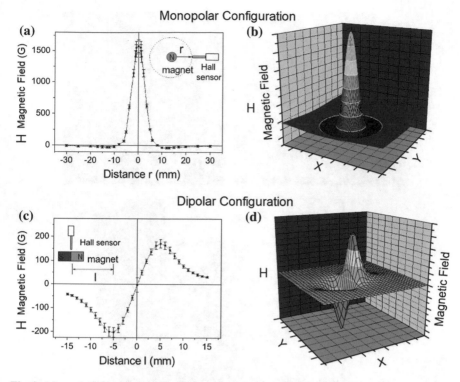

Fig. 2 Magnetic field of the permanent magnet used in the experiment of Fig. 3, for monopolar and dipolar configuration: **a** Magnetic field of monopolar configuration in one dimension, **b** the same magnetic field in two dimensions. In **c** the magnetic field of the dipolar configuration, and in **d** the same magnetic field for the two-dimensional plot

2 Magnetism and Diffraction

We have developed a magneto-optical system which simulates the stability of fixed points and the trajectories of orbits present in dynamical systems. The magnetic field is a vector field in the three-dimensional space, as it is shown in Figs. 2 and 3. In order to understand the structure of the magnetic field of a permanent magnet, we have used a gaussmeter obtaining the plots shown in Fig. 2. This field passes through the plane delimited by the Ferrocell, and in this way, we just visualize the effects of the magnetic field projected in this plane, like the two patterns of Fig. 3a, b.

One way to see directly what is happening inside the Ferrocell is obtaining the polarimetry of the Ferrocell, see Tufaile et al. [2]. We have observed that the nanoparticles create a diffracting grating following the orientation of the magnetic field, forming locally a needlelike structure. The light is diffracted by these structures following perpendicular lines to the nanoparticles array. The origin of the light line is the light source because each diffracted line is perpendicular to the ferrofluid needlelike structure.

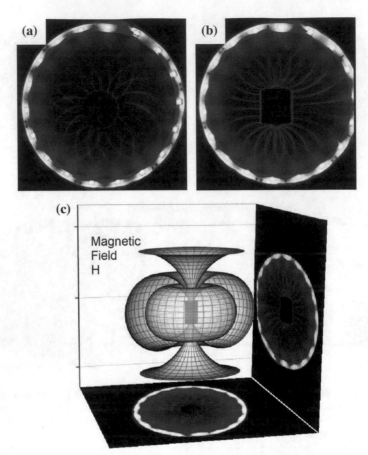

Fig. 3 The monopolar and dipolar configurations: two light patterns observed in the Ferrocell and their relative position for two different perspectives of the magnetic field, placing the source of light (red LEDs) in a circular array

Using a circular array of red LED (Light Emitting Diodes), and a magnet with the magnetic field described in Fig. 2, for the monopolar and dipolar configuration, we have directly observed the effects of the magnetic field in the Ferrocell using different planes of observation in Fig. 3a, b and the general perspective of the magnetic field of a single magnet for these two light patterns in Fig. 3c.

In this way, the transmitted light in the Ferrocell changes for different values of the magnetic field and the orbits can be observed, giving the impression that we can bend the light. The trajectories of these orbits are obtained by the diffracted light, which consists of light patterns irradiated from the liquid film, and these light patterns can be related to some magneto-optical effects, such as Faraday effect, birefringence, and dichroism. The birefringence is the difference between the refractive indices of light polarized parallel and light polarized orthogonal to the magnetic field, while

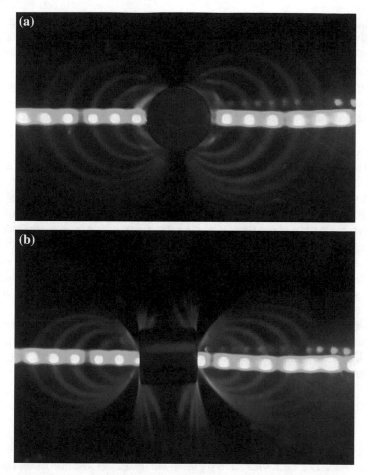

Fig. 4 **a** Image of the diffracted curves for a cylindrical magnet with one of its poles facing the Ferrocell for an array of line of LEDs. **b** The same magnet placed laterally on the array of blue LEDs

dichroism is the difference in the extinction of light between the two polarization directions. One interesting aspect of this type of diffraction is the inexistence of fringes, as one can see in Fig. 4, indicating the existence of multiple diffraction from many nanoparticles arrays, in the shape of needles.

For a volume of ferrofluid, we have the combination of wave optics and geometric optics at same time in this system, with the complex refractive index of the ferrofluid given by two components, one perpendicular (N_{perp}) and another parallel (N_{para}) to the magnetic film:

$$N^2_{perp} = (\eta_{perp} + ik_{perp})^2$$
$$N^2_{para} = (\eta_{para} + ik_{para})^2 \,,$$

where η_{perp} and η_{para} represents refraction, and k_{perp} and k_{para} represents extinction in two directions perpendicular and parallel to the magnetic field [3]. For a thin film of Ferrofluid, we observed these effects of light patterns inside the liquid film plane and transmitted light [4].

In the presence of an external magnetic field, the ferrofluids forms small needlelike structures, and these structures will align themselves with this magnetic field. When light passes through the ferrofluid, it undergoes diffraction, producing characteristic patterns. The main feature of this system is that the ferrofluid inside of the Ferrocell behaves like a variable diffracting grating, whose shape depends on the magnetic field. In this way, these line patterns are diffracted rays associated to the magnetic field.

The magneto-optical properties of these light patterns were extensively studied by Philip and Laskar [5], summarized by the statement that the formation of curved light lines can be explained by considering the scattering of light by cylinders, and evaluating the scattered electromagnetic field from a cylindrical surface. According to these authors, the reason for the formation of some ring structures on the scattered pattern can be explained by considering scattering of light by a cylinder. Some curious aspects of this type of diffraction can be summarizes by the Geometrical Theory of Diffraction [6].

Considering the ferrofluid as a diffracting grating dependent on the magnetic field, these light lines are diffracted lines. We can understand these lines algebraically if we consider the magnetic scalar potential V_m, so that the magnet can be modeled as two magnetic charges producing a magnetic dipole with [7]:

$$V_m = \frac{\mu M_c}{4\pi} \left[\frac{1}{r_n} - \frac{1}{r_s} \right].$$

The nanoparticles create the diffracting grating following the orientation of the magnetic field. The light diffracted by these nanoparticles seemed to follow isopotential lines of this scalar field having the light source as the origin of the light line, because each diffracted line is perpendicular to the ferrofluid neddlelike structure [7]. Considering the case of a monopolar configuration, with r_s tending to infinity, the representation of equipotential lines surrounding this pole of the magnet is concentric circles (see Fig. 5). The relationship between the diffracted lines and the magnetic field is clear when considering a single dimension. In the one-dimensional case, $r = x$, so that $V_m = constant/x$. Hence the vector associated with the light pattern D and the magnetic scalar potential V is:

$$D = -\frac{dV}{dx}.$$

This relation means that the strength of the diffracted lines at any point in space is the change rate of the magnetic potential over space. However, the diffracted line intersects another line from a different light source. The explanation for these

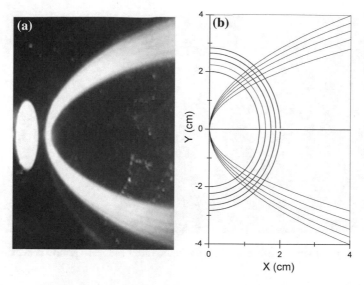

Fig. 5 Diffracted rays obtained from the experiment in (**a**). In **b** functions representing the diffracted lines with their respective orthogonal functions. These orthogonal functions are related to disposition of the nanoparticles inside the Ferrocell

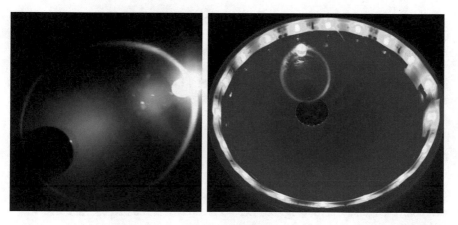

Fig. 6 The elliptical trajectory of light passing though the light source and around the face of a pole of the magnet is a limit cycle. We can see that each elliptical line is related to each LED

intersections is the overlapping projection of diffracted lines in the plane of the Ferrocell. Choosing the correct lightening configuration, these overlaps are canceled, as it is shown in Figs. 4 and 6.

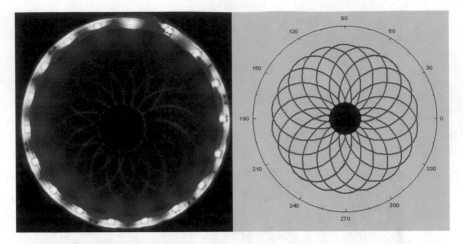

Fig. 7 Example of light pattern obtained with the Ferrocell and its respective simulation, for 18 red LEDs

3   Fixed Points and Trajectories

We start comparing the phase space of a pendulum and the diffracted lines, with two centers and one saddle point. In the Ferrocell system, the attractors are represented by the set of trajectories of light patterns (Figs. 7 and 8).

In Fig. 9 there is another example of application is the evolution from a limit cycle in (1), the separatrix is represented by the green line (2), and open trajectories (3) and (4), placing the magnet in the dipolar configuration.

4   More Attractors

We have explored some magnets configurations and observed some attractors. For example, the next attractor (Fig. 10) was obtained with three magnets and many light trajectories.

For three magnets, we have obtained three possible configurations near to the magnets, as it is shown in Fig. 11.

5   Conclusions

We have explored the possibility of creating some representations of dynamical systems using magnetic fields and optic fluidics. Our dynamical system is the direct observation of diffracted lines in the Ferrocell, a Hele-Shaw cell containing ferrofluid.

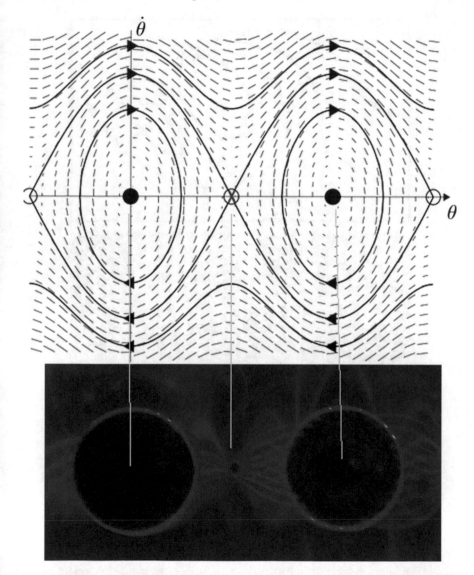

Fig. 8 Comparison between diagram of phase of a pendulum and diffracted lines: Two centers and a saddle point. We are using two cylindrical magnets with two north poles facing the Ferrocell

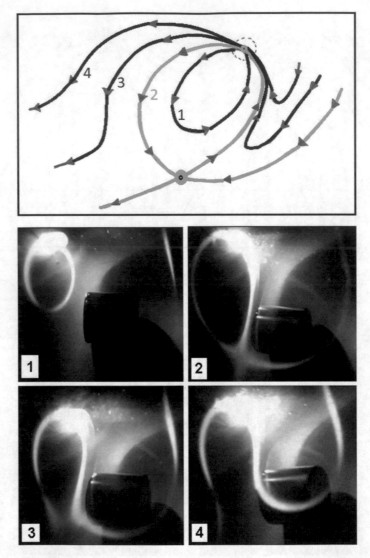

Fig. 9 Diagram of the evolution of a dynamical system, and the evolution of trajectories for different initial conditions in the Ferrocell, with a limit cycle, separatrix and non-closed trajectories

This device is an interesting tool to visualize directly the abstract phase space, and it can help the community involved with dynamical systems to explore, teach and create some types of attractors.

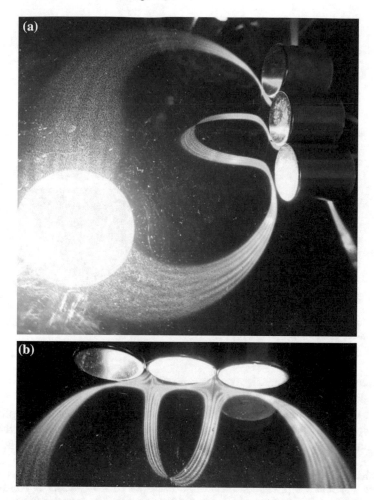

Fig. 10 Three cylindrical magnets with polarity N-S-N creates the attractor in figure (**a**). In **b** another perspective of the same attractor in the region closer to the magnets. There is a folding between two saddle points, with a behavior resembling Moebius strip

The Ferrocell presents some curious properties involving scattering of light partially confined in the plane of the ferrofluid, like a plane wave guide, as well as diffraction for the transmitted and reflected light when this system is subjected to an external magnetic field, suggesting the use of this device as light modulator or display.

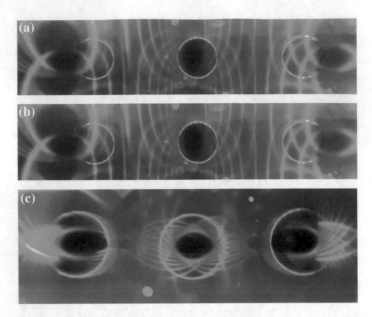

Fig. 11 Top view of the diffracted patterns of the previous attractor. The polarity of the magnets in **a** is N-S-N, in **b** N-N-S, and in **c** N-N-N

Although the comprehensive explanation of these diffracted lines is not something trivial, and still demands studies of the optical properties of light deflection by the thin film of nanoparticles, the light patterns obtained with this system is captivating, and using analogy between different areas, the intellectual intersection of the areas of dynamical systems and electromagnetism can give us interesting insights.

Acknowledgements This work was partially supported by *Conselho Nacional de Desenvolvimento Científico e Tecnológico (CNPq), Instituto Nacional de Ciência e Tecnologia de Fluidos Complexos (INCT-FCx), and by Fundação de Amparo à Pesquisa do Estado de São Paulo* (FAPESP) FAPES/CNPq #573560/2008-0.

References

1. A. Tufaile, T.A. Vanderelli, A.P.B. Tufaile, Observing the jumping laser dogs, J. Appl. Math. Phys. **4**, 1977–1988 (2016)
2. A. Tufaile, T.A. Vanderelli, A.P.B. Tufaile, Light polarization using ferrofluids and magnetic fields. Adv. Condens. Matter. Phys. article ID 2583717 (2017)
3. J. Fu, J. Li, Y.Q. Lin, X.D. Liu, H. Miao, L.H. Lin, Study of magneto-optical effects in γ-$Fe_2O_3/ZnFe_2O_4$ nanoparticle ferrofluids, using circularly polarized light. Sci China-Phys. Mech. Astronom. **55**, 1404–1411 (2012)
4. S. Wang, C. Yu, L. Du, C. Sun, J. Yang, Response sensitivity of Faraday magneto optical effects of ferrofluids. Mater. Sci. Forum **663–665**, 1221–1224 (2010)

5. J. Philip, J.M. Laskar, Optical properties and applications of ferrofluids—A review. J. Nanofluids **1**, 3–20 (2012)
6. A. Tufaile, A.P.B. Tufaile, The dynamics of diffracted rays in foams. Phys. Lett. A **379**, 3059–3068 (2015)
7. M. Snyder, J. Frederick, Photonic dipole contours of ferrofluid Hele-Shaw cell. (2009). https://arxiv.org/ftp/arxiv/papers/0805/0805.4364.pdf

Rainbows, Billiards and Chaos

Alberto Tufaile and Adriana P. B. Tufaile

Abstract Starting at the end of last century, Chaos Theory is used to explain since the dynamics of a dripping faucet to the essence of black holes. The main aspect of this ubiquity is because chaotic systems involve nonlinear systems, and most systems behave linearly only when they are close to equilibrium, far from this region, we can observe a myriad of behaviors. We studied some phenomena involving rays and waves in optics and acoustics, such as rainbows, fogbows, Glory effect, iridescent clouds, halos and sound waves in acoustic billiards from the point of view of chaotic systems. We explore the aspects of ray splitting and their relationship with Chaos Theory, based on different subjects, such as Random Matrix Theory, Caustics, Interference and Geometrical Theory of Diffraction. One interesting case in such systems is that the existence of discontinuities or singularities can lead to wave diffraction, which is related to additional contributions to the trace formula, with the presence of creeping orbits and caustics. This approach can be extended to quantum systems, such as nuclear rainbow. We will present scattering of light in open systems and compare them to the scattering of particles. We are presenting experimental results of light scattering in a cylinder and observing the "spiral rainbow" pattern.

Keywords Billiard · Wave Chaos · Random Matrix · Ray Optics · Diffracted Rays

1 Introduction

One of most beautiful atmospheric phenomena observed by many people is the rainbow, as can be seen in Fig. 1a. The rainbow is a caustic that forms perfect circular arcs with beautiful colors [1]. Other beautiful phenomenon is the Glory effect [2], which presents concentric colored halos around the shadow of the observer, as it is shown in Fig. 1b. These kinds of phenomena inspire myths, poets and physicists to

A. Tufaile (✉) · A. P. B. Tufaile
Soft Matter Laboratory, Escola de Artes, Ciencias e Humanidades, Universidade de Sao Paulo, Sao Paulo, SP, Brazil
e-mail: tufaile@usp.br

© Springer Nature Switzerland AG 2019
C. H. Skiadas and I. Lubashevsky (eds.), *11th Chaotic Modeling and Simulation International Conference*, Springer Proceedings in Complexity, https://doi.org/10.1007/978-3-030-15297-0_26

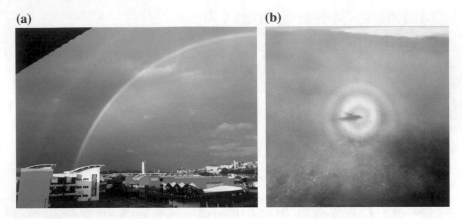

Fig. 1 Rainbows (**a**) and Glory effect (**b**)

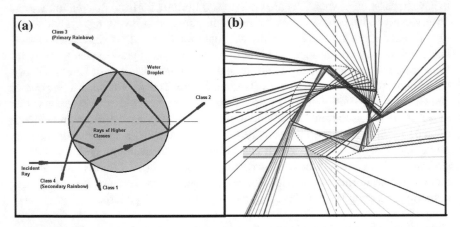

Fig. 2 Light rays in a water drop representing an open billiard system. In **a** the different class of rays, and in **b** multiple ray trajectories

study their properties. According to Nussenzveig, some of the most powerful tools of mathematical physics were created to solve the questions raised by the observation of rainbows and with closely related problems, such as the Glory effect.

The Chaos Theory is a tool used to study old problems from a new point of view, based on two main concepts: deterministic nonlinear systems with few degrees of freedom and sensitiveness to initial conditions. If we consider the Geometric Optics of an incident ray of light in a spherical drop of Fig. 2a, we can see these concepts of chaotic systems present in this system. The nonlinearity is present in the abrupt change of the refractive index at the border of the drop, with the trajectory of the light ray changing following just two laws of reflection and refraction in some points:

$$\theta_i = \theta_r;$$
$$n_i \sin \theta_i = n_t \sin \theta_t. \tag{1}$$

The signature of chaotic systems is observed in Fig. 2b, when we try to follow some trajectories of some light rays in a single drop: the light rays split and bounce back and forth following the equations (1), making the long-term prediction of the trajectories of these rays very difficult, as the same as it is observed in systems known as open billiards.

Inspired by these ideas, we have explored experimentally some open billiards using light and compared them with some systems involving rays and waves in acoustic billiards, which can be applied in quantum systems.

2 Spheres and cylinders

The drop of Fig. 2 shows some trajectories of the light rays, such as that rays of class 3 create the brighter rainbows, and the rays of class 4 create secondary rainbows, with its place higher in the sky than primary bow, reversed order of color due to the second internal reflection, and fainter than the primary bow due to another ray splitting. These features will be always present because this is a deterministic system.

However, where is the chaotic behavior for this system? In order to explore the most obvious result of chaotic system, the "butterfly effect", we have to obtain a numerous sequence of reflections and refractions. We have developed a system to observe multiples reflections and refractions using, instead a sphere, a cylinder of Fig. 3, and instead of the Sun, diode laser beams. The transversal section of the sphere and of the cylinder are circles, and in this way we have reduced our study to a two dimensional system. Even though the Sun is not a laser beam, they share some properties, for example, the sunlight that reaches us can be considered a plane wave or parallel rays. And for some peculiar conditions, such as the observed during the dawn, the Sun light is roughly a monochromatic source, creating the red rainbow of Fig. 4.

The laser beam is injected obliquely in a glass cylinder (diameter = 1.0 cm) with a refractive index around 1.4, forming the conical scattering of light of Fig. 3, which is projected in a screen. Depending on the angle formed between the cylinder axis and the laser beam, different patterns can be observed in a screen, and recorded in a picture, as it is shown in Fig. 5.

In Fig. 6 we are presenting the classes of rays obtained in our experiment, each time the laser beam hits the surface, part of light is reflected and part is refracted. The label 1 is given for rays partially reflected from the surface of the cylinder. The remaining light is transmitted into the cylinder, with a change of direction due to refraction, and those rays transmitted directly through the cylinder are designated class 2. Rays emerging after one internal reflection are labeled rays of class 3, and these rays are equivalent to the case of primary rainbow in nature. Rays of class 4 have undergone two internal reflections, and following the analogy with the light

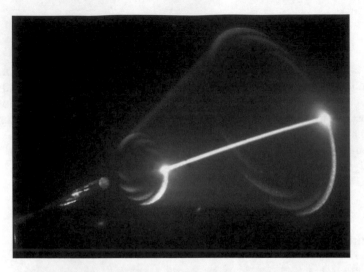

Fig. 3 A red laser beam inject obliquely in a cylinder forms a conical scattering

Fig. 4 Red rainbow observed in Denver, Colorado, USA, during the dawn

scattering in a sphere, they are associated with the secondary rainbow. The class 5 rays emerge after three internal reflections, and it is that give rise to the tertiary rainbow for the case of the sphere. The quaternary rainbow is associated with the rays of class 6. At the next boundary ray is again split into reflected and transmitted components, and the process is observed multiple times.

In Fig. 7 we have 3 spiral bows made with three laser beams.

Fig. 5 Image of the pattern obtained in a screen for a diode laser beam scattered in a glass cylinder obliquely, creating a pattern resembling a "spiral rainbow"

The reduction in size of the successive ray classes is due to the effect of the reduction of the distance from the emerging ray from the cylinder to the screen. We also observe the reduction of the light intensity of the successive rays as the effect of the multiple ray splitting. With this open billiard system, we are able to record the sequence of seventeen transformations of Eq. (1), given by the Geometric Optics of reflection and refraction. One interesting fact that this experiment revealed in Fig. 6b, c is the observation of the folding of a bow, in a process which resemble the horseshoe transformation, which is a map contracting distances horizontally and expanding distances vertically. This result resembles the Baker map or the Horseshoe map developed by Smale.

Depending on the author, the Glory effect is associated with different concepts, for example, in Ref. [2], we have the concept of light tunneling, in which most of the light seen in a Glory is the result of energy "tunneling" into water droplets from light rays passing close to them. Other authors claim that the light grazing a droplet could temporarily turn into electromagnetic surface waves. We can grasp these different perspectives with the experiment of light scattering in a Reuleaux bubble of Fig. 8.

The different explanations of the existence of the Glory effect have in common the idea of wave optics of light backscattering. In our experiment of the Eye of the Horus pattern, we can observe the effect of the interference of a curved thin film, projected into a screen, forming the center of the concentric fringes in a different position of the incident light ray, which is hitting the screen at the left of the light circle. This light circle (Fig. 8e) is known as parlaseric circle. Although the light interference phenomenon separates the colors of the sunlight in the Glory effect, while in our

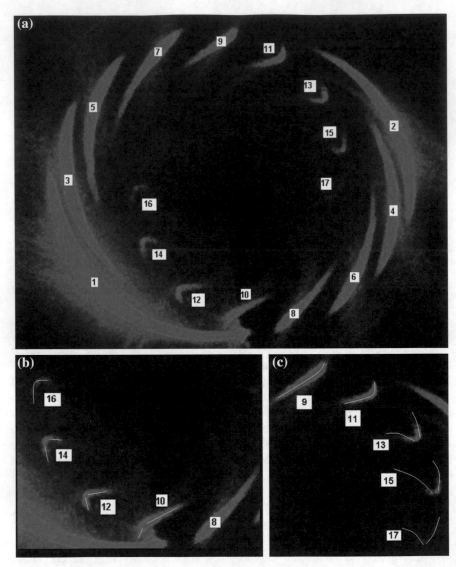

Fig. 6 Classes of rays: we can follow the sequence of class ray from 1 to 17 in this system. Pictures in (**b**) and (**c**) are zoom of two parts of the picture in (**a**)

Fig. 7 The spiral rainbow using three different diode laser beams, red, green and blue at same time

experiment we have used monochromatic blue light, in Ref. [3] is demonstrated that these fringes are sensitive to the light wavelength.

These two beautiful phenomena observed in atmospheric optics were used to stimulate the interest in the abstract concepts which involves the physical systems of light scattering ranging from Geometric Optics to Wave Optics. One interesting aspect of the ray splitting in the spiral rainbow is the mechanism of chaotization involved, because even though the process is completely deterministic, the trajectories change very abruptly due to reflection and refraction of light, different from the case of smooth separation of trajectories in an exponential evolution. In nature, ray splitting is very common in optical systems involving some kind of interface, like drops, bubbles or foams. In this way, the source of Chaos can be found more in the boundary conditions, than in the intrinsic equations of the system. In other words, you can have systems described by linear equations, or solvable equations with uniqueness of solutions, which behaves very wild when they meet an interface.

3   Wave Chaos

Several authors studied many aspects of the rainbow with different tools and techniques, such as Descartes, Newton, Young and Airy, just to cite some of them. In simplified classification, Descartes and Newton studied this atmospheric phenomenon

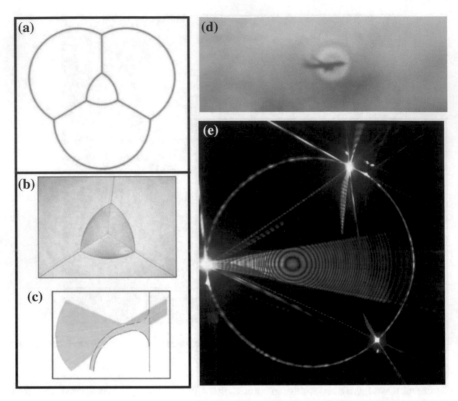

Fig. 8 Light scattering in a Reuleaux bubble compared to the Glory effect. This bubble is related to the Reuleaux triangle, which is the central bubble in (**a**) of a mathematical model of a four-bubble planar soap bubble cluster. The three-dimensional version of this triangle is shown in the real photography in (**b**). When light grazes this bubble in (**c**), we have an analog pattern of the Glory effect in (**d**), with concentric fringes due to light interference in the light pattern obtained in (**e**), known as "The Eye of Horus" explained in more detail in Ref. [3]

using Geometric Optics, Young used the idea of diffracted ray and Airy studied using Wave Optics. The complete description of the rainbow physics was given by Mie, using Maxwell's equations for a plane wave hitting a sphere in a homogeneous medium. The interesting analogy here is that these multiple studies are bridges connecting classical and quantum physics. The Geometric Optics is linked to light rays, like the trajectories of particles in billiards, and it is based in Classical Physics, the Young's method is based in semi-classics, and Wave Theory is related to the Quantum Physics. Each one of these studies is very important from the qualitative and quantitative aspects, enhancing the comprehension of the complexity of this phenomenon. One interesting point is the change of view of the Huygens principle of Optics to the Young principle, which inserts the idea of the directionality of the light wave propagation, considering the ray tracing with wave interference phenomena, creating the concept of diffracted ray [3]. In broad sense, this approach is the limit

Fig. 9 Plot of the light intensity as function of the angle for the rainbow phenomena using Geometric Optics (Descartes), diffracted rays (Young) and wave approach (Airy). The concept of Young is associated with semi-classics, the territory of Quantum Chaos

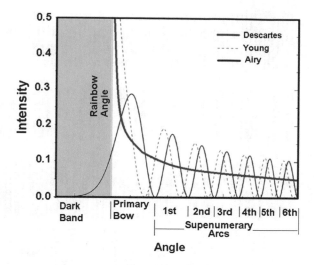

between classical physical and quantum mechanical descriptions of phenomena, in the same way as Geometrical Optics is the limit of the electromagnetism, the classical physics is the limit of quantum physics. In this frontier we found the Quantum Chaos, the branch of Physics which studies how chaotic classical dynamical systems can be described in terms of Quantum Theory. The abrupt transition between light and shadow observed in Geometric Optics presents diffraction effects that are related to the Quantum Chaos. For example, the dark region described by Descartes theory of rainbow presents the diffraction aspects of the Airy theory (see Fig. 9). The optical phenomena of rainbow and Glory have their counterparts in nuclear, atomic and molecular scattering [2].

The studies of Quantum Chaos are very elusive because they confront two different paradigms in Physics: the nonlinear determinism of classical physics and the statistics of the quantum states described by the Schrödinger wave equation. To obtain something consistent with these two descriptions of nature, is necessary to apply the Correspondence Principle, which states that Classical Mechanics is the classical limit of Quantum Mechanics, the KAM theorem (Kolmogorov-Arnold-Moser Theorem), connecting chaos and Hamiltonians, giving conditions under which chaos is restricted in extent, or BGS Conjecture (Bohigas-Gianonni-Schimit Conjecture), which reveals a very strong link that exists between Chaos and Random Matrix Theory [4, 5], which is a model for the statistical description of spectra and wave functions of quantum systems. For this conjecture, how the spectra distribute themselves in space is an important aspect of wave systems. For example, a graph of nearest-neighbor spacing distribution promptly distinguishes among the periodic, random and chaotic systems, because for periodic spectra, all the spacing are the same or located in set of points. For the case of spectra with random distribution, the distribution of nearest-neighbor spacing follows the function e^{-x}, implying that the smallest spacing between levels is very easy to be found. The chaotic systems follow the Gaussian distribution for

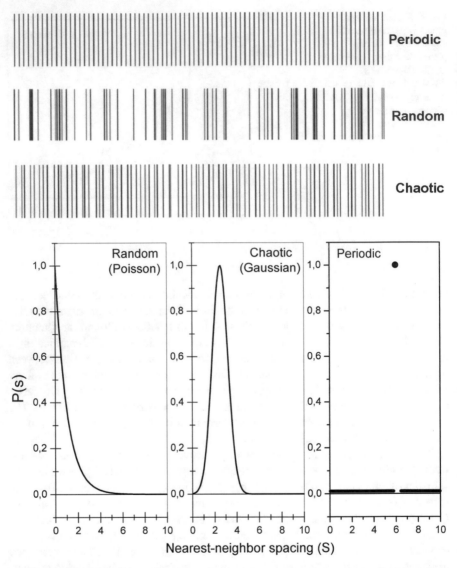

Fig. 10 Spectra and their statistics. Examples of spectrum lines and nearest-neighbor spacing distribution P(s) for the periodic, random and chaotic cases

the nearest-neighbor intervals, due to a property known as level repulsion, which is present in systems formed by "heavy nuclei" in nuclear physics, and chaotic systems fall in the class of Gaussian Ensembles, according to the jargon of this area, such as Gaussian Unitary Ensemble (GUE), Gaussian Symplectic Ensemble (GSE), and Gaussian Orthogonal Ensemble (GOE) [4] (see Fig. 10).

Fig. 11 Effect of ray splitting in the fluctuation statistics of fused quartz blocks. The curve labeled "8 GOE" corresponds to a superposition of eight independent Gaussian Orthogonal Ensembles, each one with fluctuation properties of the Gaussian distribution family

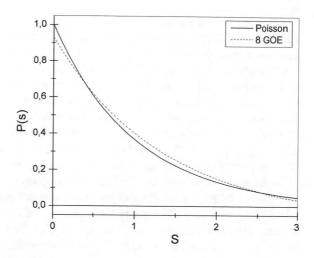

4 Elastomechanics of Quartz Blocks

The Random Matrix Theory used for these kinds of spectra, also explains the nearest-neighbor spacing distribution of the spectra of acoustic billiards, which are analogous to the quantum systems, because the Helmholtz equation is mathematically equivalent to the time independent Schrödinger equation. Besides the aspects of quantum systems, this study improves the understanding of the properties of pieces of quartz and aluminum when they are subjected to mechanical waves. For example, the spectral pattern of the aluminum fuselage of a plane can be recorded and compared with spectra after many flights, to analyze the integrity of the material without intrusive techniques. Another advantage of the use of acoustic billiards is the fact that these kinds of spectra can be obtained with high precision, enabling the analysis of many aspects of the fluctuation statistics of the resonances, including other important features, such as the spectral rigidity, also called Δ_3. The standing waves in this type of resonator, explained by the branch of Physics known as Elastomechanics, can be compared to the trajectories of classical particles bouncing inside the profile of the resonator. Again, we are facing scattering problems of semi-classical physics discussed previously for the case of the rays in spheres and cylinders.

Using the temperature as a perturbation of the quartz blocks, and measuring the ultrasound transmission in fused quartz blocks, within the frequencies ranging from 600 to 885 kHz, with low pressure (around 3 mbar), the fluctuation statistics of fused quartz blocks is compared to the computations of Random Matrix Theory, and this plot in Fig. 11 shows the existence of 8 Gaussian Orthogonal Ensembles of symmetrical real matrices. The mode conversion which occurs in the wave reflections at the block interfaces creates chaotic behavior, even for a regular a system with a regular shape. This mode conversion is related to the change of the characteristics of the sound wave which propagates inside the quartz block, and these waves can be transversal, longitudinal or mixed.

In other words, the acoustic billiards using quartz blocks also present ray-splitting with polarized waves in a similar way which occurs for electromagnetic waves. The vast majority of the resonances are mixtures of transverse and longitudinal motion, yet a small number of special resonances remain pure, if the temperature of this system is perturbed [5]. In this way, the multiple reflections in the case of the laser in the cylinder, and the multiple reflections of quartz blocks are somehow related with chaotic behavior. In order to explore in detail, the richness and the complexity of some ideas discussed in this work we encourage the reader to explore the work of Adam [6] about rainbows and glories, and the study of Legrand and Mortessagne [7] about wave chaos and the Helmholtz equation. The first work [6] explores the different approaches involving the study of the optics applied to the case of rainbows and glories, with an interesting discussion about different kinds of diffracted rays, while the second work [7] explore the aspects of diffractive orbits. In this paper, the authors claim that the nearest-neighbor spacing derived from Random Matrix Theory has never totally justified on semi-classical arguments, and in this way non-chaotic systems can mimic level repulsion, such as billiards with point scatterers, leading to level repulsion in systems that are not chaotic. Another interesting theoretical paper exploring quantum, classical and semi-classical chaos is the work of King [8] which presents beautiful simulations of patterns in the stadium billiard, with the existence of caustics.

5 Conclusions

Quantum and classical systems present an interesting frontier: semi-classical systems. The concepts of particles/rays and waves are not enough to understand this region, and the use of diffracted rays is interesting to improve our comprehension of physical systems presenting this type of duality, since from rainbows to quantum chaos, because ray theory breaks down when diffraction is present. Besides diffraction effects, we observed that the mechanism of ray splitting is one of the main features of chaotic behavior, because the trajectories diverge in an abrupt fashion, not only exponentially. This ray splitting introduces another degree of divergence in each case, and whole system has to be examined all at once, using the different points of view of classical and quantum systems, probably because ray splitting tends to destroy invariant tori and stable islands in the phase space, increasing the ergodic component of the dynamics. During our studies, we have found the interesting case of "spiral rainbow", based in the experiment of a laser scattering in a glass cylinder. In this spiral rainbow, we have observed something similar to the horseshoe map for the multiples reflections in the cylinder. The mechanism of ray splitting is also present in the elastomechanics of the fused quartz blocks, involving mode conversion as a symmetry-breaking mechanism that acts to mix transverse and longitudinal wave motion. The resonances in these quartz blocks are mixtures of transverse and longitudinal motion. Although numerical simulation shows that the classical rectangular three-dimensional ray-splitting billiard is not chaotic, the spectral fluctuations statistics of the measured eigenfrequencies follow superposed GOE spectra.

Acknowledgements This work was partially supported by Conselho Nacional de Desenvolvimento Científico e Tecnológico (CNPq), Instituto Nacional de Ciência e Tecnologia de Fluidos Complexos (INCT-FCx), and by Fundação de Amparo à Pesquisa do Estado de São Paulo (FAPESP) FAPES/CNPq #573560/2008-0. We thank professor Ângela M. M. L. Hutchison for the picture of the red rainbow.

References

1. H.M. Nussenzveig, *The Theory of the Rainbow* (Scientific American, 1977), pp. 116–127
2. H.M. Nussenzveig, The science of glory. Sci. Am. **306**, 68–73 (2012)
3. A. Tufaile, A.P.B. Tufaile, The dynamics of diffracted rays in foams. Phy. Lett. A **379**, 3059–3068 (2015)
4. M.S. Hussein, C.P. Malta, M.P. Pato, A.P.B. Tufaile, Effect of symmetry breaking on level curvature distributions. Phys. Rev. E **65**, 057203 (2002)
5. K. Schaadt, A.P.B. Tufaile, C. Ellegaard, Chaotic sound waves in a regular billiard. Phys. Rev. E **67**, 026213 (2003)
6. J. Adam, The mathematical physics of rainbows and glories. Phys. Rep. **356**, 229–265 (2002)
7. O. Legand, F. Mortessagne, Wave chaos for Helmholtz equation, in *New Directions in Linear Acoustics and Vibration,* ed. by M.Wright, R. Weaver (Cambridge University Press, 2010), pp. 24–41
8. C. King, Exploring quantum, classical and semi-classical chaos in the stadium billiard. Quanta **3**(1), 16–31 (2014)

Experimental and Numerical Study of Spectral Properties of Three-Dimensional Chaotic Microwave Cavities: The Case of Missing Levels

Vitalii Yunko, Małgorzata Białous, Szymon Bauch, Michał Ławniczak and Leszek Sirko

Abstract We present an experimental and numerical study of missing-level statistics of chaotic three-dimensional microwave cavities. The nearest-neighbor spacing distribution, the spectral rigidity, and the power spectrum of level fluctuations were investigated. We show that the theoretical approach to a problem of incomplete spectra does not work well when the incompleteness of the spectra is caused by unresolved resonances. In such a case the fraction of missing levels can be evaluated by calculations based on random matrix theory.

Keywords High-dimensional chaos · Quantum chaos · Noise and Brownian motion · Time series analysis

PACS Numbers 05.40.-a · 05.45.Jn · 05.45.Mt · 05.45.Tp

1 Introduction

Low-dimensional microwave systems are extremely useful for studying quantum chaos. One-dimensional (1D) microwave networks [1–3] can be used to simulate quantum graphs due to the equivalence of the telegraph equation describing them and the Schrodinger equation describing corresponding quantum graphs [1, 4–7]. In turn, two-dimensional (2D) microwave cavities [8–15] can be used to simulate quantum billiards due to the analogy of the Helmholtz scalar equation and the Schrödinger equation describing these systems, respectively. In the case of three-dimensional (3D) microwave cavities, there is no direct analogy between Helmholtz's vector equation and the Schrödinger equation. Thus, the cavities cannot simulate quantum systems. Nevertheless, the spectral statistics of irregular/rough 3D cavities display behavior characteristic for classically chaotic quantum systems [16–24]. Therefore, the 3D

V. Yunko · M. Białous · S. Bauch · M. Ławniczak · L. Sirko (✉)
Institute of Physics, Polish Academy of Sciences,
Aleja Lotników 32/46, 02-668 Warsaw, Poland
e-mail: sirko@ifpan.edu.pl

microwave cavities are also very interesting objects in the research of the properties of wave chaos. However, studies of such systems, both theoretical, including properties of random electromagnetic vector fields [25–28], as well as experimental are very rare.

In the spectral measurements of any systems the loss of some resonances is inevitable. It may be caused by low signal-to-noise ratio, by degeneration or overlap of resonances due to losses (absorption/system openness). In the case of 3D systems we have an additional obstacle: a large density of states. However, one should mention that in the studies of spectral statistics of acoustic resonances in 3D aluminum [29] and quartz [30] blocks, characterized by high quality factors $Q \simeq 10^4 - 10^5$, no missing resonances were reported. In billiard systems, even higher quality factors were obtained in the experiments with superconducting microwave cavities [17]. In normal conducting resonators the quality factors are much lower ($Q \simeq 10^3$) and the loss of some modes is either very likely or even inevitable, therefore, in the recent study of such a chaotic 3D microwave cavity the missing levels were explicitly taken into account [31].

The determination of the system chaoticity and symmetry class defined in the random matrix theory (RMT) using its spectral properties requires knowledge of complete series of eigenvalues, so experimentally it is generally a difficult task [32–35]. The procedures developed for microwave networks with broken [36] and preserved [37] time reversal symmetries (TRS) show that this is possible provided that several statistical measures, e.g. a short-range correlation function (the nearest-neighbor spacing distribution—NNSD), long-range correlation functions (e.g., the spectral rigidity), and the power spectrum of level fluctuations [38–40] will be analyzed. Nuclei and molecules [41–44] are examples of the real physical systems for which such procedures are crucial because in their case one always deals with incomplete spectra [45–47].

2 Experimental Setup and Measurements

The overall view of the experimental setup is shown in Fig. 1a. The 3D microwave cavity was made of polished aluminum type EN 5754 and consists of four elements. The rough semicircular element of height 60 mm (marked by (1) in Fig. 2 and visible in Fig. 1b) is closed by two flat parts: side (labeled by (2) in Fig. 2) and upper ones, both visible in Fig. 2. The bottom element is a slightly inclined and convex plate (labeled by (3) in Fig. 2) preventing the appearance of bouncing balls orbits between the upper and lower walls of the cavity. The radius function $R(\theta) = R_0 + \sum_{m=2}^{M} a_m \sin(m\theta + \Phi_m)$, where the mean radius $R_0 = 10.0$ cm, $M = 20$, a_m and Φ_m are uniformly distributed on [0.084, 0.091] cm and on [0, 2π], and $0 \leq \theta < \pi$, described the rough, semicircular element on the plane of the cross-section (Fig. 2). An aluminum scatterer inside the cavity, mounted on the metallic axel in its upper wall (see Fig. 1) was used to realize various cavity configurations. The orientation of the scatterer was changed by turning the axle around in 18 iden-

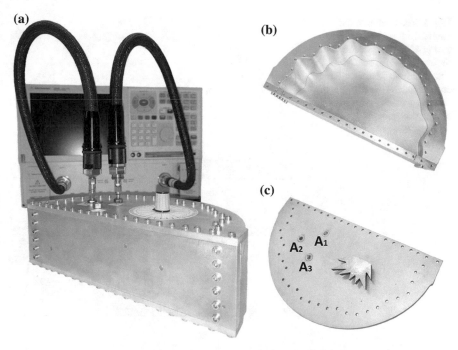

Fig. 1 **a** The 3D microwave cavity connected to the vector network analyzer. The panels **b** and **c** show the cavity without the upper cover and the inner side of the upper cover with the scatterer and marked holes for the antennas

tical steps, each equal to $\pi/9$. Besides the hole for the scatterer in the bottom wall there are three other holes A1, A2, and A3 for antennas.

An Agilent E8364B vector network analyzer (VNA) was used to perform two-port measurements of the four-element scattering matrix $\hat{S}$ [31, 48].

$$\hat{S} = \begin{bmatrix} S_{11} & S_{12} \\ S_{21} & S_{22} \end{bmatrix} \tag{1}$$

The cavity was connected to the VNA via two antennas and the flexible microwave cables HP 85133-616 and HP 85133-617. The measurements were done in the frequency range 6–11 GHz for all three combinations of the antennas positions. The antennas penetrated 6 mm into the cavity with a wire of 0.9 mm in diameter. The "third" empty hole was plugged by brass plug during the measurement. Previous measurements [23] have shown that the total absorption of the cavity is mainly related to internal absorption much greater than that associated with antennas/channels.

In Fig. 3 we present the examples of the modules of the reflected $|S_{11}|$, $|S_{22}|$, and transmitted $|S_{12}|$ signals measured in the lowest 6–7 GHz and highest 10–11 GHz frequency ranges. It should be noted that a full cross-correlation was observed between S_{12} and S_{21}, which means that TRS is preserved. The spectra were obtained in the

Fig. 2 The sketch of the cavity. The perpendicular and parallel cross-sections to the cavity height are shown. The positions of the scatterer and the antennas are marked. See the text for a detailed description

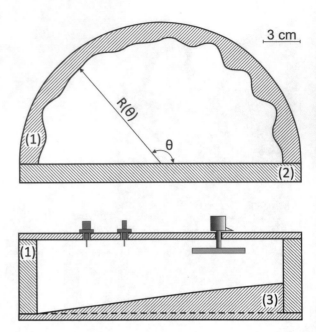

two-port measurement with the antennas at positions 2 and 3 (Fig. 1c) connected to the ports 1 and 2 of the VNA, respectively. In the low frequency range, when the resonances are well separated, comparing the spectra $|S_{11}|$ and $|S_{22}|$, it is clearly seen that the number of detected resonances may depend on the position of the antenna. The transmission signal $|S_{12}|$ can be also used in the search for the resonances, however, even for this low frequency range not all resonances are separated and visible. In turn, in the higher frequency range, due to the cubic dependence of the number of resonances on the frequency, the overlapping resonances have appeared. Therefore, losing resonances in the measurements is inevitable.

3   Statistical Measures of Experimental Spectra of Chaotic Systems

In order to analyze experimental data, it is useful to eliminate their dependence on specific features of the studied system, such as its dimensions, for example. This may be achieved by the rescaling procedure, which in the case of the 3D chaotic systems is carried out using the Weyl formula [28, 49–51]:

$$N(\nu) = A\nu^3 - B\nu + C. \tag{2}$$

The coefficient $A = \frac{8}{3}\frac{\pi}{c^3}V$, where c is speed of light in vacuum and $V = (7.267 \pm 0.012) \times 10^{-4}$ m^3 is the volume of the cavity reduced by the volume of the scatterer. The dependent on the surface of a cavity, a term proportional to ν^2 disappears due to boundary conditions of the electromagnetic field in the conducting cavity walls [49]. The coefficient B depends on the surface curvature, internal angles, and the edge length of the cavity [28]. The constant C is also associated with the shape of the cavity and in the simple case of the cubic cavity $C = 1/2$ [50]. The coefficients B and C are generally difficult to determine exactly, so fitting procedures are necessary to obtain the cumulative number of levels $N(\nu)$ for 3D irregular cavities.

In order to analyze the data we will use the short-range spectral fluctuation function, the nearest-neighbor spacing distribution, i.e. the distribution of spacings between adjacent eigenvalues $s_i = \epsilon_{i+1} - \epsilon_i$, where $\epsilon_i = N(\nu_i)$ are rescaled eigenvalues obtained by unfolding procedure will be used. Also, the integrated nearest-

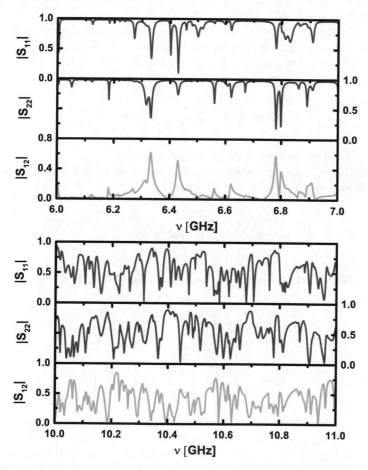

Fig. 3 Modules of the elements $|S_{11}|$ $|S_{22}|$ and $|S_{12}|$ of the two-port scattering matrix $\hat{S}$ (antennas in holes A$_2$ and A$_3$) of the 3D microwave cavity measured in the frequency range 6–7 and 10–11 GHz, respectively

neighbor spacing distribution $I(s)$, very sensitive to the symmetry class of the system (preserved or broken time reversal symmetry) will be used. The spectral rigidity of the spectrum $\Delta_3(L)$ which is the least-squares deviation of the integrated resonance density from the straight line that best fits it in the interval L [52] will be exploit as the measure of long-range spectral fluctuations. We also will take into account the power spectrum of the deviation of the qth nearest-neighbor spacing from its mean value q [42].

$$\eta_q = \sum_{i=1}^{q} (s_i - <s>) = \epsilon_{q+1} - \epsilon_1 - q \tag{3}$$

In the systems with losses, when the problem of missing levels may be very severe, the fluctuations of the scattering matrix elements can be useful. The correlation functions [53, 54], the Wigner's reaction matrix and the elastic enhancement factor [2, 7, 48] are sensitive measures of system chaoticity, however don't provide any information about missing energy levels.

In the article [47] the authors derived analytical expressions for the spectral rigidity, the number variance, and the nearest-neighbor spacing distribution describing incomplete spectra. The parameter $0 < \varphi \le 1$ denotes the fraction of observed levels. The formula for the spectral rigidity reads as follows:

$$\delta_3(L) = (1 - \varphi)\frac{L}{15} + \varphi^2 \Delta_3\left(\frac{L}{\varphi}\right). \tag{4}$$

where $\Delta_3(L)$ is the expression for the complete spectra.

The NNSD is expressed by the sum of terms of the $(n + 1)$st nearest-neighbor spacing distribution $P(n, \frac{s}{\varphi})$:

$$p(s) = \sum_{n=0}^{\infty} (1 - \varphi)^n P\left(n, \frac{s}{\varphi}\right). \tag{5}$$

For GOE systems the first and second term of Eq. (5) is approximated by

$$P\left(0, \frac{s}{\varphi}\right) = \frac{\pi}{2}\frac{s}{\varphi} \exp\left[-\frac{\pi}{4}\left(\frac{s}{\varphi}\right)^2\right]. \tag{6}$$

$$P\left(1, \frac{s}{\varphi}\right) = \frac{8}{3\pi^3}\left(\frac{4}{3}\right)^5 \left(\frac{s}{\varphi}\right)^4 \exp\left[-\frac{16}{9\pi}\left(\frac{s}{\varphi}\right)^2\right]. \tag{7}$$

When $\varphi = 1$ the above formulas reduces to Wigner surmise formula for the NNSD for $P(0, s)$ and to the NNSD of the symplectic ensemble with $\langle s \rangle = 2$ for $P(1, s)$. For the higher $n = 2, 3, \ldots$ spacing distributions $P(n, \frac{s}{\varphi})$, are well approximated by their Gaussian asymptotic forms, centered at $n + 1$.

$$P(n, \frac{s}{\varphi}) = \frac{1}{\sqrt{2\pi V^2(n)}} \exp\left[-\frac{(\frac{s}{\varphi} - n - 1)^2}{2V^2(n)} \right],$$ (8)

with the variances

$$V^2(n) \simeq \Sigma^2(L = n) - \frac{1}{6}.$$ (9)

The number variance $\Sigma^2(L)$ in Eq. (9) is the variance of the number of levels contained in an interval of length L [52].

The integrated nearest-neighbor spacing distribution $I(s)$, which is very useful for distinguishing of GOE and GUE (Gaussian unitary ensemble—systems with broken TRS) due to its high sensitivity from s in the range of small level separations, reads:

$$I(s) = \int_0^s p(s')ds'.$$ (10)

The analytical expression for the power spectrum of level fluctuations for incomplete spectra was given in Ref. [38]

$$\langle S(\tilde{k}) \rangle = \frac{\varphi}{4\pi^2} \left[\frac{K\left(\varphi\tilde{k}\right) - 1}{\tilde{k}^2} + \frac{K\left(\varphi\left(1 - \tilde{k}\right)\right) - 1}{(1 - \tilde{k})^2} \right]$$

$$+ \frac{1}{4\sin^2(\pi\tilde{k})} - \frac{\varphi^2}{12},$$ (11)

where $K(\tau) = 2\tau - \tau \log(1 + 2\tau)$ is the spectral form factor for GOE system and $0 \leq \tilde{k} = k/N \leq 1$.

The $< S(k) >$ is given in terms of the Fourier spectrum transform from "time" q to k

$$S(k) = |\tilde{\eta}_k|^2$$ (12)

with

$$\tilde{\eta}_k = \frac{1}{\sqrt{N}} \sum_{q=0}^{N-1} \eta_q \exp\left(-\frac{2\pi i k q}{N}\right)$$ (13)

In Refs. [39, 40] authors showed that for $\tilde{k} \ll 1$ the $\langle S(\tilde{k}) \rangle \propto (\tilde{k})^{-\alpha}$ with α equals 2 and 1 for regular and chaotic system, respectively, regardless the system time symmetry. The power spectrum and the power law behavior were studied numerically in [55–58]. The usefulness of the power spectrum in analyzing experimental results was confirmed in [42]—measurements of molecular resonances and in the investigations of microwave networks [36, 37, 59], and billiards [60].

4 Results

In Fig. 4 we present the experimental results obtained in the frequency range 6–11 GHz for 30 realizations of the cavity. The number of the detected resonances depends on the cavity realization, therefore, from each cavity spectrum a small number of resonances has been randomly removed to obtain the same number $\Delta N_{exp} = 208$ for all cavity configurations. The NNSD, the integrated NNSD, the spectral rigidity and the power spectrum are shown in panels (a), (b), (c) and (d), respectively. The solid (black) line denotes the theoretical predictions for the complete spectra $\varphi = 1$ of the GOE system. Ten terms of Eq. (5) was used in the numerical calculation of the NNSD. Experimental data are represented by a blue histogram and blue full diamonds in panels (a), (b) and blue full triangles and circles in panels (c), (d). Using the equations for the spectral rigidity (4) and the power spectrum (11) we found that the best agreement between the theoretical predictions for incomplete spectra (dark cyan dash-dot line) and experimental results occurs for $\varphi = 0.85$, as shown in panels (c) and (d) of Fig. 4. In turn, $\varphi = 0.85$ was inserted into equations (5) and (10) to calculate the NNSD and the INNSD which are shown in panels (a) and (b). It is easily seen that the spectral rigidity and the power spectrum are much more sensitive measures of losing states than the nearest-neighbor spacing distribution. The excellent agreement between the theoretical predictions for incomplete spectra and the experimental results shows that the investigated system, 3D irregular cavity, belongs to systems characterized by preserved TRS (GOE) and that the resonances have been randomly lost.

In order to estimate the theoretical number of resonances ΔN_w in the frequency range 6–11 GHz, which is required for the calculation of φ, we found for nine configurations of the cavity complete spectra in the frequency range 7–9 GHz [31]. This allowed us to make the fits of the experimental staircase functions to the formula (2) $N(\nu) = A\nu^3 - B\nu + C$ with the fixed coefficient $A = \frac{8}{3}\frac{\pi}{c^3}V = (0.2259 \pm 0.004) \times 10^{-27} s^3$. The fits gave the average values of the coefficients $B = (1.442 \pm 0.174) \times 10^{-9} s$ and $C = -66.0 \pm 1.4$. Then using the formula (2) we calculated the theoretical number of resonances $\Delta N_w = 245$ in the frequency range 6–11 GHz.

The fraction of the detected levels $\varphi = 0.85$ estimated from the missing-level statistics (Eqs. 4 and 11) can now be compared with the fraction $\Delta N_{exp}/\Delta N_w = 208/245 \simeq 0.849$ obtained as a ratio of the experimentally founded eigenfrequencies and those predicted from the Weyl formula in the frequency range 6–11 GHz. The agreement is excellent.

The discussed above theoretical and experimental results are additionally compared to the numerical results obtained directly by the application of the random matrix theory. We created 99 realizations of random, real, symmetric matrices of a size N = 295, representing GOE system. 25 eigenvalues of the matrix were removed from the beginning and the end of each set of eigenvalues, yielding 245 eigenvalues as for the complete spectrum. Then, 15% of the eigenvalues were randomly removed, so we finally got 208 eigenvalues as in the experiment, which were rescaled using a fifth

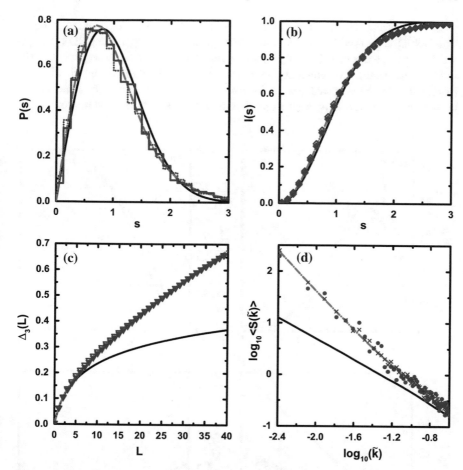

Fig. 4 Short- and long-range spectral correlation functions calculated for the 3D microwave cavity. Panel **a** shows the experimental nearest-neighbor spacing distribution NNSD (blue histogram) compared to the NNSD evaluated for eigenvalues of random matrices for the fraction of observed levels ($\varphi = 0.85$) (red doted histogram). Panel **b** shows the experimental integrated nearest-neighbor spacing distribution INNSD (blue diamonds) compared the INNSD evaluated for eigenvalues of random matrices for ($\varphi = 0.85$) (red empty diamonds). Panel **c** shows the experimental spectral rigidity (blue triangles) compared to the RMT for ($\varphi = 0.85$) (red empty triangles). Panel **d** shows the experimental average power spectrum of level fluctuations (blue circles) compared to the RMT for ($\varphi = 0.85$) (red crosses). The GOE predictions ($\varphi = 1$) and the missing level statistics predictions evaluated for the fraction of observed levels ($\varphi = 0.85$) are shown in all four panels by black solid and dark cyan dash-dot lines, respectively

order polynomial. The results for all considered measures are also presented in Fig. 4. The NNSD is marked by the red dotted histogram, the integrated NNSD by empty red diamonds, the spectral rigidity by empty red triangles and the power spectrum by red crosses. Again, the agreement with the experimental results is remarkable, confirming that the investigated system, 3D irregular cavity, belongs to the systems with preserved time reversal symmetry and that the resonances have been randomly lost.

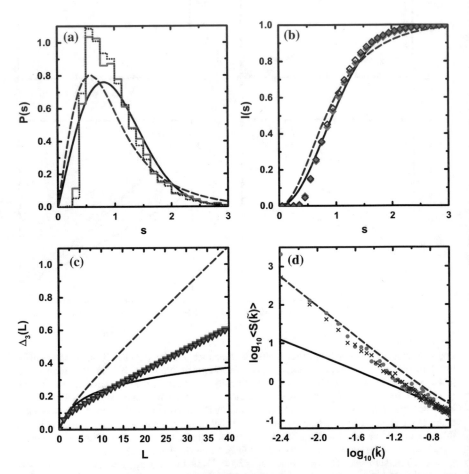

Fig. 5 Short- and long-range spectral correlation functions calculated for the modified experimental spectra. In such spectra "the closest" (see the text for details) resonances were eliminated to get the fraction of observed levels $\varphi = 0.65$. Panels **a–d** show the NNSD (dark green histogram), the INNSD (dark green diamonds), the spectral rigidity of the spectrum (dark green triangles), and the average power spectrum of level fluctuations (dark green circles), respectively, for modified experimental spectra. The results obtained from RMT calculations in panels **a–c** are marked by dashed line of vine colors, and in panel **d** by vine crosses. The GOE predictions ($\varphi = 1$) and the missing-levels statistics predictions calculated for the fraction of observed levels $\varphi = 0.65$ are shown in all panels by full black and purple broken lines, respectively

From the experimental point of view it is important to analyze a more complicated situation when some of the resonances are not randomly lost. We modified the experimental spectra, originally with 15% of randomly missing levels, using the following procedure. In the step by step procedure in an analyzed spectrum we identified the pair of resonances, the least distant from each other, and eliminated one of them until we reached $\varphi = 0.65$. In this way we additionally removed 20% resonances due to clustering. In Fig. 5 we compare the results for the modified experimental spectra with the GOE prediction for the complete spectra, predictions for the missing-level statistics [Eqs. (4), (5), (10), and (11)] calculated for $\varphi = 0.65$, and with the results of RMT calculations. It is clearly seen that the missing-level statistics (purple dashed line in all panels) fail in description of the spectra in which resonances were not lost randomly [green full line histogram—(a), green full diamonds—(b), green full triangles—(c), green dots—(d)]. This is a very important problem because the loss of energy levels due to their degeneration or overlap caused by absorption or openness of a system is very common.

The results of RMT calculation [vine doted histogram—(a), vine empty diamonds—(b), vine empty triangles—(c), vine crosses—(d)] in which the procedure of eliminating eigenvalues mimic the procedure used for the experimental data are in good agreement with experimental ones. It should be pointed out that these numerical calculation are sensitive to the order in which the eigenvalues are deleted. The results may slightly vary depending whether we start the deletion from the overlapping resonances or from random ones.

5 Conclusions

We present an experimental and numerical study of the fluctuation properties in incomplete spectra of the 3D chaotic microwave cavity. We analyzed the two important cases: the situation of randomly lost resonances and the situation when an additional fraction of resonances is omitted due to their clustering (overlapping). In the case of randomly missing resonances our results are in agreement with the level-missing statistics. However, in the case of many overlapping resonances direct random matrix theory calculations are required to properly simulate the experimental results.

Acknowledgements This work was supported in part by the National Science Centre Grants Nos. UMO-2016/23/B/ST2/03979 and 2017/01/X/ST2/00734.

References

1. O. Hul, S. Bauch, P. Pakoński, N. Savytskyy, K. Życzkowski, L. Sirko, Phys. Rev. E **69**, 056205 (2004)
2. M. Ławniczak, S. Bauch, O. Hul, L. Sirko, Phys. Rev. E **81**, 046204 (2010)

3. O. Hul, M. Ławniczak, S. Bauch, A. Sawicki, M. Kuś, L. Sirko, Phys. Rev. Lett **109**, 040402 (2012)
4. T. Kottos, U. Smilansky, Phys. Rev. Lett. **79**, 4794 (1997)
5. T. Kottos, U. Smilansky, Ann. Phys. **274**, 76 (1999)
6. P. Pakoński, K. Życzkowski, M. Kuś, J. Phys. A **34**, 9303 (2001)
7. M. Ławniczak, S. Bauch, L. Sirko, in *Handbook of Applications of Chaos Theory*, ed. by C. Skiadas, C. Skiadas (CRC Press, Boca Raton, USA, 2016), p. 559
8. H.J. Stöckmann, J. Stein, Phys. Rev. Lett. **64**, 2215 (1990)
9. S. Sridhar, Phys. Rev. Lett. **67**, 785 (1991)
10. H. Alt, H.-D. Gräf, H.L. Harner, R. Hofferbert, H. Lengeler, A. Richter, P. Schardt, A. Weidenmüller, Phys. Rev. Lett. **74**, 62 (1995)
11. P. So, S.M. Anlage, E. Ott, R.N. Oerter, Phys. Rev. Lett. **74**, 2662 (1995)
12. U. Stoffregen, J. Stein, H.-J. Stöckmann, M. Kuś, F. Haake, Phys. Rev. Lett. **74**, 2666 (1995)
13. L. Sirko, P.M. Koch, R. Blümel, Phys. Rev. Lett. **78**, 2940 (1997)
14. S. Hemmady, X. Zheng, E. Ott, T.M. Antonsen, S.M. Anlage, Phys. Rev. Lett. **94**, 014102 (2005)
15. B. Dietz, A. Richter, CHAOS **25**, 097601 (2015)
16. S. Deus, P.M. Koch, L. Sirko, Phys. Rev. E **52**, 1146 (1995)
17. H. Alt, C. Dembowski, H.-D. Gräf, R. Hofferbert, H. Rehfeld, A. Richter, R. Schuhmann, T. Weiland, Phys. Rev. Lett. **79**, 1026 (1997)
18. U. Dörr, H.-J. Stöckmann, M. Barth, U. Kuhl, Phys. Rev. Lett. **80**, 1030 (1998)
19. C. Dembowski, B. Dietz, H.-D. Gräf, A. Heine, T. Papenbrock, A. Richter, C. Richter, Phys. Rev. Lett. **89**, 064101 (2002)
20. O. Tymoshchuk, N. Savytskyy, O. Hul, S. Bauch, L. Sirko, Phys. Rev. E **75**, 037202 (2007)
21. N. Savytskyy, O. Tymoshchuk, O. Hul, S. Bauch, L. Sirko, Phys. Lett. A **372**, 1851 (2008)
22. M.R. Schröder, Acoustica **4**, 456 (1954)
23. M. Ławniczak, O. Hul, S. Bauch, L. Sirko, Acta Phys. Pol. A **116**, 749 (2009)
24. J.-H. Yeh, Z. Drikas, J. Gil, S. Hong, B.T. Taddese, E. Ott, T.M. Antonsen, T. Andreadis, S.M. Anlage, Acta Phys. Pol. A **124**, 1045 (2013)
25. H. Primack, U. Smilansky, Phys. Rev. Lett. **74**, 4831 (1995)
26. T. Prosen, Phys. Lett. A **233**, 323 (1997)
27. L.R. Arnaut, Phys. Rev. E **73**, 036604 (2006)
28. J.-B. Gros, O. Legrand, F. Mortessagne, E. Richalot, K. Selemani, Wave Motion **51**, 664 (2014)
29. R.L. Weaver, J. Acoust. Soc. Am. **85**, 1005 (1989)
30. C. Ellegaard, T. Guhr, K. Lindemann, J. Nygard, M. Oxborrow, Phys. Rev. Lett. **77**, 4918 (1996)
31. M. Ławniczak, M. Białous, V. Yunko, S. Bauch, L. Sirko, Phys. Rev. E **98**, 012206 (2018)
32. O. Bohigas, M.J. Giannoni, C. Schmit, Phys. Rev. Lett. **52**, 1 (1984)
33. O. Bohigas, R.U. Haq, A. Pandey, in *Nuclear Data for Science and Technology*, ed. by K.H. Böckhoff (Reidel, Dordrecht, 1983)
34. M. Sieber, U. Smilansky, S.C. Creagh, R.G. Littlejohn, J. Phys. A **26**, 6217 (1993)
35. B. Dietz, T. Guhr, B. Gutkin, M. Miski-Oglu, A. Richter, Phys. Rev. E **90**, 022903 (2014)
36. M. Białous, V. Yunko, S. Bauch, M. Ławniczak, B. Dietz, L. Sirko, Phys. Rev. Lett. **117**, 144101 (2016)
37. B. Dietz, V. Yunko, M. Białous, S. Bauch, M. Ławniczak, L. Sirko, Phys. Rev. E **95**, 052202 (2017)
38. R.A. Molina, J. Retamosa, L. Muñoz, A. Relaño, E. Faleiro, Phys. Lett. B **644**, 25 (2007)
39. A. Relaño, J.M.G. Gómez, R.A. Molina, J. Retamosa, E. Faleiro, Phys. Rev. Lett. **89**, 244102 (2002)
40. E. Faleiro, J.M.G. Gómez, R.A. Molina, L. Muñoz, A. Relaño, J. Retamosa, Phys. Rev. Lett. **93**, 244101 (2004)
41. A. Frisch, M. Mark, K. Aikawa, F. Ferlaino, J. Bohn, C. Makrides, A. Petrov, S. Kotochigova, Nature (London) **507**, 475 (2014)
42. J. Mur-Petit, R.A. Molina, Phys. Rev. E **92**, 042906 (2015)

43. H.I. Liou, H.S. Camarda, F. Rahn, Phys. Rev. C **5**, 131 (1972)
44. T. Zimmermann, H. Köppel, L.S. Cederbaum, G. Persch, W. Demtröder, Phys. Rev. Lett. **61**, 3 (1988)
45. J. Enders, T. Guhr, N. Huxel, P. von NeumannCosel, C. Rangacharyulu, A. Richter, Phys. Lett. B **486**, 273 (2000)
46. U. Agvaanluvsan, G.E. Mitchell, J.F. Shriner Jr., M. Pato, Phys. Rev. C **67**, 064608 (2003)
47. O. Bohigas, M.P. Pato, Phys. Lett. B **595**, 171 (2004)
48. M. Ławniczak, M. Białous, V. Yunko, S. Bauch, L. Sirko, Phys. Rev. E **91**, 032925 (2015)
49. R. Balian, C. Bloch, Ann. Phys. (N.Y.) **64**, 271 (1971)
50. H.P. Baltes, Phys. Rev. A **6**, 2252 (1972)
51. R. Balian, C. Bloch, Ann. Phys. (N.Y.) **84**, 559 (1974)
52. M.L. Mehta, *Random Matrices* (Academic Press, London, 1990)
53. B. Dietz, T. Friedrich, H.L. Harney, M. Miski-Oglu, A. Richter, F. Schäfer, J. Verbaarschot, H.A. Weidenmüller, Phys. Rev. Lett. **103**, 064101 (2009)
54. B. Dietz, T. Friedrich, H.L. Harney, M. Miski-Oglu, A. Richter, F. Schäfer, H.A. Weidenmüller, Phys. Rev. E **81**, 036205 (2010)
55. J.M.G. Gómez, A. Relaño, J. Retamosa, E. Faleiro, L. Salasnich, M. Vraničar, M. Robnik, Phys. Rev. Lett. **94**, 084101 (2005)
56. L. Salasnich, Phys. Rev. E **71**, 047202 (2005)
57. M.S. Santhanam, J.N. Bandyopadhyay, Phys. Rev. Lett. **95**, 114101 (2005)
58. A. Relaño, Phys. Rev. Lett. **100**, 224101 (2008)
59. M. Ławniczak, M. Białous, V. Yunko, S. Bauch, L. Sirko, Acta Phys. Pol. A **132**, 1672 (2017)
60. E. Faleiro, U. Kuhl, R.A. Molina, L. Muñoz, A. Relaño, J. Retamosa, Phys. Lett. A **358**, 251 (2006)

Author Index

© Springer Nature Switzerland AG 2019
C. H. Skiadas and I. Lubashevsky (eds.), *11th Chaotic Modeling
and Simulation International Conference*, Springer Proceedings
in Complexity, https://doi.org/10.1007/978-3-030-15297-0

Printed in the United States
By Bookmasters